Deformation and Failure Mechanism of Rock Tunnels under Earthquake Loading

Seismic damage to rock tunnels from recent earthquakes indicates an urgent need for seismic assessment and aseismic design of underground structures. This book offers a comprehensive account of seismic performance and the response of underground structures under earthquake loading, necessary for adequate assessment and design.

The book presents research methods for the rate-dependent mechanical behavior of rock and for the seismic behavior of underground structures. It describes analytical solutions to investigate the seismic response of tunnels subjected to seismic waves, toward an improved quantitative understanding of the seismic deformation and failure mechanism in both longitudinal and transversal aspects. A performance-based restoration design criterion and aseismic design are also proposed for future tunnel planning.

- Includes a detailed case study for the seismic performance assessment of rock tunnels under earthquake loading
- Explores the relationship between seismic damage to underground structures and ground deformation
- Covers a range of issues from mechanisms, analysis, assessment, and design of both new tunnels and restoration projects

The book is ideal for earthquake engineers and researchers, and will also be of interest to contractors, clients, researchers, lecturers, and advanced students working on tunnel engineering.

Yujing Jiang is a professor of rock mechanics and geo-environmental engineering at Nagasaki University, Japan and Shandong University of Science and Technology, China. He was elected as a fellow of the Japan Society of Civil Engineers (JSCE) in 2010 and a Foreign Associate of the Engineering Academy of Japan (EAJ) in 2018.

Xuepeng Zhang is an associate professor at Shandong University of Science and Technology, China.

Deformation and Failure Mechanism of Rock Tunnels under Earthquake Loading

Yujing Jiang and Xuepeng Zhang

CRC Press
Taylor & Francis Group
Boca Raton London

CRC Press is an imprint of the
Taylor & Francis Group, an **informa** business

First edition published 2024
by CRC Press
2385 NW Executive Center Drive, Suite 320, Boca Raton, FL 33431

and by CRC Press
4 Park Square, Milton Park, Abingdon, Oxon, OX14 4RN

CRC Press is an imprint of Taylor & Francis Group, LLC

© 2024 Yujing Jiang and Xuepeng Zhang

ISBN: 978-1-032-50921-1 (hbk)
ISBN: 978-1-032-51301-0 (pbk)
ISBN: 978-1-003-40159-9 (ebk)

DOI: 10.1201/9781003401599

Typeset in Sabon
by Apex CoVantage, LLC

Contents

Preface

Underground structures are more and more frequently constructed to facilitate different needs in a wide range of engineering applications, including subways and railways, highways, material storage, and sewage and water transport. Historically, underground structures have proved to experience a lower rate of damage than surface structures under both static and seismic loadings. However, many instances with noticeable seismic damage to rock tunnels were reported to indicate that tunnels could be damaged at different levels regarding earthquakes, such as the 1999 Chi-Chi earthquake, the 2008 Wenchuan earthquake, and the 2016 Kumamoto earthquake. Such records indicate the urgent requirement for seismic assessment and aseismic design for underground structures. Accurate seismic assessment and aseismic design require a comprehensive understanding of the seismic performance and response of underground structures subjected to an earthquake. The book presents the research method for the seismic behavior of underground structures and surrounding rock.

Firstly, a numerical simulation test based on particle flow code and laboratory direct shear test are conducted for the rate-dependent mechanical behavior of the intact rock and jointed rock. The rate-dependent behavior of the intact rock is discussed from the viewpoint of compression strength, tensile strength, failure modes and degree, strain energy rate, and acoustic emission with four loading velocities. Shear rate and cyclic shear loading are two aspects of the seismic-dependent consideration for jointed rock. Jointed rock with/without reinforcement is considered. Meanwhile, a modified shear strength criterion based on Barton's JRC-JCS criterion is developed considering the shear rate effect. On the other hand, an analytic method based on the displacement discontinuity method and dynamic centrifuge model test are conducted for the seismic behavior of the fractured rock. Analytical models considering that different types of seismic waves (P-wave, SH-wave and SV-wave) propagate across an inclined rock fracture (joint) were formulated. Then dynamic centrifuge tests on artificial rock-like models with various joint distributions were performed to investigate the seismic responses of fractured rock foundations to horizontal shear waves (simple harmonic waves, SV-waves).

Secondly, analytical solutions are described to investigate the seismic response of tunnels subjected to seismic waves, aimed at improving a quantitative understanding of the seismic deformation and failure mechanism in both longitudinal and transversal aspects. The seismic performance assessment of rock tunnels under earthquake loading is discussed by taking the Tawarayama tunnel in Kumamoto, Japan, which was subject to the 2016 Kumamoto earthquake, as one example. A possible relationship between seismic damage to underground structures and ground deformation is also explored. Finally, a performance-based restoration design criterion and aseismic design especially for the longitudinal aseismic design are proposed for future tunnel planning.

Acknowledgments

The experimental facilities and the numerical analysis platform are provided by the Geo-environment Laboratory at Graduate School of Engineering, Nagasaki University and the State Key Laboratory of Mining Disaster Prevention and Control Co-founded by Shandong Province and the Ministry of Science and Technology, Shandong University of Science and Technology. The authors appreciate their essential supports.

The authors gratefully acknowledge the support of the Kumamoto River and National Highway Office, Kyushu Regional Development Bureau, Ministry of Land, Infrastructure, Transport and Tourism in the site investigation after the 2016 Kumamoto earthquake.

We wish to thank the staff at CRC Press and Routledge (imprints of Taylor & Francis), in particular Tony Moore, Aimee Wragg and Jayachandran Rajendiran for their kind help and support.

April 2023

Yujing JIANG
Xuepeng ZHANG

Chapter 1

Introduction

1.1 BACKGROUND

An earthquake is the shaking of the surface of the Earth, resulting from the sudden release of energy in the Earth's lithosphere that creates seismic waves. Earthquakes can range in size from those that are so weak that they cannot be felt to those violent enough to toss people around and destroy whole cities. It is estimated that around 500,000 detectable earthquakes occur each year with current instrumentation (United States Geological Survey, 2010). Figure 1.1 lists a map of the earthquakes in the world with a magnitude of 6.0 and greater from 1900 to 2017. Since 1900, 11,095 earthquakes with a magnitude of 6.0 and greater have occurred in the world. Shaking and ground rupture are the main effects created by earthquakes, principally resulting in more or less severe damage to buildings and other rigid structures.

In the past few decades, underground structures such as tunnels, metro stations, and underground parking stations have become major components of any transportation system. They are more and more frequently constructed to facilitate different needs in a wide range of engineering applications, including subways and railways, highways, material storage, and sewage and water transport. Historically, underground structures have proved to experience a lower rate of damage than surface structures subjected to both static and seismic loadings. Therefore, they were assumed to be seismic resistant due to being situated deep within rock/soil layers, especially for the mountain tunnels (Dowding & Rozen, 1978; Sharma & Judd, 1991; Towhata, 2008). Nevertheless, if a tunnel located close to an earthquake fault experiences strong-magnitude shaking and has special geologic or construction conditions, the tunnel will be damaged (Yoshikawa et al., 1981). Many instances with noticeable seismic damage were reported to indicate that mountain tunnels could be damaged at different levels by earthquakes, such as the 1923 Kanto earthquake, the 1995 Kobe earthquake, the 1999 Chi-Chi earthquake, the 2004 Niigataken-Chuetu earthquake, the 2007 Niigataken Chuetu-Oki earthquake, the 2008 Wenchuan earthquake, and the 2016 Kumamoto earthquake. Table 1.1 lists the past damage to mountain tunnels

DOI: 10.1201/9781003401599-1

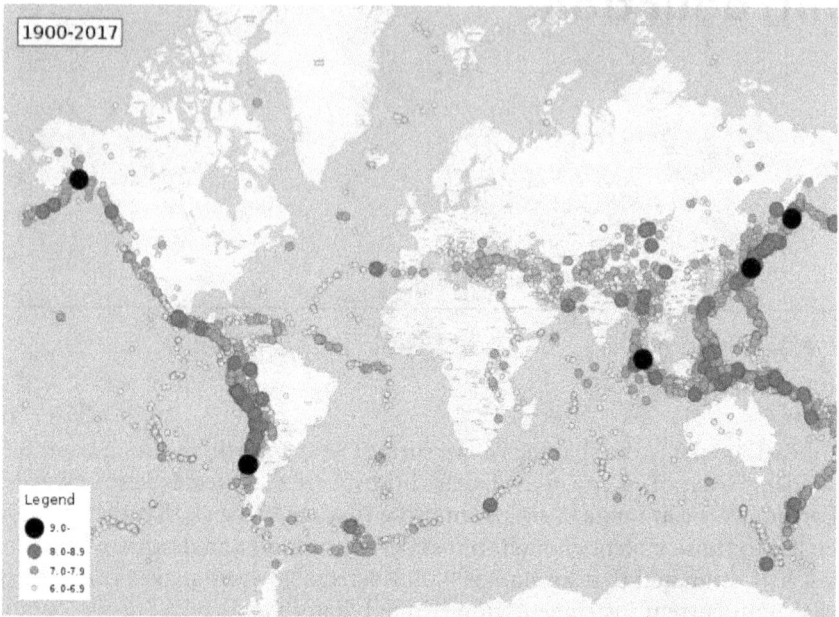

Figure 1.1 Map of earthquakes in the world with a magnitude of 6.0 and greater (1900–2017).

Source: Wikipedia, 2018. Data source: Search Earthquake Archives, USGS KML data.

Table 1.1 Past Damage to Mountain Tunnels by Earthquakes

Year	Name	Magnitude	Tunnel Performance
1906	San Francisco (United States)	8.3	Extensive; most severe damage to 2 tunnels crossing San Andreas Fault
1923	Kantō (Japan)	7.9	Extensive; most severe damage to more than 100 tunnels in the southern Kanto area
1927	Kita-Tango (Japan)	7.3	Very slight damage to 2 railway tunnels in the epicentral region
1930	Kita-Izu (Japan)	7.3	Very severe damage to 1 railway tunnel due to an earthquake fault
1948	Fukui (Japan)	7.1	Severe damage to 2 railway tunnels within 8 km of the earthquake fault
1952	Tokachi-Oki (Japan)	8.2	Slight damage to 10 railway tunnels in Hokkaido
1952	Kern (United States)	7.7	Severe damage to 4 railway tunnels
1961	Kita-Mino (Japan)	7.0	Cracking damage to a couple of aqueduct tunnels

1964	Niigata (Japan)	7.5	Extensive damage to about 20 railway tunnels and 1 road tunnel
1968	Toikchi-Oki (Japan)	7.9	Slight damage to 23 railway tunnels in Hokkaido
1970	Tonghai (China)	7.7	Severe damage to road tunnels, especially the portal collapse
1971	Los Angeles (United States)	6.6	Several damage to mountain tunnels crossing Thelma Fault; slight damage to 3 mountain tunnels
1978	Izu-Oshima-Kinkai (Japan)	7.0	Very severe damage to 9 railway and 4 road tunnels in a limited area
1978	Miyagiken-Oki (Japan)	7.4	Slight damage to 6 railway tunnels, mainly in Miyagi Prefecture
1982	Urakawa-Oki (Japan)	7.1	Slight damage to 6 railway tunnels near Urakawa
1983	Nihonkai-Cyubu (Japan)	7.7	Slight damage to 8 railway tunnels in Akita, etc.
1984	Naganoken-Seibu (Japan)	6.8	Cracking damage to 1 hydraulic power tunnel
1987	Chibaken-Toho-Oki (Japan)	6.7	Damage to the wall of 1 railway tunnel at the Kanagawa-Yamanashi border
1993	Notohanto-Oki (Japan)	6.6	Severe damage to 1 road tunnel
1993	Hokkaido-Nansei-Oki (Japan)	7.8	Severe damage to 1 road tunnel due to a direct hit of a falling rock
1995	Hyogoken-Nanbu (Japan)	7.2	Damage to over 20 tunnels, of which about 10 required repair and reinforcement
1999	Chi-Chi (Taiwan China)	7.6	Damage to about 57 tunnels, of which 14 were severely damaged, 11 were moderately damaged, and 23 were slightly damaged
2004	Niigataken-Chuetsu (Japan)	6.8	Damage to about 50 tunnels, of which about 25 needed reinforcement or repair
2007	Niigataken Chuetu-Oki (Japan)	6.8	Damage to about 21 tunnels, of which 5 were severely damaged, 13 were moderately damaged, and 4 were slightly damaged
2008	Wenchuan (China)	8.0	Damage to about 55 mountain tunnels in the seismic active area, of which 10 collapsed or were severely damaged, 11 were moderately damaged, and 17 were slightly damaged
2016	Kumamoto (Japan)	7.3	Severe damage to 1 mountain tunnel, moderate/slight damage to 3 railway and road tunnels

Source: Modified after Asakura et al., 2000; Shimizu et al., 2005, 2007; Saito et al., 2007; Konagai et al., 2009; Jiang et al., 2010b; Wang et al., 2012; Okano et al., 2018; Isago & Kusaka, 2018; Zhang et al., 2018.

by earthquakes. It can be observed that the underground structures in the seismic-active area suffered from damages that ranged from minor cracking to even failure.

Therefore, considering the significance of underground structures in modern societies, all of these records indicate the urgent requirement for the investigation of the seismic performance of underground structures subjected to seismic loadings and their seismic design for future planning.

1.2 SEISMIC WAVE

1.2.1 Characteristics of Seismic Waves

When earthquakes occur, the suddenly released energy produces seismic waves. Seismic waves are the manifestation of dynamic stress changes, and can be regarded as stress waves (Chapman, 2004). There are two basic types of seismic waves: the body waves that travel through the interior of the earth and the surface waves that travel along the ground surface. Body waves can be subdivided into P-waves (primary waves) and S-waves (secondary waves). The P-wave causes a series of compressions and dilations of the medium and induces particle movements parallel to its propagation direction. The S-wave causes shearing deformations of the medium and induces particle motions perpendicular to its propagation direction. An S-wave is classified as S_H-wave if it is polarized in the horizontal plane, or as S_V-wave if polarized in the vertical plane. Comparing with the P-wave, the S-wave typically travels more slowly in geomaterials, but has a greater impact on ground surface movements (Day, 2002). Surface waves such as Love waves and Rayleigh waves usually travel more slowly than body waves, with lower frequencies and longer durations. They diminish quickly as the depth increases.

In earthquake engineering, it is usually difficult to distinguish between the different types of seismic waves that could impact the site. Instead, the combined effect of waves in terms of producing a peak ground acceleration is of primary interest. Seismograms, which are records of ground motions during earthquakes, are commonly utilized to describe the seismic characteristics. A typical seismogram obtained during the M9.0 Great Tohoku earthquake in Japan (March 11, 2011) is shown in Figure 1.2a, indicating the variation of ground acceleration with time. It usually has three components, representing the ground motions in the directions of north-south (NS), east-west (EW), and up-down (UD), respectively.

For a specific seismic record (see Figure 1.2b), a further analysis is necessary to obtain some important properties of the seismic wave, including the maximum amplitude, the duration, the frequency (or period), the wave number, and the wavelength (Osaki, 1984). The maximum amplitude and the duration of a seismic wave can be obtained directly from the seismogram.

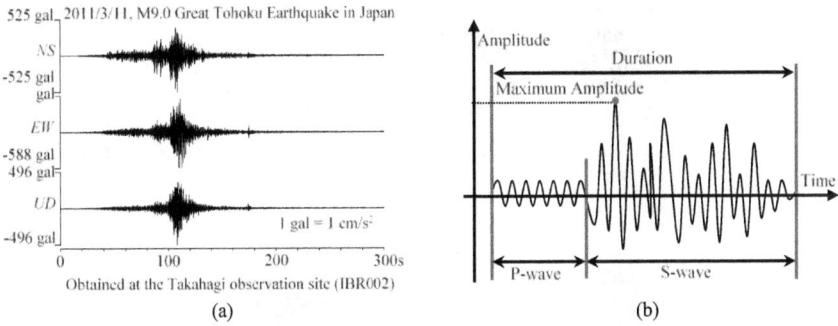

Figure 1.2 A typical seismogram and the characteristics of seismic waves. (a) A typical seismogram (from *K-NET*); (b) characteristics of seismic wave.

The frequency cannot be derived directly due to the complicated waveform of a real seismic wave. In engineering analysis, the dominant frequency f_d is usually used to represent the frequency characteristic of seismic wave, and can be obtained through the Fourier transform. The Fourier transform converts a wave function in the time domain into a form in the frequency domain, thereby, it can be used to investigate the frequency spectrum of a real seismic wave. Oppositely, the inverse Fourier transform expresses a frequency domain function in the time domain. The Fourier transform and its inverse form are given by

$$F(\omega) = \int_{-\infty}^{\infty} f(t)e^{-i\omega t}\,dt,\ f(t) = \frac{1}{2\pi}\int_{-\infty}^{\infty} F(\omega)e^{i\omega t}\,d\omega \qquad (1.1)$$

where $F(\omega)$ and $f(t)$ represent the functions of waves in the frequency and time domains, respectively; ω is the angular frequency and $\omega = 2\pi f$ (f is the frequency); and t denotes the time. After f_d is obtained, the wave number (K) and wavelength (λ) can be expressed in terms of the f_d and the wave velocity (V) as:

$$K = 2\pi / \lambda = 2\pi f_d / V = \omega / V,\ \lambda = 2\pi / K = 2\pi V / \omega = V / f_d \qquad (1.2)$$

1.2.2 Investigation of Seismic Waves at the Deep Rock Formation

Comparing with artificial seismic waves and simple harmonic waves, the real seismic wave taken from similar sites is more representative to the real situation when a rock foundation is subjected to earthquake loads. The seismic records obtained from observation sites during an earthquake represent the seismic motions of the ground surface. When earthquake occurs, seismic

Layer No.	Coordinate	Propagation direction	Properties of the rock layer	Thickness
1	u_1 ρ x_1		G_1 η_1 ρ_1	h_1
. . .	u_2 x_2			
m	u_m x_m		G_m η_m ρ_m	h_m
$m+1$	u_{m+1} x_{m+1}		G_{m+1} η_{m+1} ρ_{m+1}	h_{m+1}
. . .	u_{m+2} x_{m+2}			
N	u_N x_N		G_N η_N ρ_N	$h_N{=}\infty$

Motion of particle

Reflected wave   Incident wave

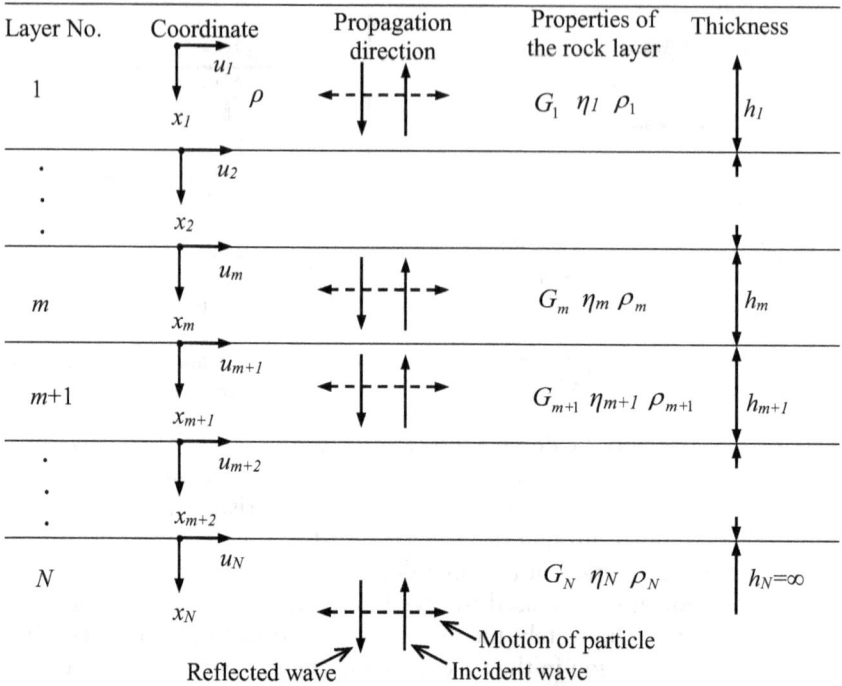

Figure 1.3 Theory of repeated seismic reflection for investigating the seismic motions at the deep rock formation.

Source: After Structure and Planning Institute of Japan, 2005a.

waves propagate from the deep rock formation to the ground surface. To reproduce such a process in numerical simulations, it is necessary to investigate the seismic motion at the deep rock formation, which can be input onto the bottom boundary of the foundation model. The theory of repeated seismic reflection assumes that rock (or soil) layers distribute horizontally in an infinite half space, which are constituted by homogenous and isotropic geomaterials, as shown in Figure 1.3. The origins of local coordinates are located at the upper surface of each layer, with the positive direction of the x-axis going downward perpendicularly. During the upward propagation of simple harmonic waves, the horizontal displacement of rock particles $u(x, t)$ induced by the shear wave, which depends on the coordinate x and the time t, should satisfy the one-dimensional equation of motion, as expressed by

$$\rho\frac{\partial^2 u(x,t)}{\partial t^2} = G\frac{\partial^2 u(x,t)}{\partial x^2} + \eta\frac{\partial^3 u(x,t)}{\partial x^2 \partial t} \tag{1.3}$$

where ρ is the density of rock; G is the shear modulus of rock; and η is the viscosity coefficient. The general solution of the equation of motion can be written as

$$u(x,t) = Ee^{i(kx+\omega t)} + Fe^{-i(kx-\omega t)} \tag{1.4}$$

where k is the wave number; ω is the angular frequency; E is the amplitude of incident wave which propagates upwards; F is the amplitude of reflected wave that propagates downwards; and i is the imaginary unit.

The velocity $v(x, t)$ and the acceleration $a(x, t)$ of particles can be obtained by calculating the first-order and second-order partial derivatives of the displacement with respect to time, as given by Equation (1.5). Additionally, the shear stress $\tau(x, t)$ and the shear strain $\gamma(x, t)$ can also be derived based on the elastic theory, in the form of Equation (1.6).

$$v(x,t) = \dot{u}(x,t) = \frac{\partial u}{\partial t}, a(x,t) = \ddot{u}(x,t) = \frac{\partial^2 u}{\partial t^2} \tag{1.5}$$

$$\tau(x,t) = G\frac{\partial u}{\partial x} + \eta\frac{\partial^2 u}{\partial x \partial t}, \gamma(x,t) = \frac{\partial u}{\partial x} \tag{1.6}$$

For any layer (e.g., the mth layer), the displacements and the shear stresses at the upper and lower boundaries ($x = 0$ and $x = h_m$) of the layer can be expressed as

$$u_m(x=0) = (E_m + F_m)e^{i\omega t}$$
$$u_m(x=h_m) = (E_m e^{ik_m h_m} + F_m e^{-ik_m h_m})e^{i\omega t} \tag{1.7}$$

$$\tau_m(x=0) = ik_m G_m^*(E_m - F_m)e^{i\omega t}$$
$$\tau_m(x=h_m) = ik_m G_m^*(E_m e^{ik_m h_m} + F_m e^{-ik_m h_m})e^{i\omega t} \tag{1.8}$$

where G_m^* is the complex shear modulus ($G_m^* = G_m + i\omega\eta_m$). The compatibility conditions of displacement and stress on interfaces between adjacent layers need to be satisfied. For example, at the interface between the mth and $(m + 1)$th layers, the boundary conditions can be given as: $u_m(x = h_m) = u_{m+1}(x = 0)$, $\tau_m(x = h_m) = \tau_{m+1}(x = 0)$. Then the relation of wave amplitudes between two adjacent layers can be given by

$$\begin{bmatrix} E_{m+1} \\ F_{m+1} \end{bmatrix} = \frac{1}{2}\begin{bmatrix} (1+\alpha_m)e^{ik_m h_m} & (1-\alpha_m)e^{-ik_m h_m} \\ (1-\alpha_m)e^{ik_m h_m} & (1+\alpha_m)e^{-ik_m h_m} \end{bmatrix}\begin{bmatrix} E_m \\ F_m \end{bmatrix} \tag{1.9}$$

where α_m represents the impedance of the mth layer, with the complex form of

$$\alpha_m = \frac{k_m G_m^*}{k_{m+1} G_{m+1}^*} = \left(\frac{\rho_m G_m^*}{\rho_{m+1} G_{m+1}^*} \right)^{1/2} \quad (1.10)$$

The seismic amplitudes at any layer can be derived from those of the adjacent layers. Therefore, if the wave amplitudes at the ground surface is known ($E_1 = F_1$, due to the fact that the shear stress on the ground surface is zero), the amplitudes in all layers can be derived, and the displacement function can be solved.

It should be noted that the expressions mentioned above are derived with the assumption of harmonic waves. When considering the real seismic waves, the situation will become much more complicated. In this case, the real seismic waves need to be firstly translated into the superposition form of multiple harmonic waves by performing the Fourier transform. Seismic waves after the Fourier transform are in the frequency-domain form, which can be translated back into the time-domain form through the inverse Fourier transform. The seismic motions at every layer caused by the real seismic waves can be obtained by superposing all the motions caused by the componential harmonic waves. The code of k-SHAKE, developed by the Structure and Planning Institute of Japan (2005a) on the basis of the theory of repeated seismic reflection, is one convenient way to investigate the seismic wave at the deep rock formation with a certain depth. The numerical model using the code of k-SHAKE is shown in Figure 1.4. The symbols $2E_0$

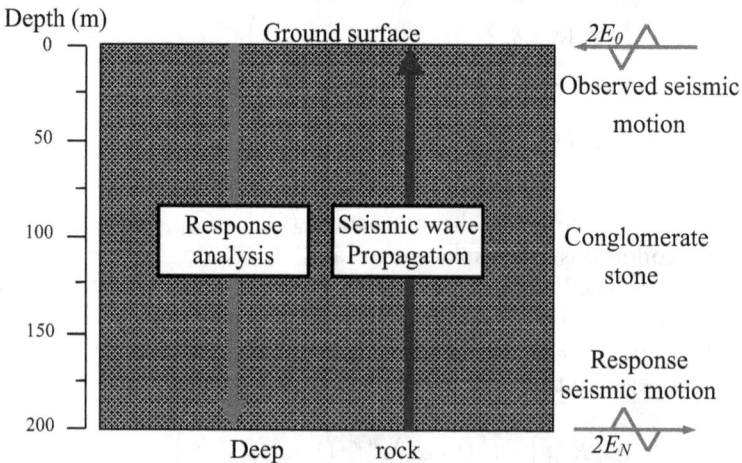

Figure 1.4 Numerical model for investigating the seismic wave at the deep rock formation with a depth of 200 m.

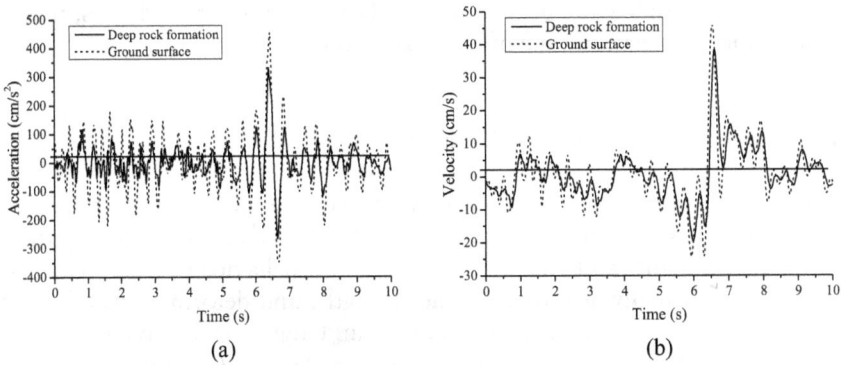

Figure 1.5 Seismic waves at the deep rock formation with a depth of 200 m. (a) Acceleration wave for FEM simulation; (b) velocity wave for DEM simulation.

and $2E_N$ represent the seismic motions on the ground surface and at the deep rock formation, respectively. In the model, $2E_0$ was assigned with the ground acceleration-time data of the observed ground seismic wave. Then the seismic waves (including the acceleration and velocity waves) at the deep rock formation $2E_N$ can be back-calculated by using the theory of repeated seismic reflection mentioned above.

Figure 1.5 illustrates the seismic waves at the deep rock formation with a depth of 200 m. The rock formation as deep as 200 m was chosen as the bottom boundary of the model, according to the dimensions of the nuclear buildings. The lateral boundaries of model were assumed to be semi-infinite rock regions. In this model, the physico-mechanical properties of the rock mass were set as follows: the density of rock mass was 2660 kg/m^3; and the propagation velocity of shear wave was 3363 m/s. For geomaterials, the damping ratio generally falls into the range from 2% to 5% (Biggs, 1964). The damping ratio of the rock mass here was set as 3%, according to the results of in-situ tests. According to the calculation principles of different numerical methods, the acceleration wave was adopted in the finite element method (FEM) simulations, while the velocity wave was used in the distinct element method (DEM) simulations. For the purpose of comparison, the acceleration and velocity waves on the ground surface are also shown in Figure 1.5. It can be found that the acceleration and velocity waves at the deep rock formation (200 m) have similar waveforms to those observed on the ground surface. However, the amplitudes at peaks and troughs of seismic waves increase during the wave propagation process, due to the amplification effect of the intact bedrock. For example, the maximum and minimum amplitudes of the acceleration wave at deep rock formation are 330.4 cm/s^2 and –266.3 cm/s^2, respectively, which are less than those observed on the ground surface (454.7 cm/s^2 and -347.13 cm/s^2, respectively). Additionally, there is a little time gap between waves obtained at different depths, representing the propagation time of the seismic wave.

1.3 MECHANICAL BEHAVIOR OF ROCK MASS

Fractured rock masses usually involved in engineering practices are composed of an assemblage of intact rock blocks, which are separated by fractures such as joints and cracks. The mechanical behavior of fractured rock masses is principally governed by the strengths and deformations of rocks and fractures. To investigate the failure and dynamic behavior of rock masses, it is necessary to begin with a general discussion about the properties of rocks and fractures as well as the basic mechanical performances of rock mass under various loading conditions. In this chapter, a general description of rock mass is provided. Properties of rocks and fractures as well as some related mechanical models are briefly depicted. Then the failure behavior of rock masses and some failure theories are described. The fundamental issues concerning the seismic behavior of rock masses are discussed in Chapters 2–5.

1.3.1 General Description of Rock Mass

A rock mass is a heterogeneous and discontinuous material commonly encountered in geotechnical engineering. The rock mass is composed of a system of rock blocks and rock fractures (e.g., joints, faults, bedding planes, cracks, cavities), which behave in mutual dependence as a unit in the material (Matula & Holzer, 1978). Due to the great diversity in the composition of intact rocks and in the properties and extent of fractures, the rock mass exhibits a high level of complexity and uncertainty in its structure, and a high level of anisotropy and nonlinearity in its mechanical properties (Palmström, 1995, 1996). To estimate the mechanical behavior of a rock mass, a comprehensive understanding of the mechanical structure of rock mass is required, involving the properties of intact rocks as well as the distribution and properties of fractures.

Intact rocks are generally made up of an aggregate of crystals of different minerals and amorphous particles joined by varying amounts of cementing materials (Jaeger et al., 2007). Boundaries between crystals are weaknesses in the rock material and lead to the discreteness of rocks. Besides the crystal boundaries, there are also some other types of defects existing in rocks, such as micro-cracks and micro-cavities, which may be generated by environmental stresses (Sang-Eun et al., 2006). Under external loads, the total deformation of rocks consists of the deformations of crystals and defects. Since defects have much lower strengths than crystals and can induce rock failure, they play a significant role in the mechanical behavior of rocks.

The mechanical performance of fractured rock masses is much more complicated than that of intact rocks due to the involvement of fractures. Fractures divide the rock mass into a large number of sub-domains or blocks, whose sizes and interactions dominate the overall behavior of the rock mass (Jing, 2003). The existence of rock fractures causes the stress redistribution and localized stress concentration in the host rock mass when subjected to external loads, which may subsequently lead to large plastic deformations and the failure of the rock mass. Besides, fractures can open or close in the normal direction to their initial planes, and slide along their planes, making a major contribution to the overall deformation of the rock mass. Due to the controlling effects of fractures on the structure and mechanical behavior of rock masses, their geometrical and mechanical properties need to be investigated sufficiently in engineering practices.

The mechanical behavior of fractured rock masses also depends upon the scale and the detail involved in analysis, which is generally referred to as the scale effect (Da Cunha, 1993; Bamant, 2000; Fardin et al., 2001). At various scales, rock masses have different structures and contain different number of fractures with various shapes and extents. On a large scale of most concern in geotechnical engineering with dimensions from several meters to kilometers, the rock mass contains a great number of fractures; thereby, it usually has a lower strength than the small-scale rock masses such as samples used in laboratory tests (see Figure 1.6). In engineering analysis, the proper scale

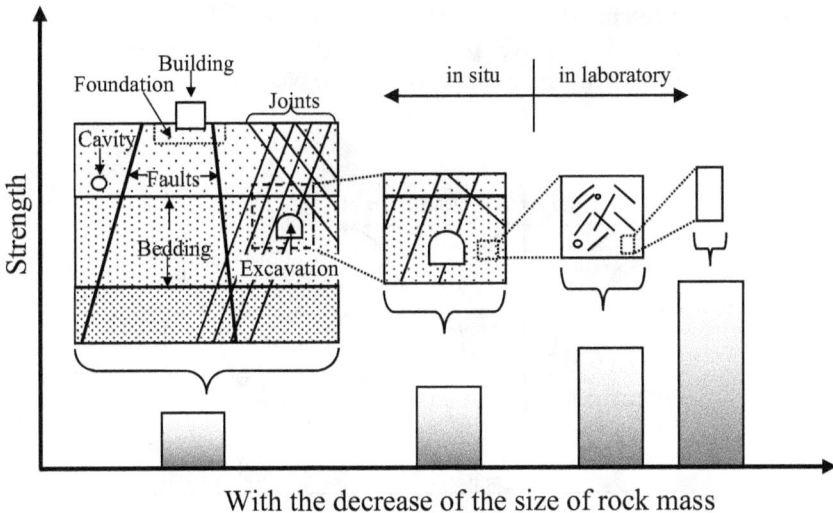

Figure 1.6 Scale effect of rock mass and the variation of strength.

of a mechanical model for a specific problem is important to perform a reliable assessment. To establish a reliable model, the rock mass structure needs to be investigated by means of geological and geophysical surveys, including lithologies, boundaries of rock types, and the distribution of fractures (especially the major joints and faults). Then mechanical properties of intact rocks and fractures need to be measured through laboratory tests on intact or fractured rock samples to provide reliable parameters for analysis.

1.3.2 Properties of Rock Mass

1. Mechanical properties of intact rocks

Mechanical properties of intact rocks are usually investigated by performing laboratory tests on samples under various loading conditions. The stress-strain curves of rocks obtained from tests represent the load-deformation process of rocks and remain as the most fundamental information for understanding the mechanical behavior of rocks. Here the mechanical properties of intact rocks will be briefly discussed by analyzing the typical stress-strain curves of rocks under uniaxial and triaxial compression.

Figure 1.7 shows the typical complete stress-strain curve for a rock sample under uniaxial compression. The load-deformation process can be divided into four stages (Cai et al., 2002). In the first stage (oa portion), the curve has an initial portion which is concave upwards with an increasing slope. This is due to the fact that at the beginning of loading, pores and micro-cracks in the rock material are closed under the compressive stress. The stress at the point a (σ_a) represents the closure stress for micro-defects. The volumetric strain in this stage is positive, indicating that the rock volume decreases as stress increases. With the increase of compressive stress, the stress-strain

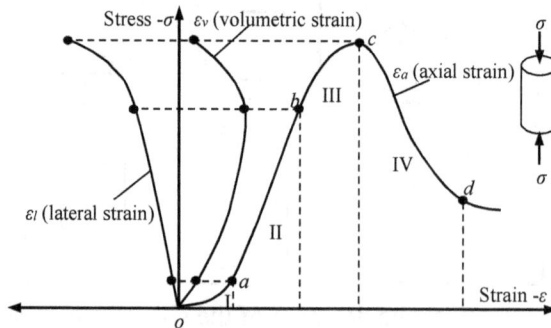

Figure 1.7 Complete stress-strain curve for a rock under uniaxial compression.

curve enters its second stage (*ab* portion). The rock behaves approximately in a linear manner, and the slope of the curve represents the elastic modulus of the rock. The rock volume decreases continuously in this stage while its compression ratio gradually declines, which can be found from the slope variation in the stress–volumetric strain curve. The stress at point *b* (σ_b) represents the ultimate elastic strength of the rock. After that, the stress-strain curve enters the third stage of plastic deformation (*bc* portion). With the increase of stress, the failure of rock occurs in this stage, in terms of crack initiation and propagation, producing large plastic deformation, which leads to the decrease of the curve slope. The volumetric strain turns to decrease in this stage, indicating that the volume of rock changes from compression to dilation. Finally, the stress reaches its maximum at the end of this stage (σ_c), which represents the uniaxial compressive strength of the rock. After the peak strength, the stress-strain curve enters the fourth stage (*cd* portion), namely the post–deformation-failure stage. In this stage, the rock fracture becomes more severe, and the load-bearing capability of rock decreases continuously. After undergoing a large deformation, the stress gradually reaches a constant value at point *d* (σ_c), which represents the residual strength of the rock.

Rocks in nature are generally subjected to three-dimensional stresses. The mechanical properties of rocks under triaxial stress conditions can be studied by performing triaxial compression tests. There are generally two types of triaxial compression tests: the first type uses cylindrical samples with the confining stress applied on their side surfaces, and the stress state of samples is $\sigma_1 > \sigma_2 = \sigma_3$; the second type uses rectangular samples with different normal stresses applied on each pair of opposite faces of samples, and the stress state is $\sigma_1 > \sigma_2 > \sigma_3$, which represents a more general stress circumstance.

Typical stress-strain curves of rocks in triaxial compression tests ($\sigma_1 > \sigma_2 = \sigma_3$) are shown in Figure 1.8. At the beginning of loading, the stress-strain curve of the hard rock exhibits a linear elastic portion, and its slope that represents the elastic modulus is nearly independent of the confining stress. After the axial stress exceeds a critical value (before the peak strength), the rock starts to dilate as a result of internal fracturing (Brady & Brown, 2004; Elliott, 1982). The confining stress has significant effects on the mechanical and deformational behavior of rocks, which can be summarized as follows: (1) as the confining stress increases, the peak strength increases and the volumetric dilation decreases; (2) the increasing confining stress can lead to a transition from typically brittle to fully ductile behavior of rocks with the introduction of large plastic deformations; (3) the portion incorporating the peak of the stress-strain curve becomes flat and wide as the confining pressure increases; and (4) the stress drop from the peak strength to the residual strength in the post-peak stage decreases and even disappears at very high confining stresses.

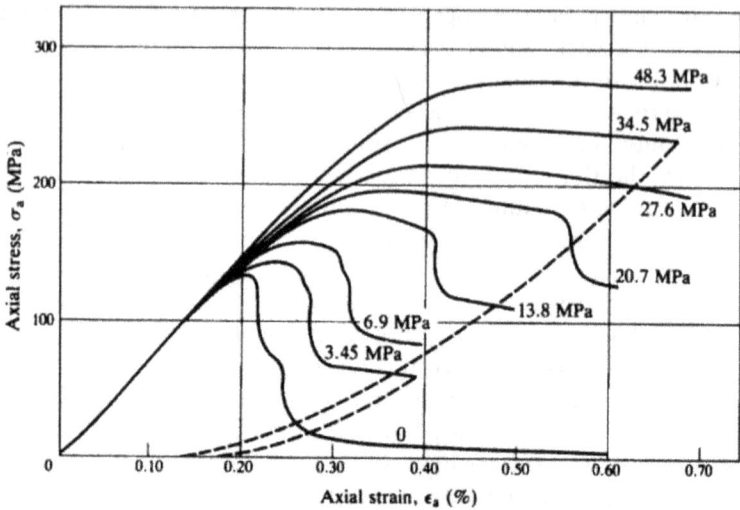

Figure 1.8 Axial stress-axial strain curves obtained in triaxial compression tests on Tennes-
see marble at various confining pressures.

Source: Brady & Brown, 2004.

Triaxial compression tests considering $\sigma_1 > \sigma_2 > \sigma_3$ have received less attention due to the difficulties caused by the end effects when applying stresses on the sample faces, which may reduce the reliability and accuracy of test results. The current results have indicated the influence of σ_2 (the intermediate principal stress) on the load-deformation behavior of rock (Paterson, 1978; Brady & Brown, 2004). Generally, the peak strength of a rock increases as the σ_2 increases with a constant σ_3.

2. Geometrical and mechanical properties of fractured rock mass

Rock fractures refer to any interruption in the rock mass with effectively zero tensile strength (Farmer, 1983). Those fractures, especially some large and persistent ones, influence greatly the strength and deformability of the rock mass. Therefore, in the engineering analysis, a thorough understanding of the geometrical and mechanical properties of fractures is necessary.

The geometrical properties of fractures mainly include the orientation, the spacing and the persistence which govern the distribution of fractures in the rock mass, and the fracture roughness and fracture aperture which affect the strength, deformation, and permeability of fractures. Based on the description given by Hudson and Harrison (1997), those geometrical properties are described as follows. (1) The fracture orientation determines the shape of separated rock blocks and controls greatly the deformation of engineering structures. It can be represented by the strike angle and dip angle of a fracture. In engineering surveys, the strike and dip angles of all fractures are usually plotted on a circular plane by using the method of stereographic projection to give a preferred or 'mean' orientation of fractures (Phillips, 1971; Priest, 1985, 1993). (2) The fracture spacing refers to the distance between adjacent fractures. It defines the size of rock blocks in a rock mass, and can be used to evaluate the quality of a rock mass, which is usually accompanied by using the RQD method (Deere, 1963). (3) The fracture persistence denotes the areal extent or size of a fracture in its own plane and can be characterized by using the classification scheme of ISRM (1978). The persistence of joints sets has a controlling effect on the large-scale sliding or the 'down-stepping' failure of rock masses. (4) The fracture roughness refers to the deviation of a fracture surface relative to the perfect planarity, and it principally governs the shear deformation and shear strength of fractures. At present, the fracture roughness is usually described by using the joint rough-ness coefficient (JRC) proposed by Barton and Choubey (1977). (5) The fracture aperture refers to the openness between two surfaces of a fracture, which concerns greatly the mechanical and hydraulic characteristics of the fracture.

Mechanical properties of fractures can be investigated via the stress-displacement curves obtained in various loading conditions. Figure 1.9 shows the typical curves of fractures obtained from normal loading and direct shear tests on samples containing single fractures (Hudson & Harrison, 1997). Under compressive stress (Figure 1.9a), the fracture surfaces are gradually pushed together, with an obvious limit when the two surfaces are completely closed. During the loading process, the curve rises up continuously until the stress reaches a limit associated with the strength of the intact rock. Normal stiffness of the fracture (k_n), as defined by the slope of stress-displacement

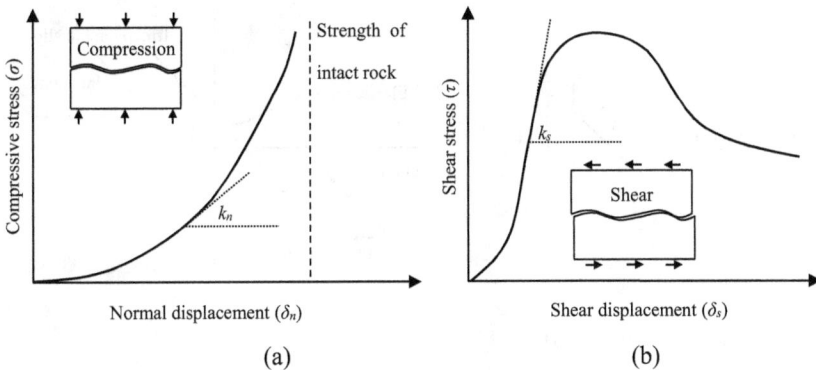

Figure 1.9 Stress-displacement curves of fractures under compressive and shear loads. (a) Normal stress vs. normal displacement; (b) shear stress vs. shear displacement.

curve, also increases with the closure of fracture surfaces due to the increased contact area. When a fracture is subjected to shear stress (Figure 1.9b), the corresponding curve has a similar shape to the stress-strain curve for intact rocks under compression, except that all failure is localized along the fracture. The curve goes up firstly with a certain slope, which represents the shear stiffness (k_s). After undergoing some degree of shear displacement, the shear stress reaches its peak that represents the shear strength of the fracture. Then some asperities on the fracture surface are sheared off, leading to the decrease of shear stress. Finally, the curve goes into the post–peak failure stage, with a residual stress caused by the frictional stress and the interlocking effect of asperities.

3. Mechanical models of fractured rock mass

In theoretical and numerical studies, the mechanical behavior of fractures can be represented by some constitutive models, such as the linear Coulomb model, the nonlinear Barton-Bandis model (Bandis et al., 1983, 1985; Barton et al., 1985), and the continuous-yielding joint model (Cundall & Hart, 1984; Cundall & Lemos, 1990), which describe the relationship between the stress and the fracture deformation. Here, the well-known Coulomb model and the Barton-Bandis model, which were commonly adopted in previous studies, are presented.

a. Linear Coulomb model

The Coulomb friction and linear deformation joint model (Figure 1.10) is the simplest model for describing the mechanical and deformation behavior of rock fractures, but it has the widest applications in rock mechanics. It is appropriate for smooth fractures like faults at residual strength, which are

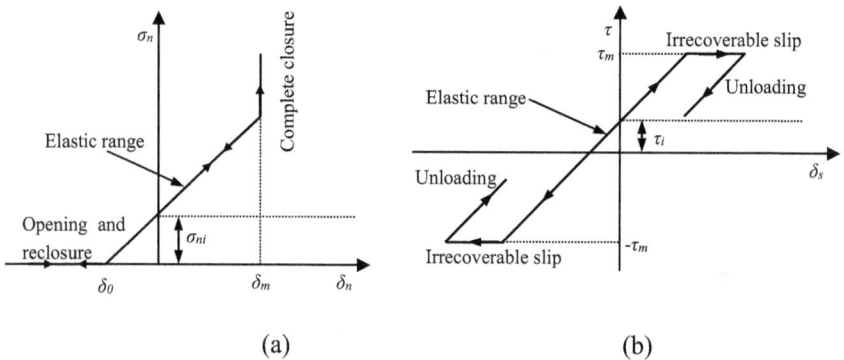

(a)   (b)

Figure 1.10 The linear Coulomb model for fractures. (a) Normal stress vs. normal displacement; (b) shear stress vs. shear displacement.

Source: Brady & Brown, 2004.

non-dilatant in shear, and it can provide a useful and reliable estimate for the static and dynamic fracture responses (Brady & Brown, 2004). This model has two different expressions according to the loading conditions (normal loading or shear loading), as given by

$$\sigma_n = k_n \delta_n + \sigma_{ni}, \, \delta_0 < \delta < \delta_m \qquad (1.11)$$

$$\tau = k_s \delta_s + \tau_i, \, -\tau_m < \tau < \tau_m \qquad (1.12)$$

In the case of normal loading (Figure 1.10a), the rock fracture is subjected to an initial compressive stress σ_{ni}. With the increase of compressive stress, the fracture undergoes linear elastic closure with a constant normal stiffness k_n, up to the maximum closure of fracture δ_m. After the fracture is completely closed, the fractured rock behaves just like intact rocks, with the same compressive strength. When the fracture is subjected to an increasing tensile stress, the tensile stress will firstly counteract the effect of σ_{ni}. After the total stress reaches the tensile strength of fracture (0), which corresponds to the displacement denoted by δ_0 ($\delta_0 = -\sigma_{ni}/k_n$), the fracture will be separated. In the case of shear loading (Figure 1.10b, the initial shear stress is τ_i), the shear stress–shear displacement curve is linear with the constant shear stiffness of k_s and reversible up to a limiting shear stress τ_m that depends on the normal stress, and then turns to be perfectly plastic with the occurrence of irrecoverable slip along the fracture plane. The shear load reversal after plastic yield is accompanied by permanent shear displacement and hysteresis.

b. Nonlinear Barton-Bandis model

The Barton-Bandis model can be used to describe the nonlinear deformation behavior of fractures. This model was established on the basis of a large number of laboratory experiments on matched fractures, where a hyperbolic function was used to predict the relationship between normal stress (σ_n) and fracture closure (δ_n), as expressed by

$$\sigma_n = \delta_n / (a - b\delta_n) \qquad (1.13)$$

where a and b are constants. The inverse of a is equal to the initial normal stiffness of the fracture (k_{ni}), and the asymptote a/b in the σ_n-δ_n plane represents the maximum allowable closure (δ_m) of the fracture. In this model, the normal stiffness of the fracture (k_n) increases as the normal stress increases, as given by

$$k_n = k_{ni} \left[1 - \sigma_n / (\delta_m k_{ni} + \sigma_n) \right]^{-2} \qquad (1.14)$$

Barton et al. (1985) proposed the empirical expressions to estimate the initial normal stiffness and the maximum allowable closure of fractures, as expressed by

$$k_{ni} = 0.02\left(JCS / E_i\right) + 2JRC - 10, \ \delta_m = A + B(JRC) + C\left(JRS / E_i\right)^D \qquad (1.15)$$

where JRC is the joint roughness coefficient; JCS is the compressive strength of fracture wall; E_i is the initial aperture of fracture; and A, B, C, and D are constants that depend on the previous stress history.

The Barton-Bandis model is capable of modeling the nonlinear behavior of rock fractures and involves some geometrical and mechanical properties of fractures, such as the JRC and JCS. However, because the parameters in this model are difficult to be obtained and are experience dependent to some extent, this model may lead to some irregularities in numerical simulations (Brady & Brown, 2004).

1.3.3 Failure of Rock Mass

Failure of rocks, as a great threat to the safety of engineering structures, generally results from the initiation and coalescence of micro-defects as well as the growth of macro-fractures existing in rock masses. Under different stress conditions, rocks fail in different patterns corresponding to different mechanical behavior of rocks (Hallbauer et al., 1973). To date, a number of failure criteria have been proposed to predict rock failure, such as the Mohr-Coulomb criterion (Coulomb, 1773), the Griffith criterion (Griffith, 1921, 1924), and the Hoek-Brown empirical criterion (Hoek & Brown, 1980a, 1980b; Hoek, 1990; Hoek et al., 1992). Besides, the theories of rock fracture mechanics have been well developed to describe the crack growth in rock masses. Here, the basic patterns of rock failure, the Mohr-Coulomb and Griffith failure criteria, are presented.

I. Failure patterns of rocks

Based on the descriptions of Jaeger et al. (2007) and Cai et al. (2002), the failure of rocks falls into seven basic patterns depending upon the loading conditions (Figure 1.11). Under uniaxial compression, three major failure patterns can be observed: (1) somewhat irregular longitudinal splitting produced by the lateral tensile stress in the sample (Figure 1.11a); (2) the failure occurs along two conjugate shear planes located symmetrically with respect to the axial direction (Figure 1.11b); and (3) the fracture along a single inclined plane produced by shear stress (Figure 1.11c). The inclination of the shear failure plane to the horizontal direction in Figure 1.11b and Figure 1.11c can be given by $\alpha = \pi/4 + \phi/2$, where ϕ is the angle of internal

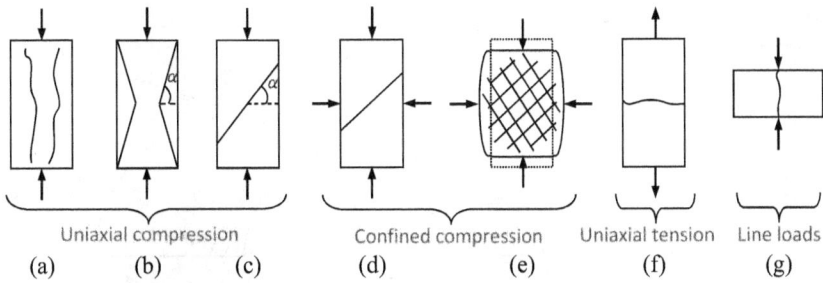

Figure 1.11 Failure patterns of rocks under various loading conditions. (a) Longitudinal splitting; (b) conjugate shear fracture; (c) shear fracture along a single plane; (d) shear fracture along a single plane; (e) multiple shear fractures; (f) extension fracture; and (g) extension fracture under opposing line loads.

Source: After Jaeger et al., 2007.

friction. According to the explanation of Paterson (1978), the shear failure under uniaxial compression seems to be an experimental artifact due to the sample ends being constrained against rotation. In the confined compression test with a moderate confining pressure, the confining stress suppresses the longitudinal fracturing, and failure occurs along an inclined plane produced by shear stress (Figure 1.11d). As the confining pressure increases, the rock becomes fully ductile, and a network of small shear fractures appears, accompanied by large plastic deformations (Figure 1.11e). For a rock under uniaxial tension, an extension fracture typically occurs with a clean separation of the two halves of the sample (Figure 1.11f). Due to the low tensile capability of rocks, the tensile failure of rocks usually occurs abruptly. To obtain the tensile strength of rocks, the common method is to compress a slab of rock or a thin cylindrical sample by applying two opposing line loads. In this case, an extension fracture due to the lateral tensile stress appears (Figure 1.11g).

2. Mohr-Coulomb failure criterion

The Mohr-Coulomb failure criterion (Figure 1.12) is the simplest criterion to predict the rock failure, which is widely used in rock mechanics and engineering. This criterion assumes that the failure of a rock occurs along an inclined plane due to the shear stress acting along that plane. This criterion can be expressed by Equation (1.16), where σ_n and τ represent respectively the normal stress and the shear stress on the failure plane; c is the cohesion of rock; and φ is the angle of internal friction.

$$\tau \geq \sigma_n \tan \varphi + c \tag{1.16}$$

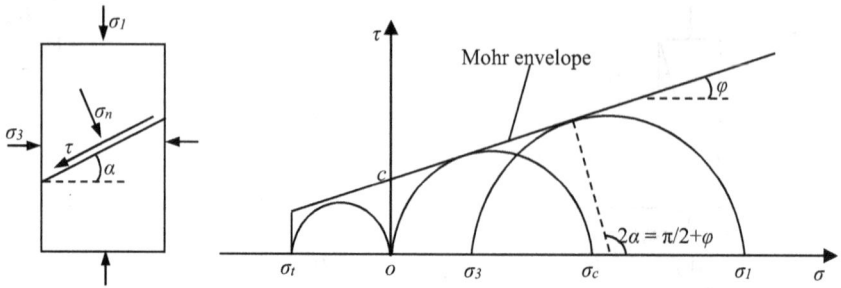

Figure 1.12 The Mohr-Coulomb failure criterion and the stress state.

The normal and shear stress on the failure plane can be derived from the principal stresses (σ_1 and σ_3) acting on the sample surface, therefore, the Mohr-Coulomb failure criterion can also be written in the form of principal stresses

$$\sigma_1 - \sigma_3 = (\sigma_1 + \sigma_3)\sin\varphi + 2c\cos\varphi \tag{1.17}$$

The failure/safety status of a rock under various stress conditions can be conveniently described by using Mohr's circles (see Figure 1.12). The envelope of Mohr's circles represents the critical stress state defined by the Mohr-Coulomb criterion. The σ-τ coordinates below the envelope imply that the rock is stable and no failure occurs. The σ-τ coordinates on the envelope represent limiting equilibrium, while the σ-τ coordinates above the envelope represent the failure conditions of rocks.

3. Griffith failure criterion

In rocks, a plenty of micro-cracks exist and control the failure behavior of rocks. To account for the effects of cracks, Griffith (1921, 1924) proposed a failure criterion to describe the failure of brittle materials with the presence of a collection of randomly oriented thin elliptical cracks. According to Griffith's theory, cracks start to propagate when the tensile strength is exceeded by the concentrated tensile stress which is induced at the tips of micro-cracks. For the simplest case considering a rock under remote tensile stresses (Figure 1.13a), the critical tensile stress (σ_f) to cause the rock failure can be given by

$$\sigma_f = \sqrt{keE/a} \tag{1.18}$$

where $k = 2/\pi$ is for plane stress or $k = 2(1 - v^2)/\pi$ is for plane strain (v is Poisson's ratio); e is the unit crack surface energy; E is the elastic modulus; and a is half the length of the crack.

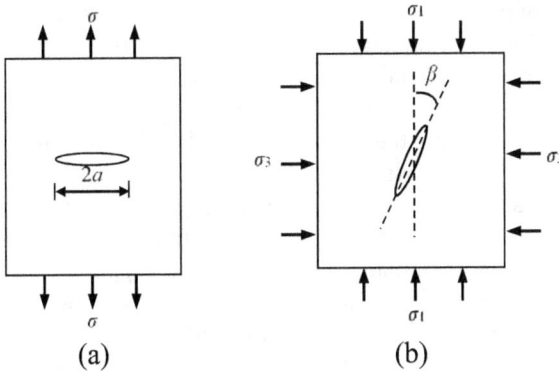

Figure 1.13 The Griffith failure criterion. (a) Under uniaxial tension; (b) under confined compression.

For the relatively complicated case that a plate containing an inclined crack is subjected to confined compression (Figure 1.13b), the Griffith criterion can be given by

$$\begin{cases} \sigma_3 = -\sigma_t & \text{if } \sigma_1 + 3\sigma_3 \geq 0 \\ (\sigma_1 - \sigma_3)^2 - 8\sigma_t(\sigma_1 + \sigma_3) = 0 & \text{if } \sigma_1 + 3\sigma_3 \leq 0 \end{cases} \qquad (1.19)$$

The crack will propagate along the direction of the maximum tensile stress, and the propagation angle to the axis of sample can be expressed by

$$\cos 2\beta = (\sigma_1 - \sigma_3) / [2(\sigma_1 + \sigma_3)] \qquad (1.20)$$

1.4 SEISMIC PERFORMANCE OF MOUNTAIN TUNNEL UNDER EARTHQUAKE FORCE

1.4.1 Performance of Mountain Tunnel Subjected to Seismic Events

Several researchers with efforts of site investigation have reported the seismic performance of tunnels (e.g., Duke & Leeds, 1959; Dowding & Rozen, 1978; Owen & Scholl, 1981; Sharma & Judd, 1991; Power et al., 1998; Wang et al., 2001; Chen et al., 2012; Wang & Zhang, 2013). Therefore, global databases are available currently. Dowding and Rozen (1978) created a database with 71 cases of seismic damage to tunnels in Japan and California and Alaska in the United States. Owen and Scholl (1981) extended the aforementioned database to a total of 127 cases. Sharma and Judd (1991) compiled a database with 192 cases of underground structure damage during 85 earthquakes. With a detailed description of damage characteristics and mechanism, Power et al. (1998) used the previous database, adding cases from more recent earthquakes (e.g., the 1995 Kobe earthquake) to further study the performance of bored tunnels. After investigating ten strong earthquakes, Chen et al. (2012) created a database for the damage situation of 81 mountain tunnels.

Even though such progress has been done for the seismic analysis, the classification criteria of seismic damages to tunnels have not been unified. Three forms of seismic damage to tunnels were suggested by Dowing and Rozen (1978): damage from earthquake-induced ground failure, damage by fault deformation, and damage by ground shaking or vibration. Wang et al. (2001) summarized damage patterns based on the characteristics and the distribution of the lining cracks, which include sheared-off lining, slope failure-induced tunnel collapse, longitudinal crack, transverse crack, inclined crack, extended cross crack, pavement or bottom cracks, wall deformation, and cracks that develop near the opening. Yashiro et al. (2007) classified the damage patterns into three types: damage to shallow tunnels, damage to tunnels in poor geological conditions, and damage to tunnels by fault slide. Wang et al. (2009b) illustrated eight major patterns of seismic damages: portal failure, longitudinal crack, transverse crack, inclined crack, shear failure of lining, pavement cracks, lining spalling, and groundwater inrush. Li (2012) classified the damage characteristics into six types: opening collapse, portal cracks, damage to the lining and surrounding rock, lining cracks and dislocation, invert damage and uplift, and primary supporting cracks and deformation. Based on ten well-documented earthquakes, Chen et al. (2012) summarized the six most common damage characteristics regarding the seismic performance of mountain tunnels: lining cracks (Figure 1.14a), shear failure of the lining (Figure 1.14b), collapse caused by slope failure (Figure 1.14c), portal cracking (Figure 1.14d), leakage (Figure 1.14e), and wall/invert damage (Figure 1.14f). Shen et al. (2014) analyzed typical seismic damage characteristics and mechanisms of mountain tunnels based on three different damage patterns: damage to the shallow tunnel, damage to the deep-buried tunnel structure, and damage to the pavement.

1.4.2 Influential Factors for Seismic Damages to the Mountain Tunnels

An explicit description of the seismic performance could contribute to an overall understanding and accurate grasp of the influential factors behind the damage. Earthquake effects on mountain tunnels can be grouped into two categories: ground shaking due to wave propagation and ground failure due to lateral spreading, landslides, and fault rupture (Dowding & Rozen, 1978; St. John & Zahrah, 1987; Hashash et al., 2001). The main factors

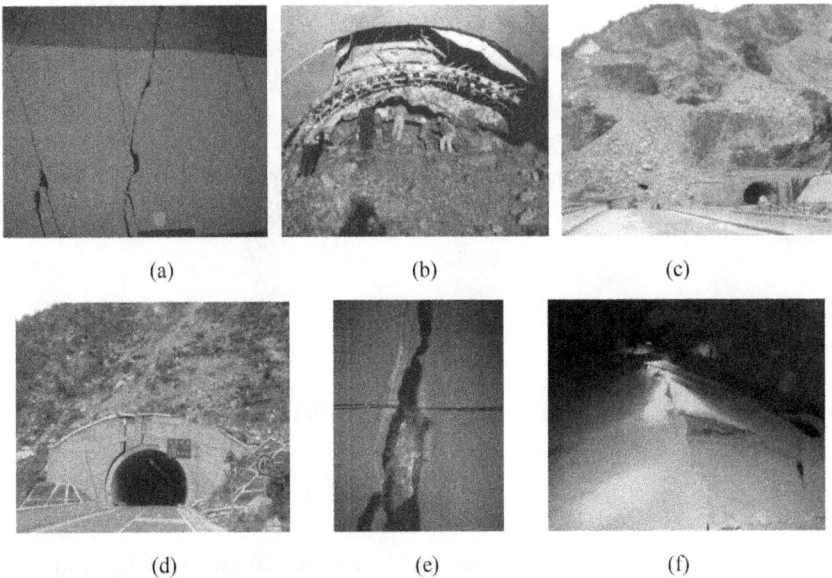

Figure 1.14 Six most common seismic damage characteristics of mountain tunnels subject to the earthquake. (a) Lining cracks (Shen et al., 2014); (b) shear failure of the lining (Li, 2012); (c) collapse caused by slope failure (Li, 2012); (d) portal cracking (Shen et al., 2014); (e) leakage (Chen et al., 2012); (f) invert damage (Yu et al., 2016b).

affecting the behavior of mountain tunnels during an earthquake can be summarized as three types: earthquake parameters, structural form, and geographic conditions on site (Chen et al., 2012).

Four factors of the earthquake parameters are mainly taken into consideration, including magnitude, focal depth, epicentral distance, and wave propagation direction (Sharma & Judd, 1991; Chen et al., 2012; Li, 2012; Wang & Zhang, 2013). The first three factors jointly determine the earthquake intensity of a particular area. The earthquake will be much more intense and its influence will be much stronger in the area with higher magnitude, shallower focal depth, and shorter epicentral distance (Chen et al., 2012; Roy & Sarkar, 2017). The deformation mode and resulting damage to tunnels due to earthquakes are significantly influenced by the propagation direction of the seismic waves (Li, 2012). Generally, three types of deformations express the effect of propagation of seismic waves on the response of underground structure (Owen & Scholl, 1981; Pitilakis & Tsinidis, 2014): axial compression and extension (Figures 1.15a and 1.15b); longitudinal bending (Figures 1.15c and 1.15d); and ovaling/racking (Figures 1.15e and 1.15f).

Structural influential factors mainly include tunnel depth, lining condition, construction method, loading form, and a section of a sudden change in the

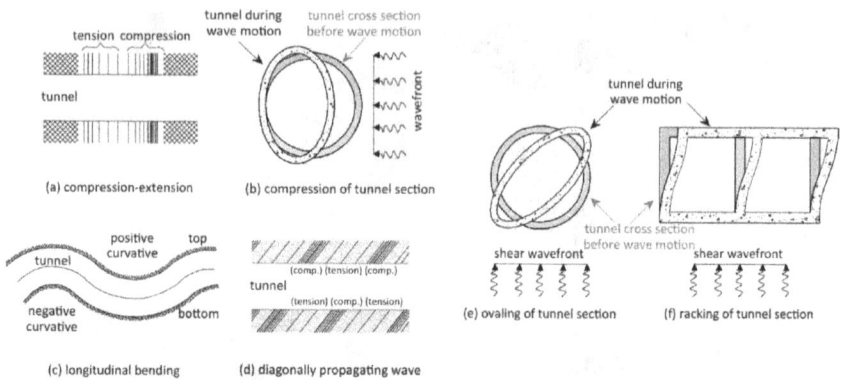

Figure 1.15 Simplified deformation modes of tunnels due to seismic waves.

Source: Owen & Scholl, 1981; Pitilakis & Tsinidis, 2014.

tunnel structure (Yashiro et al., 2007; Chen et al., 2012; Li, 2012; Wang & Zhang, 2013). One significant characteristic of the underground structure compared with the aboveground structure is that the underground structure interacts with the surrounding rock/soil medium. Therefore, for the interaction between the structures and surrounding rock/soil medium, two site conditions regarding the geographic conditions are considered: permanent ground deformation and deterioration of site conditions (Chen et al., 2012). Fault movement and deformation of surrounding rock/soil are considered for permanent ground deformation (Li, 2012; Roy & Sarkar, 2017). On one hand, seismic motions often induce a large shear movement of the fault. There is a high probability that seismic damages due to collapse, squeezing, and pulling occur when the tunnel passes through shear areas (Wang et al., 2001; Chen et al., 2012; Li, 2012). On the other hand, several aspects are mainly taken into consideration for the deterioration of site conditions. They are liquefaction and degradation of seismic subsidence for the soft-soil area and high weathering and decompression for the hard-rock area (Yakovlevich & Borisovna, 1978; Chen et al., 2012; Li, 2012). Slope failures often occur due to the poor rock mass quality of the slopes at the tunnel opening.

1.5 SEISMIC DESIGN AND ANALYSIS OF UNDERGROUND STRUCTURES UNDER EARTHQUAKE FORCE

The seismic design theory for underground structures was developed with the development of the seismic theory of surface structures. However, underground structures have features that make their seismic behavior distinct from most surface structures, most notably their complete enclosure in soil or rock, and their significant length (e.g., tunnels). Therefore, the design of underground structures to withstand seismic loading has aspects that are very different from the seismic design of surface structures (Hashash et al., 2001). The seismic design consists of the following three major aspects: (1) definition of the seismic environment, (2) evaluation of ground response to shaking, and (3) assessment of the seismic response of underground structures (Hashash et al., 2001; Pitilakis & Tsinidis, 2014).

1.5.1 Definition of Seismic Environment

There are three parts to the definition of the seismic environment: seismic hazard analysis, earthquake criteria design, and ground motion parameters. Two accepted methods, namely the deterministic seismic hazard analysis (DSHA) and the probabilistic seismic hazard analysis (PSHA), are available for the seismic hazard analysis to characterize the potential strong ground motion (Reiter, 1991; Krinitzsky, 1995, 1998). DSHA provides a straightforward framework for the evaluation of worst-case scenarios at a site, although it is unable to provide information about the likelihood or frequency of occurrence of the controlling earthquake. PSHA incorporates uncertainties in the source-to-site distance, magnitude, rate of recurrence, and variation of ground motion characteristics into the analyses (Reiter, 1991; Hashash et al., 2001). After the characterization of the seismic hazard at the site, the level of design earthquake or seismicity needs to be defined. For the current seismic design philosophy, two-level design criteria are available for many critical facilities. They are maximum design earthquake (MDE) and operating design earthquake (ODE) (Wang, 1993; Hashash et al., 2001). The aforementioned procedures provide the necessary ground motion parameters to characterize the design event, such as ground acceleration, velocity, and displacement amplitudes (Hashash et al., 2001; Pitilakis & Tsinidis, 2014). Experience has shown that effective, rather than peak, ground motion parameters tend to be better indicators of structural response, as they are more representative of the damage potential of a given ground motion (Nuttli, 1979; Hashash et al., 2001).

1.5.2 Evaluation of Ground Response to Shaking

The evaluation of ground response to shaking consists of ground failure, and ground shaking and deformation. Ground failure refers to the deformation due to liquefaction, lateral spreading, slope instability, and fault rupture (Hashash et al., 2001; Pitilakis & Tsinidis, 2014). Ground shaking and deformation refer to the deformation due to seismic waves. It may be demonstrated in terms of three principal types of deformations (Owen & Scholl, 1981; Wang, 1993), as illustrated in Figure 1.15.

1.5.3 Assessment of Seismic Response of Underground Structures

Two aspects are undertaken for the assessment of the seismic response of underground structures: seismic design loading criteria and seismic response of underground structures (Hashash et al., 2001; Pitilakis & Tsinidis, 2014). Once the ground motion parameters for the maximum and operational design earthquakes have been determined, load criteria are developed for the underground structure using the load factor design method (Wang, 1993). The transversal and longitudinal directions of the underground structure are usually undertaken separately to analyze the seismic response of underground structures, and then the computed internal forces are combined (e.g., Hashash et al., 2001; Hung et al., 2009; Pitilakis & Tsinidis, 2014). Several methods are available in the literature for seismic design in both directions. A short but comprehensive review of the research methods for the seismic response of underground structures is presented independently in Chapter 6.

Current seismic design for mountain tunnels usually considers portals and sections near slope surfaces, with seldom consideration for deeper mined parts and areas near intersections (Wang et al., 2001; Yu et al., 2016b). However, this is in contrast to some site observations, such as the Wenchuan earthquake and the Kumamoto earthquake. The earthquakes inflicted seismic damage that ranged from minor cracking to even failure along the entire tunnel. The damages include lining cracks, spalling and collapse of concrete lining, construction joint damage, pavement damage, and groundwater leakage. The observation provides sufficient evidence for further research of tunnel seismic damage mechanisms and better design for the safety of tunnels and other underground structures in seismically active areas.

Chapter 2

Rate-Dependent Mechanical Behavior of Intact Rock

Numerical Test

2.1 NUMERICAL TEST METHODOLOGY

2.1.1 Principle of Particle Flow Code (Itasca Consulting Group Inc., 2004)

The mechanical behavior of rock is usually characterized using the traditional stress-strain/displacement relationship, which can be divided into four periods including the pre-peak linear period, pre-peak nonlinear period, post-peak period, and residual strength period. However, this traditional relationship provides only limited information regarding the mechanical response of rock. Deformation and failure of rock are an evolution process with the exchange of energy and mass between rocks and the outer environment (Xie et al., 2004). A better picture of rock behavior can be obtained by considering its energy-related mechanism (Wasantha et al., 2014). Currently, a combination of physical tests, theoretical analysis, and numerical simulation, has become a much more efficient way to conduct research in the area of rock mechanics and other areas. In terms of discontinuity of rock masses, the distinct element method (DEM) is a better choice for the numerical analysis of rock mass with advantages as followings (Itasca Consulting Group, Inc., 2004): the failure and fracture development can be simulated free from the limits of the mesh deformation, and the distributions of actual discontinuities can be considered, which is different from continuum analysis, and it provides a reasonable failure mode that meets the physical state. Particle flow code (PFC), which simulates the mechanical behavior of a collection of non-uniform-sized circular rigid particles that displace independently of one another and interact only at each contact point, is currently a widely used one.

The calculation cycle in PFC is a time-stepping algorithm that requires the repeated application of the law of motion to each particle, a force-displacement law to each contact, and a constant updating of wall positions. Contacts, which may exist between two balls or between a ball and a wall, are formed and broken automatically during the course of a simulation. The calculation cycle is illustrated in Figure 2.1. At the start of each timestep, the set of contacts is

DOI: 10.1201/9781003401599-2

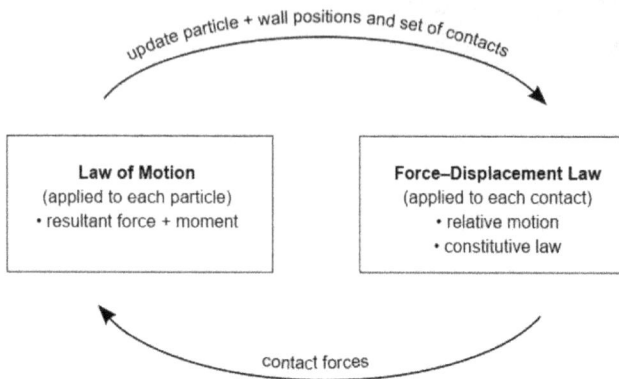

Figure 2.1 Calculation cycle in particle flow theory.

Source: Itasca Consulting Group Inc., 2004.

updated from the known particle and wall positions. The force-displacement law is then applied to each contact to update the contact forces based on the relative motion between the two entities at the contact and the contact constitutive model. Next, the law of motion is applied to each particle to update its velocity and position based on the resultant force and moment arising from the contact forces and anybody forces acting on the particle. Also, the wall positions are updated based on the specified wall velocities. The calculations performed in each of the two boxes of Figure 2.1 can be done effectively in parallel.

The overall constitutive behavior of a material is simulated in PFC by associating a simple constitutive model with each contact. The constitutive model acting at a particular contact consists of three parts: a stiffness model, a slip model, and a bonding model. The stiffness model provides an elastic relation between the contact force and relative displacement. The slip model enforces a relation between shear and normal contact forces such that the two contacting balls may slip relative to one another. The bonding model serves to limit the normal and shear forces that the contact can carry by enforcing bond-strength limits. Two bonding models are supported: a contact-bond model and a parallel-bond model. For the simulation of rock material, a bond model is necessary to characterize the presence of cement between grains. A contact bond can be envisioned as a pair of elastic springs (or point glue) with constant normal and shear stiffness acting at the contact point. A parallel bond can be envisioned as a set of elastic springs with constant normal and shear stiffness, uniformly distributed over a circular disk lying on the contact plane and centered at the contact joint, as illustrated in Figure 2.2. In the contact-bond model, because contact stiffness is still active even after bond breakage as long as particles are kept in contact, bond breakage may not significantly affect the macro stiffness, which is unlikely

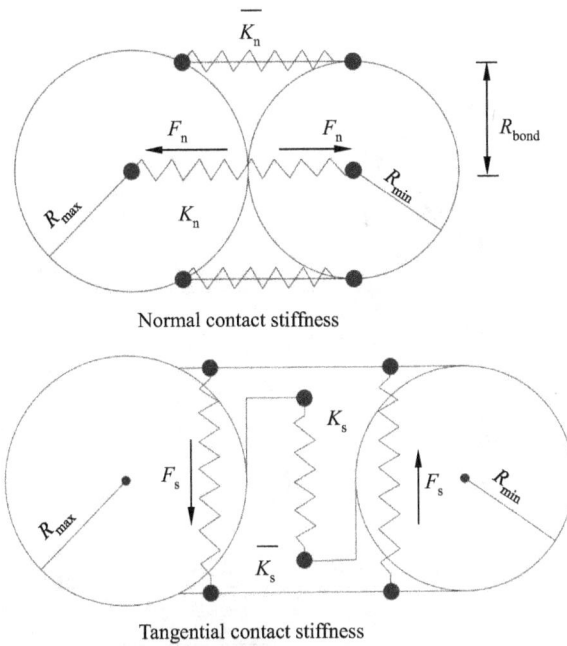

Figure 2.2 Sketch of parallel bond model.

for rock. In the parallel-bond model, however, stiffness is contributed by both contact and bond stiffness. Thus, bond breakage immediately results in stiffness reduction. In this sense, the parallel-bond model is seen as a more realistic model for rock where bonds may break in either tension or shearing with an associated reduction in stiffness (Potyondy & Cundall, 2004; Park & Song, 2009). Once the bond's maximum tensile force exceeds the bond's ultimate tensile strength, the bond breaks with the presence of a tensile crack. Once the bond's maximum shear force exceeds the bond's ultimate shear strength, the bond breaks with the presence of a shear crack. Figure 2.3 shows the yielding process for a parallel bond. Crack monitoring during the calculation process can be realized through the built-in FISH language.

2.1.2 Numerical Experimental Procedure

1. General approach using PFC

A general approach for treating a numerical model with PFC as if it were a laboratory (e.g., compression test, Brazilian test) is recommended. Table 2.1 lists the steps recommended to perform a successful numerical experiment.

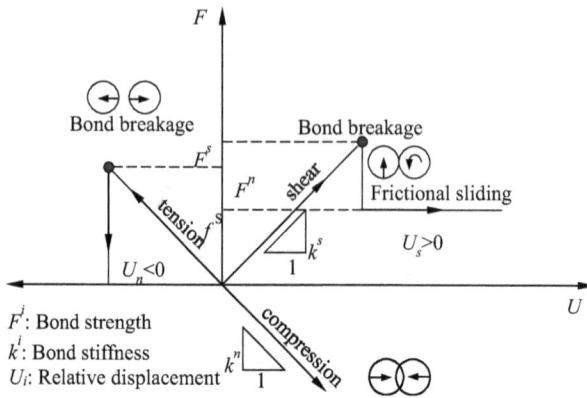

Figure 2.3 Illustration of yielding process for parallel bond.
Source: Itasca Consulting Group Inc., 2004.

Table 2.1 Recommended Steps for Numerical Experiment

Step No.	Content
1	Define the objectives for the model analysis
2	Create a conceptual picture of the physical system
3	Construct and run simple idealized models
4	Assemble problem-specific data
5	Prepare a series of detailed model runs
6	Perform the model calculation

Source: Itasca Consulting Group Inc., 2004.

For simplicity and ease of understanding, the seven steps above can be summarized into five aspects: (1) particle generation; (2) boundary and initial conditions; (3) choice of contact model and material properties; (4) loading, solution, and sequential modeling; and (5) interpretation of results.

2. Numerical model calibration

One aspect distinctive from the continuum simulation is the choice of material properties. Micro-properties of particles and contact cannot be obtained directly from the physical tests. These unknown properties require extensive calibration until the macro-response of the sample matches the results of the physical test or field investigation. In this case, we must use 'inverse modeling'; that is, perform several tests on samples with assumed properties and compare the results with the desired response of the real intact material. When a match has been found, the corresponding set of properties may be used in the full simulation. For rock materials, it is common to calibrate the micro-properties against the macro-properties involving uniaxial compressive strength (UCS), elastic modulus (E), and Poisson's ratio (v) of the intact rock (Potyondy & Cundall, 2004; Cho et al., 2007; Asadi et al., 2012; Park & Song, 2009; Xia et al., 2012; Zhou et al., 2012; Bahaaddini et al., 2013, 2014; Lambert & Coll, 2014). The macroscopic mechanical parameters of the material are determined through laboratory experiments, that is, elastic modulus E, Poisson's ratio v, compressive strength σ_c, cohesion c, and angle of internal friction ϕ. By analyzing macroscopic mechanical parameters, the particle contact modulus and parallel bond modulus of particles are preliminarily determined. The contact modulus here is different from the macroscopic elastic modulus and is often larger than the macroscopic elastic modulus (Xu et al., 2010; Itasca Consulting Group Inc., 2004).

The calibration process of micro-properties of particle and contact is shown in Figure 2.4. The first step of the calibration process is to determine the initial micro-properties of the particle and bond. Micro-parameters of the model particles mainly include particle contact modulus E_c, the ratio of normal stiffness to tangential stiffness k_n / k_s, and friction coefficient f. Micro-parameters of the parallel bond model include parallel bond radius multiplier λ, bond modulus $\bar{E}_c$, the ratio of normal stiffness to tangential stiffness $\bar{k}_n / \bar{k}_s$, and normal and tangential bond strength $\bar{\sigma}_c, \bar{\tau}_c$. The initial particle stiffness can be given by

$$k_n = 2E_c \begin{cases} t, & (PFC2D) \\ 2\tilde{R}, & (PFC3D) \end{cases} \tag{2.1}$$

$$k_s = \frac{k_n}{(k_n / k_s)} \tag{2.2}$$

$$\tilde{R} = \frac{R^{[A]} + R^{[B]}}{2} \tag{2.3}$$

where k_n and k_s are the particle normal stiffness and tangential stiffness, t is the disk thickness, and $R^{[A]}$ and $R^{[B]}$ are the radii of two particles in contact. The initial parallel bond stiffness can be given by

$$\bar{k}^n = \frac{\bar{E}_c}{R^{[A]} + R^{[B]}} \tag{2.4}$$

$$\bar{k}^s = \frac{\bar{k}_n}{(\bar{k}^n / \bar{k}^s)} \tag{2.5}$$

where $\bar{k}^n$ and $\bar{k}^s$ are the parallel bond normal stiffness and tangential stiffness.

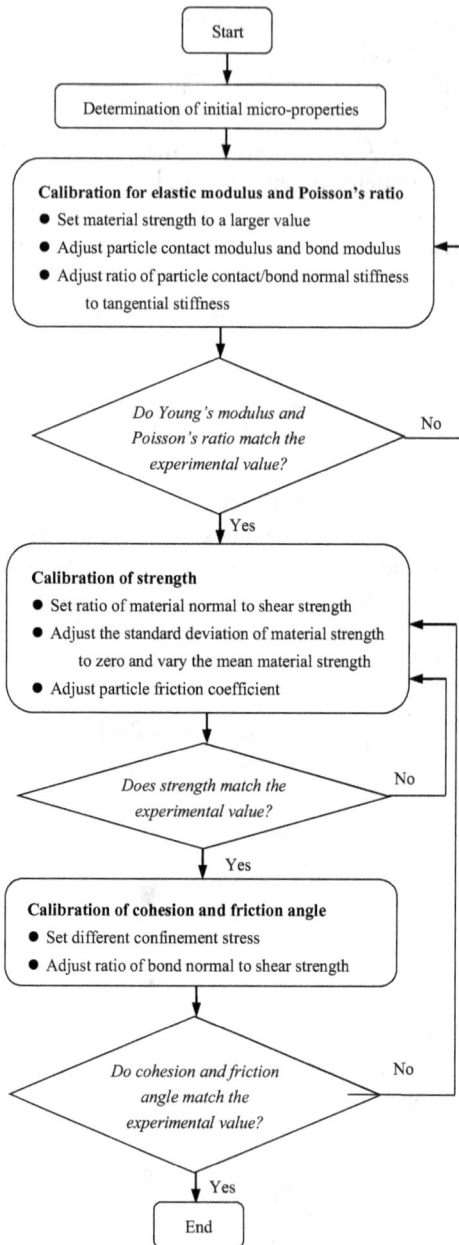

Figure 2.4 Calibration process of micro-properties of particle and bond in PFC.
Source: Modified after Potyondy and Cundall, 2004.

With initial micro-properties, match Young's modulus by setting material strengths to a large value and varying E_c (and $\bar{E}_c$ for a parallel-bonded material). Then match the Poisson's ratio by varying k_n / k_s (and $\bar{k}_n / \bar{k}_s$ for a parallel-bonded material). It may be necessary to perform a few iterations to match both values. Once the desired elastic response has been obtained, match the peak strength at zero confinement by setting the standard deviation of material strengths to zero and varying the mean material strengths. Note that the ratio of material normal to shear strength will affect the behavior; thus, keep this ratio fixed. If one desires to reproduce post-peak behavior, vary the particle friction coefficient. At this point, one can obtain the strength envelope by performing a set of triaxial tests at different confinements to calibrate the cohesion and internal friction angle.

Based on the procedure listed in Figure 2.4, 2D numerical samples with 50 mm width and 100 mm height (same as the size of the cylinder samples in the physical compression test; Zhang et al., 2014) were generated to conduct numerical compression tests, as illustrated in Figure 2.5. In the current chapter, triaxial compression test results are selected, and the rock triaxial compressive strength is selected as the compressive strength. After a sequence of 'trial-and-error' calibrations, physico-mechanical parameters in PFC match the laboratory results as shown in Table 2.2. The final micro-properties for particle contact and bond are listed in Table 2.3.

2.1.3 Realization of the Compression and Brazilian Test

A 2D numerical rectangular box with a 50 mm width and 100 mm height and a circular box with a 25 mm radius was created using the discrete element code PFC2D as illustrated in Figure 2.6. The sample genesis procedure is described in detail by Potyondy and Cundall (2004). For the compression

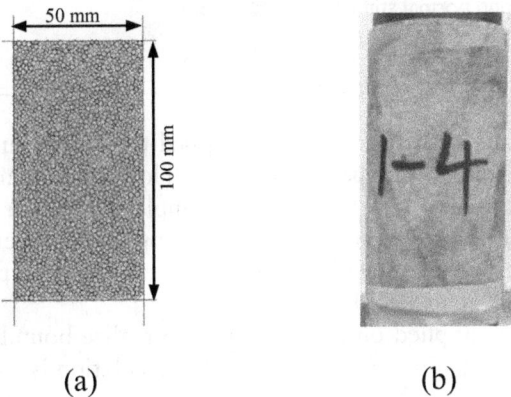

(a)   (b)

Figure 2.5 PFC model and granite sample. (a) PFC numerical model; (b) rock sample in physical.

Table 2.2 Physico-Mechanical Parameters and PFC Results

Macro Property	Value	Remark
Young's modulus (GPa)	28.7	Physical test
	28.4	Numerical test
Poisson's ratio	0.2300	Physical test
	0.2285	Numerical test
Compression strength (MPa)	130.5	Physical test
(under confinement stress of 6MPa)	132.8	Numerical test
Cohesion (MPa)	15.90	Physical test
	20.87	Numerical test
Internal friction angle (°)	49.86	Physical test
	43.68	Numerical test

Table 2.3 Mesoscopic Physico-Mechanical Parameters of Granite

Micro-Property	Value	Remark
Minimum radius/mm	0.3	
Radius multiplier	1.66	
Density/kg·m⁻³	2800	Uniform distribution
Particle contact modulus (GPa)	5.0	
Ratio of particle contact normal stiffness to tangential stiffness	3.0	
Friction coefficient	0.8	
Parallel bond modulus (GPa)	43.0	
Parallel bond normal stiffness (MPa)	88 ± 10	Gaussian distribution
Parallel bond normal stiffness (MPa)	160 ± 10	Gaussian distribution
Ratio of parallel bond normal stiffness to tangential stiffness	3.0	
Parallel bond radius multiplier	1.0	

sample, two walls with IDs 1 and 3 wrapped the side boundary and two walls with IDs 2 and 4 wrapped the upper and lower boundary. For the Brazilian sample, a circular wall wrapped the sample area which is deleted after the particle generation and two walls on the upper and lower side with IDs 1 and 2 represent the loading plate. Micro-properties for the particle contact and parallel bond refer to Table 2.3.

Velocities were applied on the rigid wall to realize boundary conditions using FISH language. In PFC simulation, the calculation is based on a time-stepping algorithm in which it is assumed that the velocities and accelerations are constant within each timestep (Δt). For the uniaxial compression test, the two side walls are deleted while the upper and lower walls remain

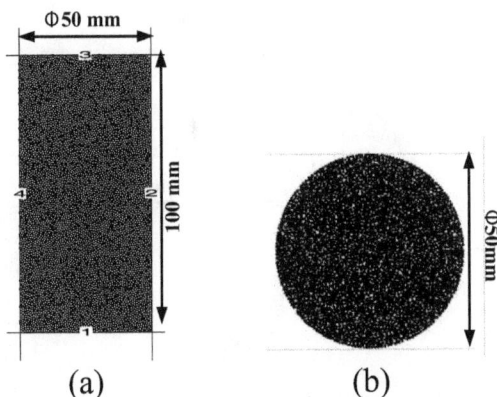

Figure 2.6 PFC models of compression and Brazilian test. (a) Compression test; (b) Brazilian test.

for compression loading. For the triaxial compression test, the horizontal velocity applied on the side walls for the compression test was converted from the corresponding confinement stress using the servo mechanism (Itasca Consulting Group Inc., 2004). For the Brazilian test, the upper and lower walls are for compression loading. The compression and tensile stresses are calculated by dividing the average force on the upper and lower walls.

2.2 QUASI-STATIC MECHANICAL BEHAVIOR OF INTACT ROCK

Quasi-static mechanical behavior for triaxial compression test under 10 MPa confinement stress is discussed from the viewpoint of energy-related aspects with consideration of micro-crack rupture evolution and energy variation law.

2.2.1 Energy-Related Mechanism Based on Micro-Crack Rupture Evolution

In the PFC numerical simulation, when the bonding strength between particles is less than the strength transmitted between particles, the particle bond will fracture, which corresponds to the micro-crack inside the rock (Hazzard et al., 2000). In the actual process of crack propagation, the strain energy will be released rapidly in the form of an elastic wave, resulting in infrasound, sound wave, or ultrasonic wave, namely acoustic emission phenomenon (AE). The general AE process is shown in Figure 2.7. Figure 2.8 shows

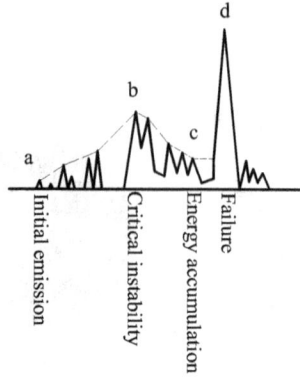

Figure 2.7 Curves of general acoustic emission process.

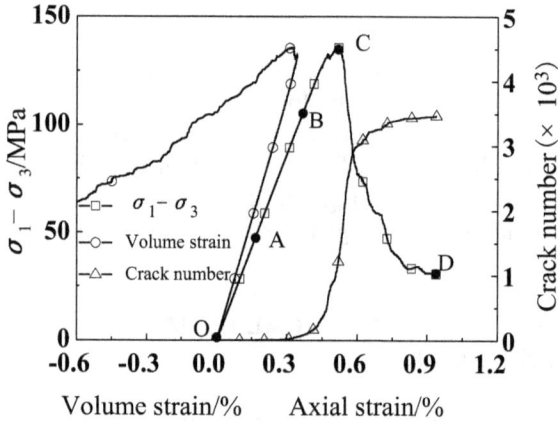

Figure 2.8 Axial stress-strain curves and mechanical behavior of compressed sample.

the axial stress-strain curve, the evolution curve of the crack number, and the relationship between axial stress and volume strain of the sample. Figure 2.9 shows the AE frequency curve. To study the evolution of crack distribution, five monitoring points noted with symbols from A to E were set in the stress-strain curve, corresponding to the axial strains of 0.0%, 0.169%, 0.388%, 0.507%, and 0.939%, respectively. Figure 2.10 shows the crack distribution of samples at each monitoring point.

As can be seen from Figures 2.8–2.10, after being subjected to an external load, the sample begins to generate micro-cracks at the point B (34%σ_c is the macroscopic crack initiation strength and σ_c is the peak strength) and emits the first acoustic emission, which is defined as the initial emission point *a* on the AE process line (point *a* in Figures 2.7 and 2.9). From the initial

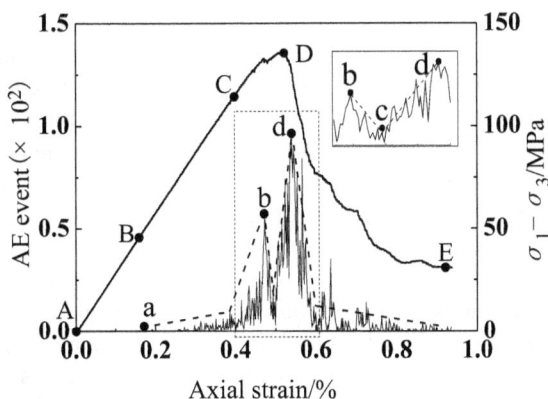

Figure 2.9 Curves of axial stress-strain and acoustic emission frequency of compressed sample.

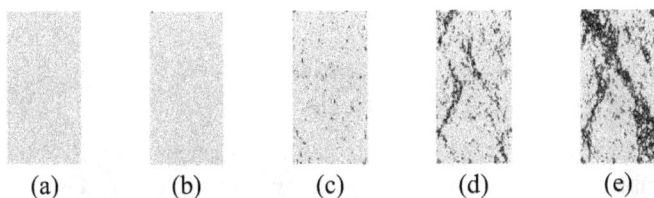

(a)   (b)   (c)   (d)   (e)

Figure 2.10 Meso-crack development state of BPM model in process of deformation and failure. (a) Point A; (b) point B; (c) point C; (d) point D; (e) point E.

emission to the stress yield point C (78.6%), the AE events are relatively sparse, and the amplitude increases with the crack opening and propagation, mainly in the range of 0–8 events, indicating the transition from the elastic stage to the stable fracture propagation stage. At this stage, the volume strain shows a linear change and the volume gradually decreased with the increase of load, and the sample shows a state of compression.

When the load exceeds the yield stress point C, the sample shifts from the stable fracture propagation stage to the unstable fracture propagation stage. The stress-strain curve presents a downward concave shape and the deformation increases. After a short unchanged phase, the volume strain curve appears to reverse bending, and the volume dilation begins, resulting in irreversible plastic deformation. The AE shows a rapid growth trend to 57 events following crack further initiation and expansion. The sample has reached critical instability point b (point b in Figures 2.7 and 2.9). When point b is exceeded, AE events will attenuate somewhat, mainly because part of the energy is consumed for the sharp increase of the new free surface generated by crack propagation. On the AE process line, there is an energy absorption valley, the bottom point of which is defined as the energy accumulation point (point c in Figures 2.7 and 2.9). After the temporary equilibrium of crack propagation, the energy begins to accumulate again with the crack compaction due to continuous compression. Volume dilation at this stage occurs since the internal pores of the sample continue to increase due to micro-crack development.

After the axial stress exceeds the peak strength σ_c (point D), the sample shows the characteristics of brittle development with a significant decrease of axial stress and a serious increase of micro-cracks. The micro-cracks develop into macroscopic cracks, and the acoustic emission events reach a maximum of 92 events (point d in Figures 2.7 and 2.9). Subsequently, the axial stress curve develops steadily, and the number of cracks increases at a constant rate. The AE decreases to a low level with the change range of 0–10 events. The sample is resumed to a stable fracture phase. At this stage, the failure of the sample mainly results from frictional sliding along the macroscopic cracks with a macroscopic fracture surface (point E in Figure 2.10).

2.2.2 Energy-Related Mechanism Based on Energy Dissipation Law

Figure 2.11 shows the energy dissipation curve of the sample under the triaxial compression test. Before the yield stress (point C in Figure 2.9), the sum of friction energy, kinetic energy, strain energy, and bond energy is equal to the boundary energy. The proportion of bond energy and strain energy is large, which is related to the generation and driving of cracks, and their development is related to the deterioration of the sample. Vice versa for the friction energy representing crack development. The crack is generated with the dissipation of the bond energy and then propagates under the drive of

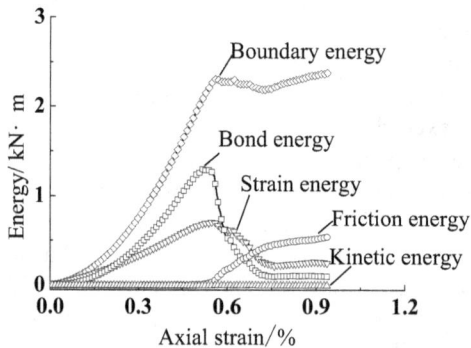

Figure 2.11 Energy dissipation laws during sample fracture process.

strain energy, the friction energy does not begin to function until the crack is generated. Therefore, there is an alternating growth and decline relation between them. When the sample reaches the peak strength, the bond energy and strain energy decrease sharply, while the friction energy increases in a large magnitude. The proportion of friction energy increases gradually with the crack's further development. It can be seen that friction is the main provider of residual strength. The kinetic energy occupies a small proportion in the whole deformation process of the sample, which is related to the loading process and the dynamic balance of the sample, indicating that the deformation of the sample is not very severe, and the crack extends through the whole sample steadily.

Based on the evolution law of micro-cracks and the energy dissipation law of samples through particle flow code, the energy-related deformation and failure mechanism of compressed rock can be revealed. Under the action of axial and confining stress, the rock's internal energy begins to accumulate. The particles first dissipate the bond energy to produce micro-cracks, and then the micro-cracks further develop with the dissipation of strain energy. The friction effect gradually begins to play a role and its proportion gradually increases with the further development of cracks. When the peak strength is reached, the strain energy and bond energy are released largely. The friction energy increases in a large magnitude with the crack development and plays a dominant role. In the residual stage, the bond energy is mostly converted to other forms of energy driven by boundary energy and strain energy, and it is difficult for particles to continuously overcome the bond strength to initiate new cracks. The final failure mode of rock is mainly sliding failure along macroscopic cracks.

2.3 RATE-DEPENDENT MECHANICAL BEHAVIOR OF INTACT ROCK

Rate-dependent mechanical behavior of rock for uniaxial compression and the Brazilian test are discussed from the viewpoint of energy-related aspects. Four rate cases with 0.001 m/s, 0.005 m/s, 0.050 m/s, and 0.500 m/s are considered.

2.3.1 Rate-Dependent Behavior of Stress versus Strain

Figure 2.12 shows the stress-strain curve of granite under uniaxial compression and Brazilian splitting at different loading rates. Under compression or splitting conditions, the sample goes through three stages: initial compaction stage, hardening dilatancy stage, and softening dilatancy stage. The loading rate does not affect the overall trend while only affects the peak strain. In the

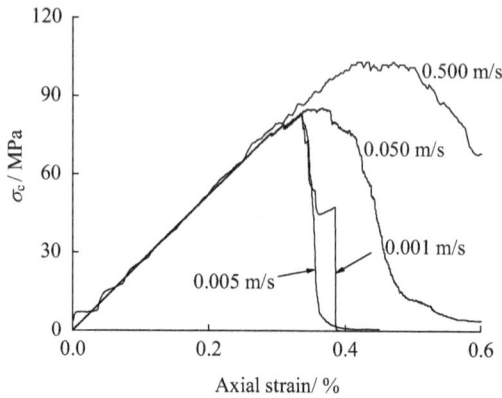

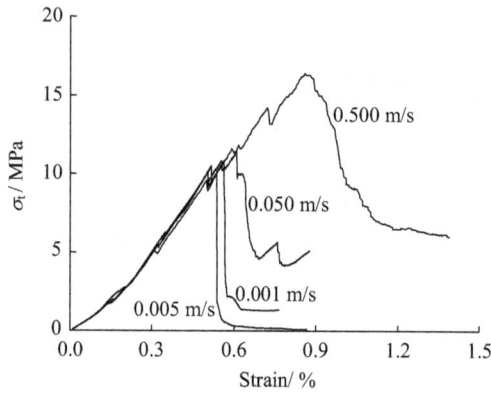

Figure 2.12 Curve of stress vs. strain of uniaxial compression test and Brazilian test under different loading rates. (a) Uniaxial compression numerical test; (b) Brazilian numerical test.

post-peak stage, the low loading rate provides sufficient time for the development of the initial inner damage and cracks, resulting in the stepwise drop of the stress-strain curve after failure. The stress-strain curve of the sample under a higher loading rate is smooth and continuous in the post-peak stage.

Table 2.4 lists the numerical results of the uniaxial compression test the and Brazilian test under different loading rates. Figure 2.13 shows the relationship between uniaxial compressive strength (*UCS*), tensile strength, and peak strain of rock with loading rate. The uniaxial compressive strength and tensile strength of rock and their corresponding peak strain show a nonlinear increase trend with the increase of loading rate. When the loading rate increases from 0.001 m/s to 0.500 m/s, the uniaxial compressive strength

Table 2.4 Numerical Results of Uniaxial Compression Test and Brazilian Test under Different Loading Rates

Loading Rate / (m/s)	Uniaxial Compression Strength/MPa	Peak Strain for UCS/%	Tensile Strength / MPa	Peak Strain for Tensile Strength/%
0.001	82.77	0.336	10.13	0.504
0.005	82.89	0.338	10.77	0.562
0.050	85.51	0.366	11.39	0.608
0.500	103.21	0.418	16.37	8.600

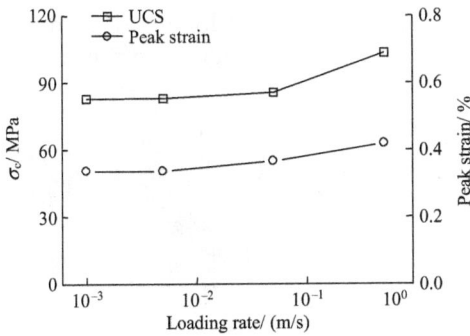

(a)

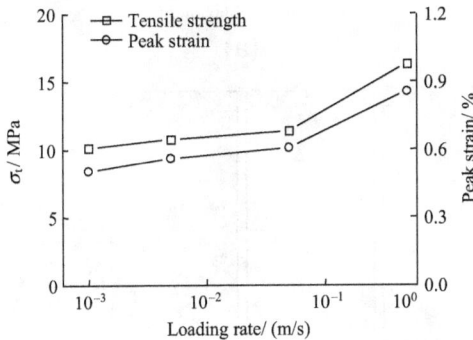

(b)

Figure 2.13 Relationships between UCS, tensile strength, peak strain, and loading rate. (a) Uniaxial compression numerical test; (b) Brazilian numerical test.

and tensile strength increase by 24.69% and 61.60%, respectively, and the corresponding peak strain increases by 24.40% and 70.63%, respectively. However, with the increase in the loading rate, the growth rate will be higher. When the loading rate increases from 0.001 m/s to 0.005 m/s, the uniaxial compressive strength and tensile strength only increase by 0.145% and 6.32%, and the corresponding peak strain increases by 0.595% and 11.51%.

When the loading rate increases from 0.050 m/s to 0.500 m/s, the growth rates of uniaxial compressive strength and tensile strength are 20.70% and 43.72%, respectively, and the corresponding peak strain growth rates are 14.21% and 41.45%, respectively.

2.3.2 Rate-Dependent Behavior of Failure Mode

To analyze the influence of loading rate on the sample fracture morphology, cracks were monitored in the numerical test, and the initiation time sequence of cracks was monitored.

Figure 2.14 shows the damage evolution mechanism of rock under different loading rates for the uniaxial compression numerical test. Four colors

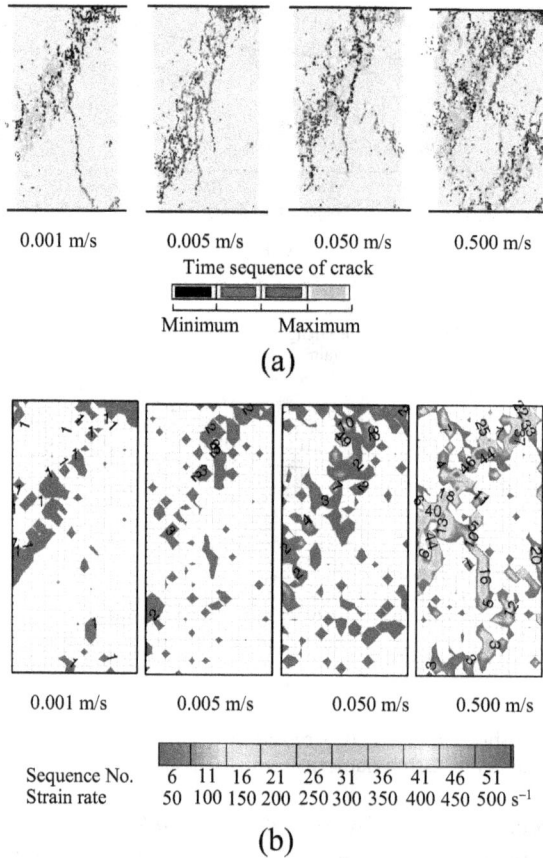

0.001 m/s   0.005 m/s   0.050 m/s   0.500 m/s

Time sequence of crack

Minimum   Maximum

(a)

0.001 m/s   0.005 m/s   0.050 m/s   0.500 m/s

Sequence No.   6   11   16   21   26   31   36   41   46   51
Strain rate   50   100   150   200   250   300   350   400   450   500 s^{-1}

(b)

Figure 2.14 Damage evolution mechanism of rock under different loading rates for uniaxial compression numerical test. (a) Micro-crack evolution; (b) strain rate in horizontal x direction.

were used to distinguish the cracks. The time sequence of crack initiation represents the relative time sequence of crack initiation and propagation. Rock under uniaxial compression experiences an elastic stage, stable fracture propagation stage, unstable fracture propagation stage, and stable fracture propagation stage in residual, as discussed in Section 2.2.1. When the rock is subjected to uniaxial compression at different loading rates, the sample shows different failure modes. With the increase in loading rate, the sample under uniaxial compression fails gradually from single inclined plane mode to multi-inclined plane mode, and the width of the main fracture zone increases largely.

It can be seen from the development of the crack number in the sample in Figure 2.15 that the failure degree of the sample gradually intensifies with the increase in loading rate. When the loading rate increases from 0.001 m/s to 0.500 m/s, the number of cracks increases from 13,650 to 34,080. On the other hand, the loading rate also affects the strain rate seriously. When the loading rate increases from 0.001 m/s to 0.500 m/s, the horizontal maximum strain rate increases from 9.220 s^{-1} to 1 940 s^{-1} in the post-peak stage, as shown in Figure 2.14(b). In addition, Figure 2.14(b) also shows that the region of high strain rate increases with the increase of loading rate indicating that the increase of loading rate will aggravate the failure degree of the rock.

Figure 2.16 shows the damage evolution mechanism of rock under different loading rates for the Brazilian numerical test. When the rock is subjected to radial splitting force, the force concentration between the loading plate and the sample due to the plate's rigidity results in the first crack initiation at this edge (corresponding to the black crack in Figure 2.16(a)). With the increase of the splitting force, the damage in the rock sample basically concentrates at the upper and lower edges, and a vertically radial micro-crack is formed

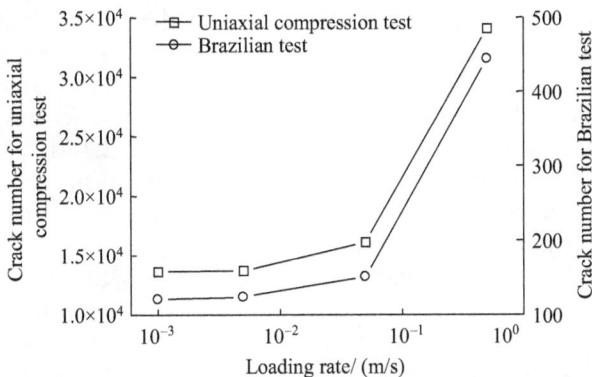

Figure 2.15 Statistical number of cracks affected by loading rate for rock in uniaxial compression and Brazilian tests.

0.001 m/s   0.005 m/s   0.050 m/s   0.500 m/s

Time sequence of crack

Minimum   Maximum

(a)

0.001 m/s   0.005 m/s   0.050 m/s   0.500 m/s

| Sequence No. | 6 | 11 | 16 | 21 | 26 | 31 | 36 | 41 | 46 | 51 |
| Strain rate | 50 | 100 | 150 | 200 | 250 | 300 | 350 | 400 | 450 | 500 s⁻¹ |

(b)

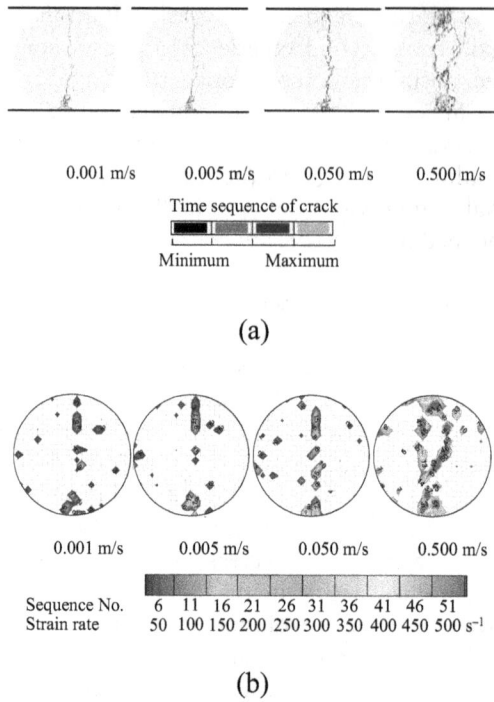

Figure 2.16 Damage evolution mechanism of rock under different loading rates for Brazilian numerical test. (a) Micro-crack evolution; (b) strain rate in horizontal x direction.

(corresponding to the crack with sequence No. of 50 in Figure 2.16(b)). With the force continuous increase, the granite damage develops centrally along the previously formed micro-crack tip to form an obvious macroscopic crack. When the macroscopic crack develops beyond the center of the sample, the crack quickly develops into a thorough fracture along the radial direction under the external force action, resulting in rock failure (corresponding to the cracks with sequence No. from 1 to 30 in Figure 2.16(b)). Figure 2.16 also shows that the main fracture zones of the sample formed in the residual stage under different loading rates correspond to the high strain rate regions of the samples in the horizontal direction.

When the granite is split with different loading rates, the sample also presents different failure modes. The sample fails in the modes from one main fracture to multiple main fractures. The cracks extend to the edges of both the right and left sides of the sample with an increase in loading rate. Meanwhile, the failure degree gradually intensifies. When the loading rate increases from 0.001 m/s to 0.500 m/s, the number of cracks increased from 121 to 444 (Figure 2.15). The effect of the loading rate on the strain rate is also obvious. When the loading rate increases from 0.001 m/s to 0.500 m/s, the horizontal maximum strain rate in the post-peak stage increases from 5.970 s^{-1} to 2420 s^{-1}, as shown in Figure 2.16(b). In addition, the increase in loading rate leads to the increase of the horizontal high strain rate region in the post-peak stage leading to the aggravation of the sample failure degree, which is consistent with the effect of loading rate on the sample under uniaxial compression.

2.3.3 Rate-Dependent Behavior of Acoustic Emission

Figures 2.17 and 2.18 show the relationship between acoustic emission (AE) events and strain energy rate of rock and axial strain and time step at different loading rates. Regardless of loading rate, the rock experiences similar micro-crack development including stable development stage, rapid increase stage, and stable development stage as discussed in Section 2.2.1.

However, the magnitudes of the strain energy rate and acoustic emission events are different due to the loading rate. Both AE events and strain energy rate show a nonlinear increasing trend with the increase of loading rate, as shown in Figure 2.19. In the uniaxial compression test, the corresponding

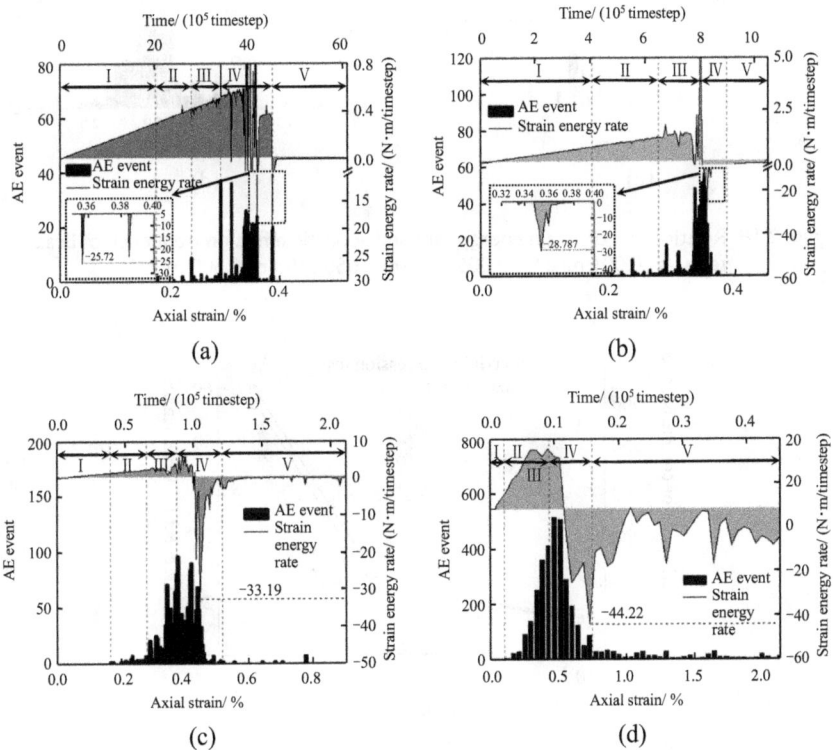

Figure 2.17 Relationship of strain energy rate and acoustic emission frequency of uniaxial compression tests. (a) 0.001 m/s; (b) 0.005 m/s; (c) 0.050 m/s; (d) 0.500 m/s.

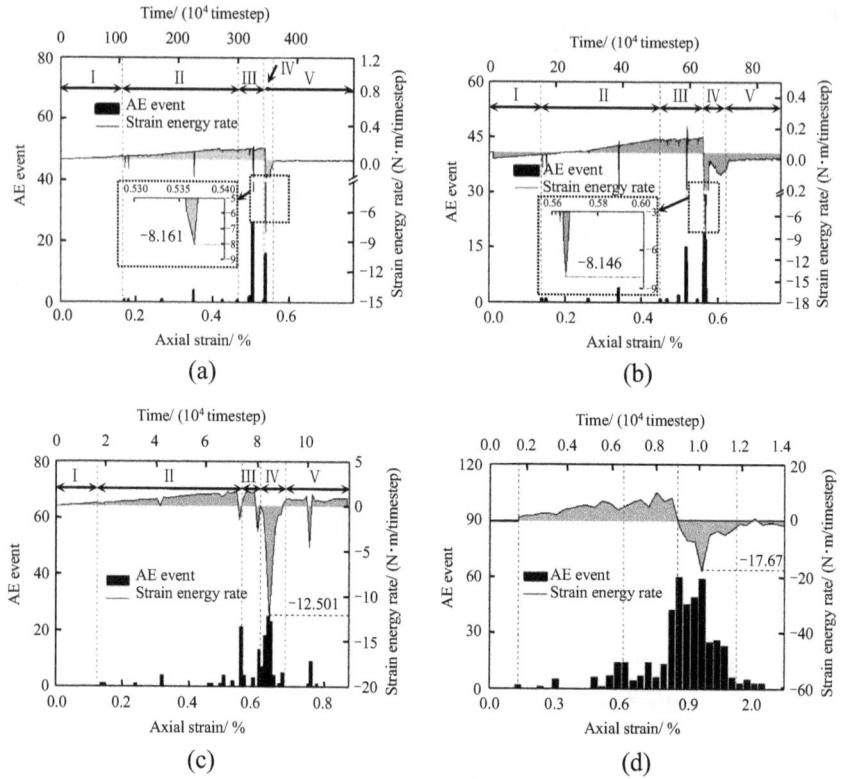

Figure 2.18 Relationship of strain energy rate and acoustic emission frequency of Brazilian tests. (a) 0.001 m/s; (b) 0.005 m/s; (c) 0.050 m/s; (d) 0.500 m/s.

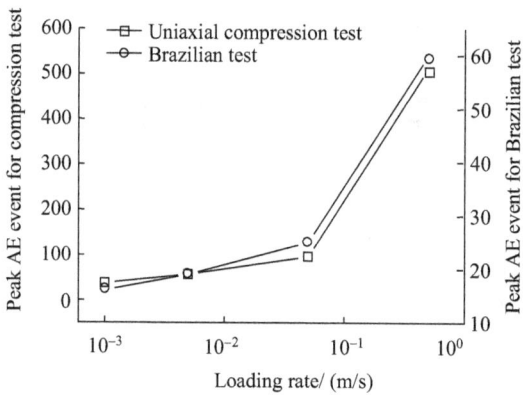

Figure 2.19 Relationships between maximum acoustic emission count and loading rate.

maximum strain energy rate in the residual stage (Stage IV) increases from 25.72 N·m per timestep to 44.22 N·m per timestep and the peak AE event number increases from 37 to 513. When the loading rate increases from 0.001 m/s to 0.500 m/s. In the Brazilian splitting test, the maximum strain energy rate in the residual stage (Stage IV) increases from 8.16 N·m per timestep to 17.67 N·m per timestep, and the peak AE event number increases from 16 to 60 when the loading rate increases from 0.001 m/s to 0.500 m/s. As the loading rate increases, AE events tend to be group seismic (stage IV in Figures 2.17 and 2.18). Also, the loading rate will affect the timestep of samples entering each stage. The larger the loading rate is, the less time it takes to enter the peak stage of AE events.

2.4 CONCLUSION

This chapter provides a numerical test method for the rate-dependent mechanical behavior of rock. Compression tests and Brazilian tests for granite were virtually simulated using the bonded particle model (BPM) in particle flow code (PFC2D). The simulated failure pattern of samples is shear failure along a single inclined plane, which agrees with the failure pattern of indoor compression tests. With the calibrated meso-mechanical parameters, the micro-crack development and energy dissipation of granite were further studied. The results show that there are three main stages during micro-crack development, which are the stable development stage, rapid increase stage, and stable development stage. The changing of boundary energy, strain energy, bond energy, frictional energy, and kinetic energy has given an excellent explanation of mesoscopic mechanics.

To quantify the rate-dependent behavior of compression strength, tensile strength, failure modes, degree, strain energy rate, and acoustic emission, four loading rates were chosen from 0.001 m/s to 0.5 m/s. The uniaxial compression strength (UCS) and its corresponding strain, as well as the tensile strength and its corresponding strain, increase non-linearly with the loading rate. Loading rate also affects the failure pattern and damage degree. Samples in compression fail with a single fracture plane under a lower loading rate, while they fail with several fracture planes forms under a higher loading rate. Crack numbers and high horizontal strain rate distributions, in addition, show that failure zone area and damage degree increase with increasing loading rate. For samples in Brazilian tests, its failure modes change from one master fracture to several master fractures as the loading rate increases. And cracks extend to the edge of the circle sample. The loading rate exacerbates the damage degree, too. Additionally, acoustic emission count and strain energy rate for samples in both tests increase in a non-linear form with increasing loading rate.

Chapter 3

Rate-Dependent Mechanical Behavior of Jointed Rock

Laboratory Test

3.1 METHODOLOGY FOR SEISMIC-DEPENDENT LABORATORY TEST

Underground openings are not safe when subjected to strong earthquakes (Dowding et al.1978; Owen et al., 1981). When a rock joint is subjected to loading and/or unloading during a seismic event, it will undergo a sequence of cyclic shearing loadings with different shear rates (Mirzaghorbanali et al., 2014). The dynamic loading during earthquake activity could seriously threaten the stability of the engineering structure and surrounding rock mass (Lin & Wang, 2006; Yu et al., 2016a; Chen et al., 2017; Liu et al., 2018). On the other hand, rock bolts and cables are widely used in rock engineering for stabilizing jointed rock mass (Wu et al., 2018a). The properties of bolted joints will affect the stability of the surrounding rock mass significantly. In engineering practice, the bolted joints could be subjected to opening movement in a direction perpendicular to the plane or shearing movement in the plane (Wang et al., 2013, 2018; Tistel et al., 2018; Srivastava & Singh, 2015; Li et al., 2016a). The research by McHugh and Signer (1999), Li et al. (2016b) indicated that a high proportion of rock bolts were damaged due to the shearing movement of rock mass (Li, 2010). Therefore, it is an important issue in terms of assessing the stability of rock mass with/without reinforcement during and after earthquakes. In response to these findings, the shear test of jointed rock is an effective way for dynamic loading consideration including shear rate and cyclic shear loading.

3.1.1 Rock Joint Morphology Description

Joint roughness is one of the most important factors for understanding and estimating the shear behavior of a rock joint. The JRC (joint roughness coefficient)-JCS (joint compressions strength) criterion proposed by Barton in 1977 is the most well-known criterion and is used in that paper to select the joints and develop the new shear strength criterion (Barton, 1977). The JRC of a joint is typically evaluated via the visual comparison of

DOI: 10.1201/9781003401599-3

measured profiles against a set of standard *JRC* profiles produced by Barton and Choubey (Barton, 1977). However, the visual evaluation of the *JRC* is subjective, and the results vary according to the opinion of the engineer performing the comparison. Tse and Cruden (1979), Bandis (1980), and Milne (1990) attempted to develop alternative methods to estimate the *JRC*. Here, the z_2 value proposed by Tse and Cruden in Equation (3.1) was used to estimate the *JRC* value using the empirical relation in Equation (3.2):

$$z_2 = \sqrt{\frac{1}{n-1} \sum_{i=1}^{n-1} \left(\frac{z_{i+1} - z_i}{\Delta} \right)} \qquad (3.1)$$

$$JRC = 60.32 \times z_2 - 4.51 \ (SI = 0.25 \ mm) \qquad (3.2)$$

where z_i and z_{i+1} are the spatial height of the *i*th and *i* + *1*th surface points, respectively; Δ is the horizontal interval in the *x* direction; and *n* is the number of spatial points in one profile.

The rough surface must be quantified before calculating its *JRC*. In this study, the rock joint surface data were acquired using a three-dimensional laser-scanning rock surface instrument. The advantages of this type of system are its high precision and good repeatability. Furthermore, it has low time and computational requirements because it can digitize the entire surface at once. The system consists of a measurement head containing a central projector unit and two high-resolution charge-coupled device (CCD) cameras. The resolution of the spatial location of each point in the three-dimensional space along the *x*, *y*, and *z* directions is ±50 μm (±50 μm in Grasselli's work [Grasselli, 2001; Grasselli et al., 2002]; ±20 μm in Xia's work [Xia et al., 2014]). In this study, the point spacing of the triangle mesh is selected as 0.25 mm (0.30 mm, in Grasselli's work [Grasselli, 2001, 2006; Grasselli et al., 2002]; 0.30 mm, in Xia's work [Xia et al., 2014]) when reconstructing the joint surface and calculating *JRC*.

To obtain the average *JRC*, a comparison of *JRC* calculation under different profile interval were conducted to determine a better profile interval, as indicated in Figure 3.1. The value of the JRC_i for the *i*th joint profile obtained using Equation (3.2) was substituted into Equation (3.3) to obtain the overall average *JRC* for each group of joints.

$$JRC = \frac{1}{n} \sum_{i=1}^{n} JRC_i \qquad (3.3)$$

The average value obtained using an interval of 0.25 mm was selected for the analysis of the rock joint roughness. To be more consistent with engineering practice, the replicas of natural rough joints were made from granite joint planes. To study the influence of the shear rate on the different

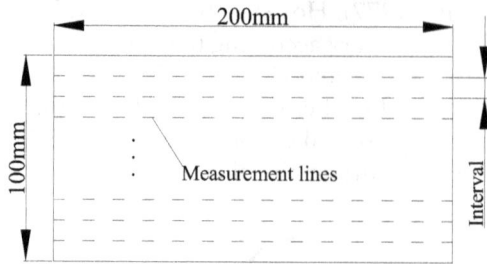

Figure 3.1 Method used to determine the JRC of rough rock joints.

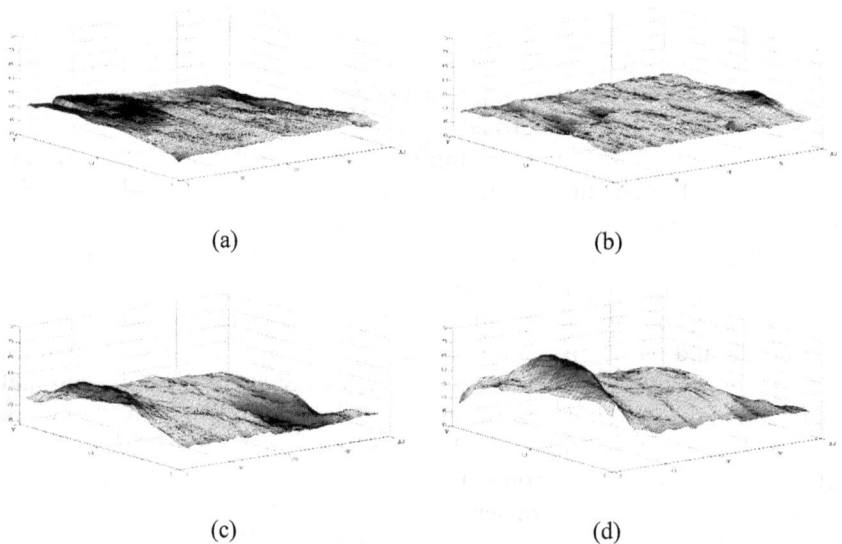

(a)   (b)

(c)   (d)

Figure 3.2 Scanning graphs of four natural rough joints for the rate-dependent test. (a) Sample S1, JRC = 3.61; (b) sample S2, JRC = 8.16; (c) sample S3, JRC = 10.61; (d) sample S4, JRC = 19.05.

joints, four rough rock joints (S1, S2, S3, and S4) were selected. These rock joints have average *JRC* values of 3.61, 8.16, 10.61, and 19.05, respectively. Figure 3.2 presents their corresponding scanning graphs. To study the influence of the cyclic shear loadings on the different joints, two natural rough joints (Figure 3.3) with a larger JRC (X2, JRC = 12.3) and a smaller JRC (X1, JRC = 3.5) were used to demonstrate the shear behavior of bolted rock joint under different roughness conditions. All the rock samples used in the shear tests were 100 mm in width, 200 mm in length, and 100 mm in height. The long side direction in Figures 3.2 and 3.3 is the orientation of the joint and the shear direction.

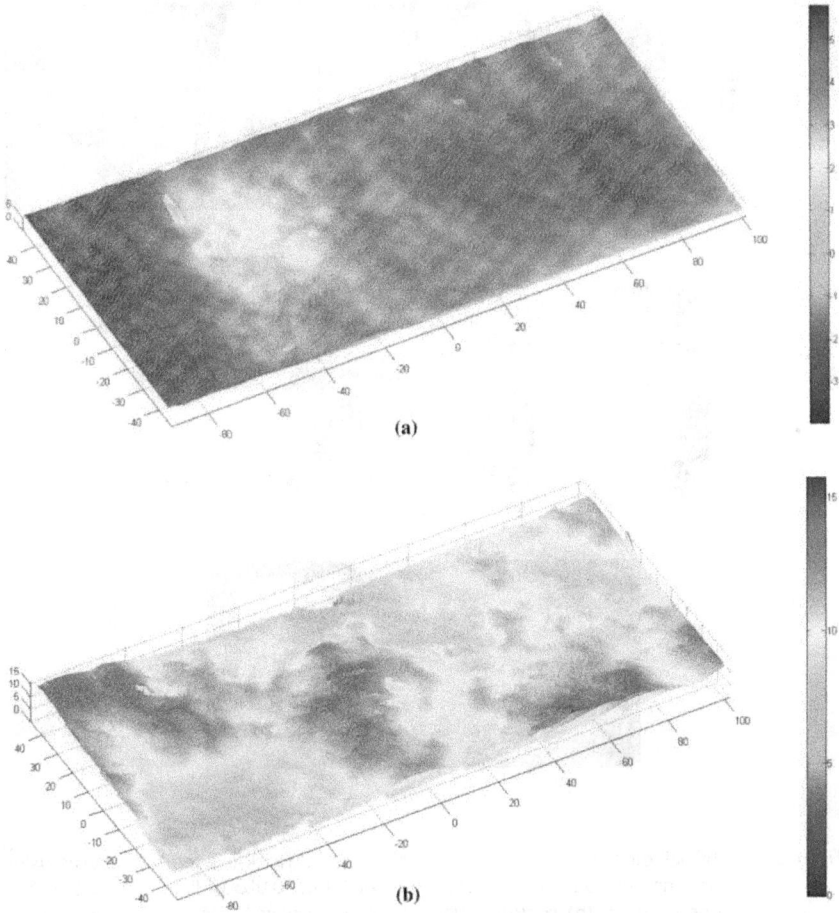

Figure 3.3 Scanning graphs of two natural rough joints for cyclic shear loading-dependent test. (a) Sample X1, JRC = 3.5; (b) sample X2, JRC = 12.3.

3.1.2 Jointed Rock Sample Preparation

1. Sample preparation for rate-dependent consideration

To conduct an experimental study on rough rock joints, it is necessary to prepare identical samples using a molding method. The artificial rock-like joints are illustrated in Figure 3.4a, with a mass ratio of 1:1:0.3:0.005 of lime, sand, water, and water reducer determined by the orthogonal experiment design. Additionally, three more groups of Barton's joints (Figure 3.4b) were prepared to validate the new shear strength criterion. The joint surfaces

(a)

(b)

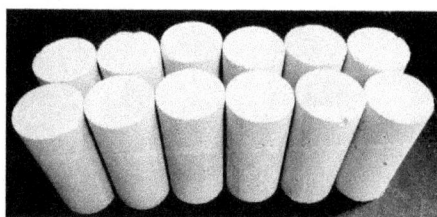

(c)

Figure 3.4 Artificial joint samples and cylinder samples. (a) Rock-like joints prepared using
a molding method with a dosage ratio of 1:1:0.3:0.005 of lime, sand, water, and
water reducer; (b) rock-like Barton joints prepared using a molding method
with a dosage ratio of 1:1:0.3:0.005 of lime, sand, water, and water reducer to
verify the modified JRC-JCS criterion; (c) cylinder samples with the same dosage
ratio as the rock-like joints.

were produced with a size of 200 mm by 100 mm. All of the samples were
held at a constant temperature of 25°C in a chamber for approximately 28
days. Each group consisted of five samples with identical morphology.

Additionally, 12 cylindrical samples (Figure 3.4c) with a diameter of 50 mm
and a height of 100 mm were used to obtain the main mechanical parameters, such as Young's modulus, Poisson's ratio, uniaxial compression strength
(UCS), cohesion and peak friction angle, which were 10.35 GPa, 0.166, 41.80
MPa, 8.78 MPa, and 44.01°, respectively. All tests were performed in triplicate
to achieve the required accuracy for the mechanical parameters. The basic
friction angle of the joint surface, φ_b, was measured by performing four direct

shear tests on the smooth joint under normal stresses of 1, 2, 3, and 4 MPa with a shear rate of 0.5 mm/min to reduce the possibility of rate effect on the results, which provided a basic friction angle, φ_b, of 36°.

2. Sample preparation for cyclic shear loading-dependent consideration

Two rock blocks with rough joint surfaces as illustrated in Figure 3.3 were used to simulate the rock joints. A fully encapsulated rock bolt was inserted at the center of the rock joint, as shown in Figure 3.5. Same to the rate-dependent test, identical samples using the molding method are also needed to conduct cyclic shear tests under different boundary conditions. All the rock samples used in the shear tests were 100 mm in width, 200 mm in length, and 100 mm in height. They were made of mixtures of plaster and water with a weight ratio of 1:0.2. The physical properties of these rock-like samples are shown in Table 3.1.

The fully encapsulated rebar bolt was a thread bar made of iron. Since the samples used in these tests were much smaller than the actual rock bolts, we have to reduce the strength of the bolt material to consider similar effects. Therefore, iron was chosen as the experimental material instead of steel, as shown in Figure 3.5b. Although there is a certain difference with the actual situation, it is possible to explore the evolution of the shear properties of bolted rock joints under different conditions through the small sample tests. The bolt diameter was 6 mm with an ultimate tensile strength of 16 kN. The

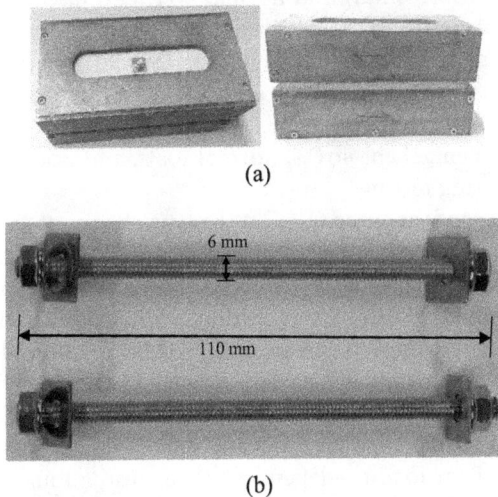

(a)

(b)

Figure 3.5 Artificial joint samples with rock bolt. (a) Joint samples with rock bolt; (b) rock bolt samples.

Table 3.1 Physico-Mechanical Properties of Samples

Physico-Mechanical Properties	Index	Unit	Value
Density	ρ	g/cm^3	2.066
Compressive strength	σ_c	MPa	38.5
Modulus of elasticity	E_s	MPa	28,700
Poisson's ratio	v	–	0.23
Tensile strength	σ_t	MPa	2.5
Cohesion	c	MPa	5.3
Internal friction angle	φ	°	60

total length of the bolt sample was 110 mm, with 5 mm exposed at both ends. A borehole with a diameter of 10 mm was drilled to install the rock bolt sample. The grout was also made of mixtures of plaster and water, with a weight ratio of 1:0.26.

3.1.3 Shear Test Methodology

Tests are performed using direct shear testing apparatus (Jiang et al., 2004; Wang et al., 2009a) under normal loading conditions. The arrangement of the shear test for seismic-dependent consideration is shown in Figure 3.6. The upper block was fixed, and a shear force was applied to the lower block. The shear and normal loads are applied using hydraulic actuators equipped with servo valves. Furthermore, the machine has a design capacity of 600 kN in both the shear (horizontal) and normal (vertical) directions and can test for both artificial and natural discontinuities. The shear and normal displacements are measured using linear variable differential transformers (LVDTs). A maximum shear displacement of 25 mm and a shear rate in the range of 0.01–100 mm/min are obtained. The shear parameters, such as shear displacement, shear stress, normal load, and normal displacement, were recorded during testing.

For the rate-dependent shear test, 2 MPa was chosen as the normal stress, which is approximately 5% of the UCS for the rock-like material, and each sample was subjected to a shear rate of 0.6, 1.2, 6, 12, or 24 mm/min. The AE of the rock-like joints was also monitored by a PAC acoustic emission system. This property has previously been studied by Moradian et al. (2010). Here, four R3a sensors with a dimension of 0.75 in. OD and 0.88 in. in height were selected. The frequency range was set between 5 and 100 kHz, the sampling rate was set to 106 samples per second, and the preamplifier was set to 40 dB.

For the cyclic shear loading-dependent shear test, a long hole was drilled at the center of the shear box to protect the bolt end when the normal stress was applied. During the shear test, a normal stress of 1 MPa was applied on the top of the upper shear box. The shear loading rate is set as 3 mm/min

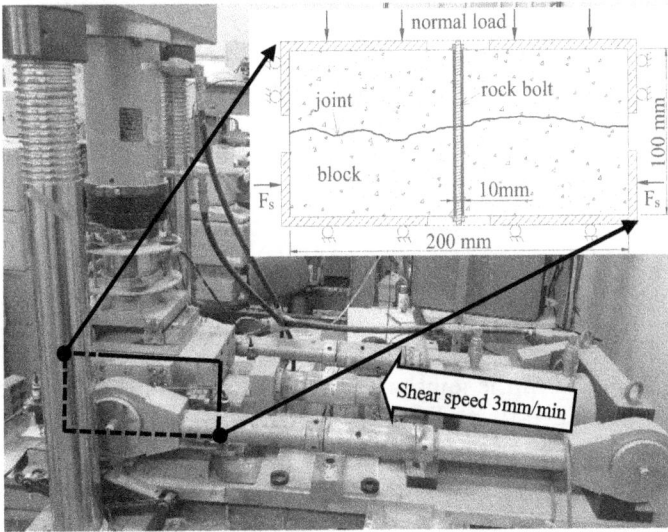

Figure 3.6 Arrangement of the shear test for seismic-dependent consideration.

to simulate the pseudo-static condition (Lee et al., 2001). As the maximum shear displacement of the rock bolt is approximately 9 mm (JRC = 3.5) and 7 mm (JRC = 12.3) in normal direct shear tests. To carry out the cyclic shear tests, the maximum cyclic displacement was determined to be 6 mm. The cyclic displacements were set as 2 mm, 4 mm, and 6 mm. Since the cyclic displacements are arithmetic series, it is relatively easy to analyze the evolution of the shear characteristic. The number of cycles was determined based on the evolution of the shear strength at different cycles. When the number of cycles is greater than five, the shear strength would not substantially change. Therefore, the number of loading cycles was set as 5. When the cyclic shear loading on samples was finished, we further carried out shear tests on the same samples to study the effect of cyclic shear loading on joint properties. The shear test after five cycles was called termination cyclic. As the shear displacement of the termination cyclic is much larger than the cyclic shear test, it is not called the sixth cycle.

3.2 RATE-DEPENDENT MECHANICAL BEHAVIOR OF JOINTED ROCK

Twenty direct shear tests were performed on four groups of rough rock joints, considering the shear rate and joint roughness. The rate-dependent behavior of the rock joints was then investigated based on an analysis of

Table 3.2 Experimental Results of Artificial Joints under Different Shear Rates

Group	Peak Shear Strength/MPa					Decrease Rate/%
	0.6 mm/min	1.2 mm/min	6.0 mm/min	12.0 mm/min	24.0 mm/min	
S1	2.679	2.525	2.458	2.283	2.244	16.237
S2	3.280	3.087	3.041	2.736	2.597	20.823
S3	3.582	3.257	3.210	2.839	2.642	26.242
S4	5.053	4.975	4.514	4.212	4.101	18.405
Increase Rate/%	88.615	97.030	83.645	84.494	82.754	

the shear strength of the joint under different shear rates. The experimental results were used to modify Barton's JRC-JCS criterion (Barton, 1977), which is widely used in practice. The modified criterion was verified by an additional shear test for groups of Barton rock joints and rough joints.

3.2.1 Rate-Dependent Behavior of Shear Strength

Table 3.2 lists the measured peak shear strength values of 20 direct shear tests for the four groups of rough rock-like joints. The relationships between shear strength and shear rate, and between shear strength and *JRC*, are presented in Figure 3.7.

The peak shear strength for the same joints decreases in a nonlinear manner with increased shear rate, and this decreasing rate of shear strength is observed to increase with increased shear rate (Figure 3.7a). As the shear rate increases from 0.6 to 24.0 mm/min, the shear strength decreases by 16.2% (S1), 20.8% (S2), 26.2% (S3), and 18.4% (S4). These values are slightly lower compared to that (30%) of black quartz syenite obtained in Curran's experiments (1983). Based on the experimental results in Table 3.2 and Figure 3.7b, under the same loading conditions, the peak shear strength is controlled by the joint roughness. The shear strength increases with the *JRC*, which increases by 88.6% (0.6 mm/min), 97.0% (1.2 mm/min), 83.6% (6.0 mm/min), 84.5% (12.0 mm/min), and 82.8% (24.0 mm/min). This increase indicates good linearity with a high correlation coefficient (>0.97 in Figure 3.7b, except for the results under a shear rate of 24.0 mm/min), which is consistent with Park's observations (2009).

3.2.2 Rate-Dependent Behavior of Acoustic Emission and Joint Surface Damage

Figures 3.8 and 3.9 provide the typical results for the AE count and AE energy rate for groups S3 and S4. The curves for the corresponding cumulative AE count and AE energy are presented in Figures 3.10 and 3.11, respectively.

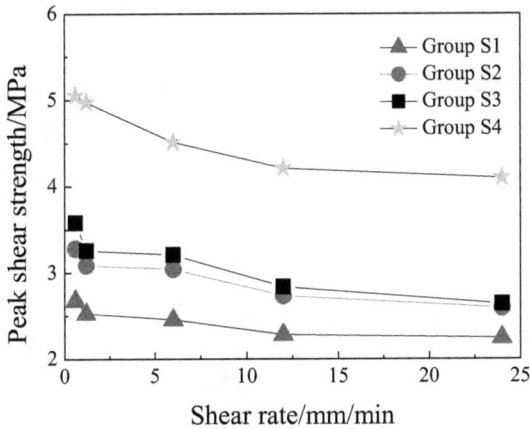

(a)

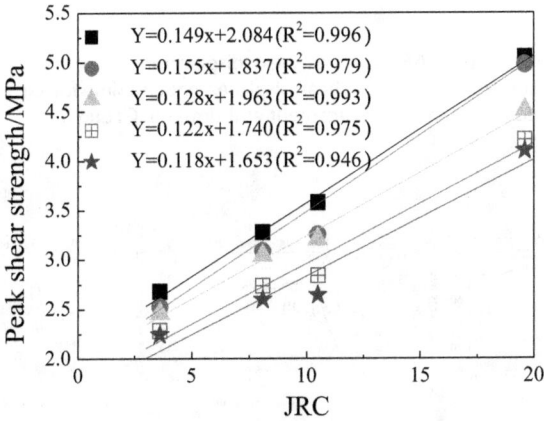

(b)

Figure 3.7 Shear strength vs. shear rate and JRC. (a) Shear strength vs. shear rate; (b) shear strength vs. JRC.

The trends in the AE for different rough joints under different shear rates are consistent. The entire shear process is accompanied by AE monitoring. In the pre-peak I stage, as illustrated in Figures 3.8 and 3.9, the AE count and AE energy rate are relatively low. In the pre-peak nonlinear stage (stage II), due to the initiation of micro-cracks, a few AEs are generated and increase proportionally with the loading. The cumulative AE count and AE energy increase linearly in this stage. In the post-peak stage (stage III), the AE count and AE energy rate increase dramatically and reach their maximum when the major asperities break. Following this peak, several smaller peaks are

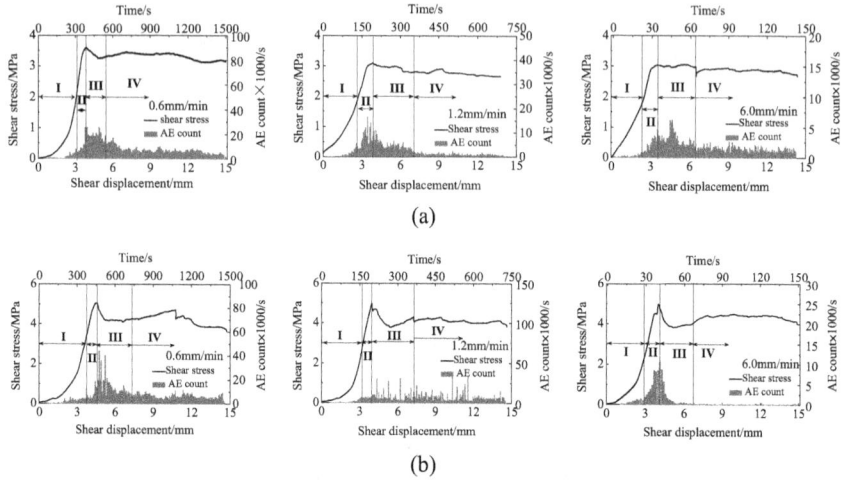

Figure 3.8 Typical results of AE count vs. shear displacement and time for artificial joints. (a) Results of AE count vs. shear displacement and time for Group S3; (b) results of AE count vs. shear displacement and time for Group S4.

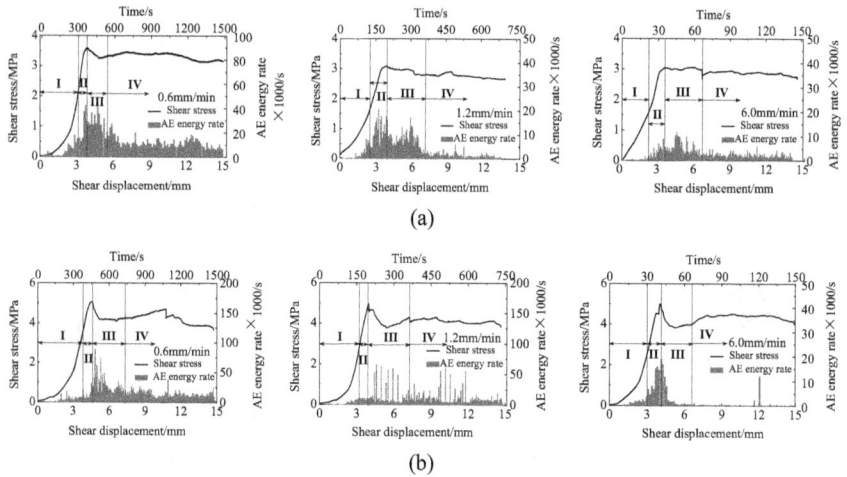

Figure 3.9 Typical results of AE energy rate vs. shear displacement and time for artificial joints. (a) Results of AE energy rate vs. shear displacement and time for Group S3; (b) results of AE energy rate vs. shear displacement and time for Group S4.

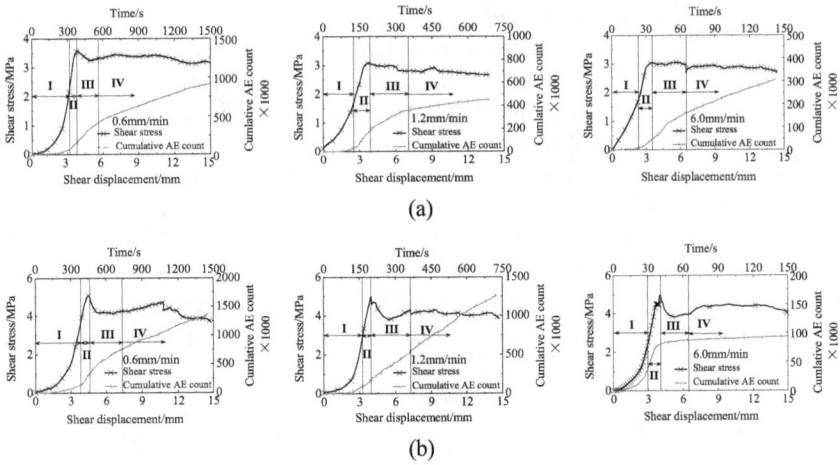

Figure 3.10 Typical results of cumulative AE count vs. shear displacement and time for artificial joints. (a) Results of cumulative AE count vs. shear displacement and time for Group S3; (b) results of cumulative AE count vs. shear displacement and time for Group S4.

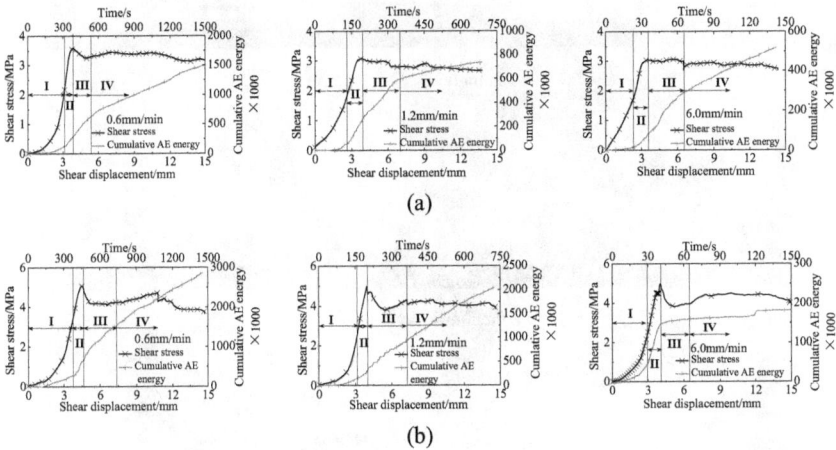

Figure 3.11 Typical results of cumulative AE energy vs. shear displacement and time for artificial joints. (a) Results of cumulative AE energy vs. shear displacement and time for Group S3; (b) results of cumulative AE energy vs. shear displacement and time for Group S4.

generated from the breakage of primary and secondary asperities. Further-more, the cumulative AE count and AE energy increase in a convex form. Because the primary and secondary asperities have been sheared off, the AE count and energy rate in stage IV reach their minimum, and thus the cumulative AE count and AE energy continue to increase extremely slowly. Sliding of the joint surface is the primary movement in this period, which is consistent with the results of Moradian's investigation (Moradian et al., 2010). The curves of the AE count and AE energy rate versus time in Figures 3.8 and 3.9 illustrate that cracks in the rock joints occur earlier under a larger shear rate, and so their corresponding AE count and AE energy rate reach their peak values earlier. Under higher shear rates, the macro-cracks are generated more rapidly with insufficient initiation and propagation of micro-cracks. There-fore, there is less cumulative joint damage with a lower cumulative AE count and AE energy (Figures 3.10 and 3.11). This damage trend can be observed in the approximated area of damage (the area surrounded by red curves) of the rock joints depicted in Figure 3.12. The area of damage decreases with

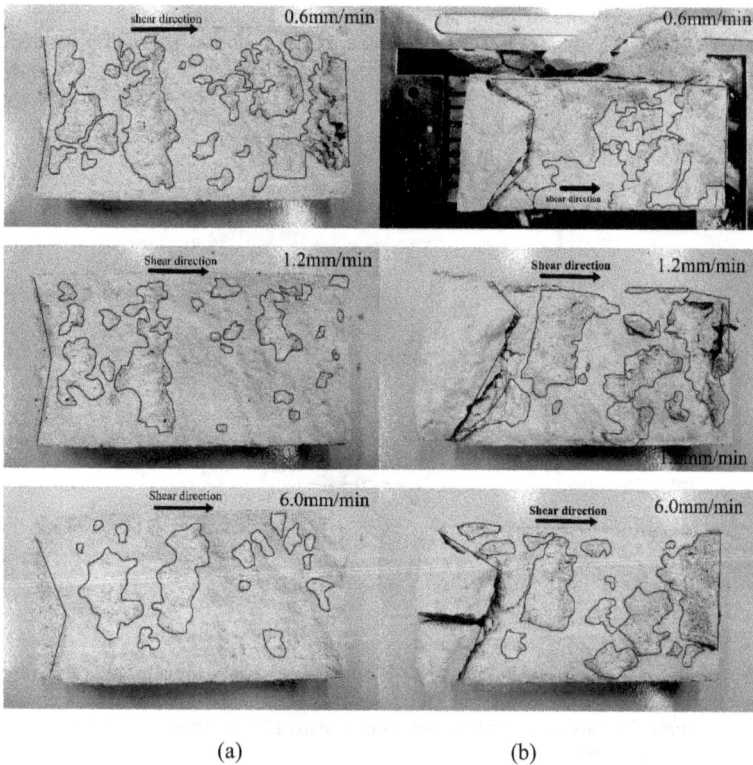

(a)   (b)

Figure 3.12 Area of damage of the joint surface for Groups S3 and S4 sheared under different shear rates. (a) Area of damage of the joint surface for Group S3; (b) area of damage of the joint surface for Group S4.

increasing shear rate. Hence, the corresponding cumulative AE count and AE energy are inversely proportional to the shear rate.

3.2.3 Modified Rate-Dependent Barton's JRC-JCS Criterion

Several criteria exist to estimate the shear strength of rock joints. Among all, Barton's *JRC-JCS* criterion is the most widely used in practice (Barton, 1977) and can be defined as

$$\tau = \sigma_n \tan\left[\varphi_b + JRC_0 \log_{10}\left(\frac{JCS}{\sigma_n}\right)\right] \qquad (3.4)$$

where σ_n, JRC_0, and φ_b represent the normal stress, initial *JRC*, and basic frictional angle, respectively. *JCS* is the joint compression strength, which is related to the degree of weathering of the joint. The *UCS* of the intact rock can be selected as the non-weathered joint compression strength (Zhao, 1998).

However, the criteria discussed above are used for static or quasi-static conditions, which do not consider the shear rate. The large deviation (average value is 32.87%) between the static shear strength calculated using Equation (3.4) and the experimental shear strength under a shear rate of 0.6 mm/min (0.01 mm/s is the static critical loading rate for rough non-weathered joints proposed by Sun [1983]) for each group of joints illustrated in Figure 3.13 reveals that the static Barton's criterion underestimates the rate-dependent shear strength. Therefore, based on the experimental data, we propose a new

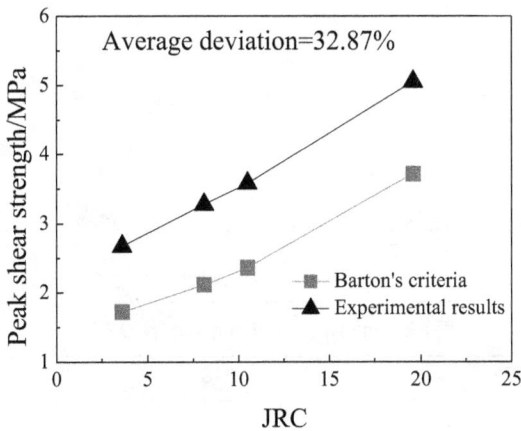

Figure 3.13 Estimate calculated using Barton's criterion in Equation (3.1) and experimental results under a shear rate of 0.6 mm/min.

shear strength criterion. Based on the previous studies (Jing, 1993; Li et al., 2006), the shear strength of a joint simply has a linear relationship with the normal stress, and the normal stress is assumed to be independent of the shear rate and remains constant due to the constant normal load condition. In addition, the friction angle is assumed to be constant. The modified equation can be expressed as

$$\tau = \sigma_n \tan\left[\varphi_b + g\left(JRC_0\right)\log_{10}\left(\frac{JCS}{\sigma_n}\right)\right] f(v) \qquad (3.5)$$

where $g(\cdot)$ is a function of the initial morphology, JRC_0, $f(v)$ is a function of the shear rate, and v is the shear rate.

To eliminate the JRC effect when obtaining the relationship $f(v)$ between the shear strength and shear rate, Equation (3.5) can be simplified as

$$\tau = G\left(\sigma_n, JRC_0\right) f(v) \qquad (3.6)$$

where $G\left(\sigma_n, JRC_0\right) = \sigma_n \tan\left[\varphi_b + g\left(JRC_0\right)\log_{10}\left(JCS / \sigma_n\right)\right]$.

Table 3.3 provides the normalized shear strength values for the four groups of rock-like joints, compared with the values obtained using a shear rate of 0.6 mm/min. Based on the exponential relationship between normalized shear strength and shear rate in Figure 3.14, the function $f(v)$ may be written as

$$f(v) = 0.982v^{-0.06} \qquad (3.7)$$

The relationship between the shear rate and JRC can be calibrated by substituting Equation (3.7) into Equation (3.5) and expressing it as T [T represents the relation $g(JRC_0)$], as defined in Equation (3.8).

$$T = arctan\left(\frac{\tau}{0.982 \cdot \sigma_n \cdot v^{-0.06}} - \varphi_b\right) / \log_{10}\left(\frac{JCS}{\sigma_n}\right) \qquad (3.8)$$

Table 3.3 Normalized Shear Strength for Four Groups of Artificial Joints

Group	Normalized Shear Strength				
	0.6 mm/min	1.2 mm/min	6.0 mm/min	12.0 mm/min	24.0 mm/min
S1	1.000	0.943	0.918	0.852	0.838
S2	1.000	0.941	0.927	0.834	0.792
S3	1.000	0.909	0.896	0.792	0.738
S4	1.000	0.985	0.893	0.834	0.812

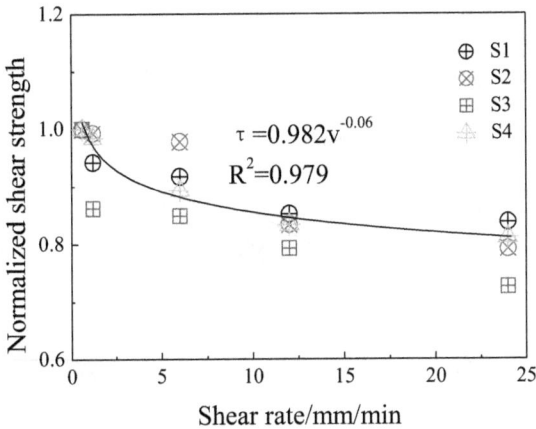

Figure 3.14 Relationship between shear rate and normalized shear strength.

Table 3.4 T Values under Different JRC Conditions

Group	T				
	0.6 mm/min	1.2 mm/min	6.0 mm/min	12.0 mm/min	24.0 mm/min
S1	12.883	11.632	11.059	9.463	9.095
S2	16.997	15.801	16.500	13.324	12.227
S3	18.681	16.859	16.573	14.094	12.593
S4	24.508	24.273	22.739	21.585	21.216

No more normalization should be performed because of the elimination of the rate effect on T while calibrating $f(v)$, and the results of T are provided in Table 3.4 and Figure 3.15. T can be expressed as follows:

$$T = 4.970 (JRC_0)^{0.475} \tag{3.9}$$

After substituting Equations (3.7) and (3.9) into Equation (3.5), a modified rate-dependent shear strength criterion can be expressed as follows:

$$\tau = 0.982 \sigma_n \tan \left[\varphi_b + 4.970 (JRC_0)^{0.475} \log_{10} \left(\frac{JCS}{\sigma_n} \right) \right] \cdot v^{-0.06} \tag{3.10}$$

An additional three groups of rock-like joints proposed by Barton and Choubey were tested to verify the proposed shear strength criterion. The

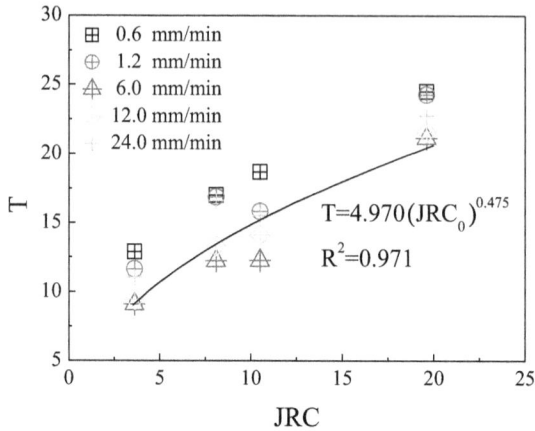

Figure 3.15 Relationship between *JRC* and *T*.

calculated *JRC* of these joints is listed in Table 3.5. Table 3.5 summarizes the experimental and calculated results and provides the estimated error of the *JRC-JCS* criterion and modified proposed criterion. The estimated values using Equation (3.10) indicate good agreement with the experimental results with an average estimation error of 11.0%, which is lower than that of Barton's *JRC-JCS* criterion (28.2%). Therefore, the modified criterion is capable of providing an accurate evaluation of the shear strength for rough rock joints under different shear rates.

3.3 CYCLIC LOADING-DEPENDENT MECHANICAL BEHAVIOR OF JOINTED ROCK

A total of 16 shear tests were conducted for two types of rough joints. Four normal direct shear tests were conducted initially to confirm the performance of the bolted joint in direct shear tests. They served as comparative tests to show the effect of cyclic loading on the shear properties of bolted joints. The effect of the rock bolts lies in the difference in shear strength between the cases with and without rock bolts.

3.3.1 Cyclic Loading-Dependent Effect on the Failure Characteristic of Rock Bolts

The rock bolt samples used in the cyclic shear tests are shown in Figure 3.16a. The positions where the rock bolts were broken are marked by short lines.

Table 3.5 Experimental Results and Criterion Estimation by Barton's Criterion and the Modified Criterion of Rock-Like Joints with Different Shear Rates

Group		Theoretical JRC	Calculated JRC	Shear Rate, mm/min	Shear Strength, MPa			Error,%	
					Experiment Result	JRC-JCS Criterion	Modified Criterion	JRC-JCS Criterion	Modified Criterion
B1	B1–1	2–4	3.67	0.6	2.547	1.727	2.257	32.2	11.4
	B1–2			1.2	2.380		2.165		9.0
	B1–3			6.0	2.174		1.966		9.6
	B1–4			12.0	2.041		1.886		7.6
	B1–5			24.0	1.653		1.809		9.5
B2	B2–1	10–12	11.05	0.6	3.352	2.427	3.050	27.6	10.4
	B2–2			1.2	2.864		2.925		12.1
	B2–3			6.0	2.632		2.656		15.2
	B2–4			12.0	2.507		2.547		16.6
	B2–5			24.0	2.184		2.444		5.5
B3	B3–1	14–16	15.10	0.6	3.864	2.945	3.462	17.2	9.0
	B3–2			1.2	3.779		3.321		2.1
	B3–3			6.0	3.555		3.015		0.9
	B3–4			12.0	3.469		2.892		1.6
	B3–5			24.0	2.935		2.774		11.9
S1	S1–1		3.61	0.6	2.679	1.723	2.249	35.7	16.0
	S1–2			1.2	2.525		2.157		14.6
	S1–3			6.0	2.458		1.959		20.3
	S1–4			12.0	2.283		1.879		17.7
	S1–5			24.0	2.244		1.802		19.7
Average								28.2	11.0

When the cyclic displacement is small (the first two bolts in Figure 3.16a), the rock bolts would be broken into two sections. When the cyclic displacement is relatively large (the last two bolts in Figure 3.16a), the rock bolts would be broken into three sections.

In the shear tests, the failure points usually do not fall precisely on the position of the rock joint as shown in Figure 3.16b. The failure point is about 5 mm from the joint surface. This is because the blocks in the 5 mm range were damaged. Under the combined action of tension and shear forces, the rock bolts broke at the edge of the damage zone. As the failure points usually do not fall precisely on the position of the rock joint, one of the bolt ends must protrude from the joint surface after the bolt broke for the first time. The protruding ends were marked in Figure 3.17. Due to the normal stress, the joint surfaces were still closely in contact with each other during the continuous shearing process. The protruding ends of rock bolts would be further sheared by the rock samples, which could result in secondary

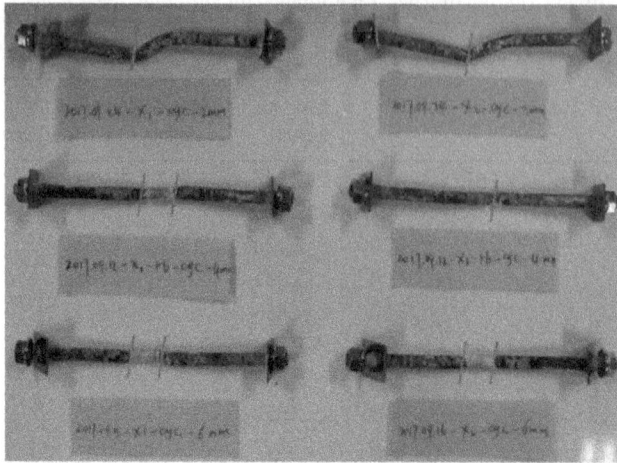

(a)

(b)

Figure 3.16 Samples used in the cyclic shear tests. (a) Rock bolts; (b) damage of rock block.

damage to the rock bolts, especially for the large cyclic displacement conditions. The rock samples used in the shear test are shown in Figure 3.17. The results showed that the asperity damage at the joint surface was relatively minor under the normal direct shear condition (Figure 3.17a). However, the asperity damage increased significantly under the cyclic loading condition (Figure 3.17b–d). The range of damaged area around the rock bolt was usually less than 30 mm.

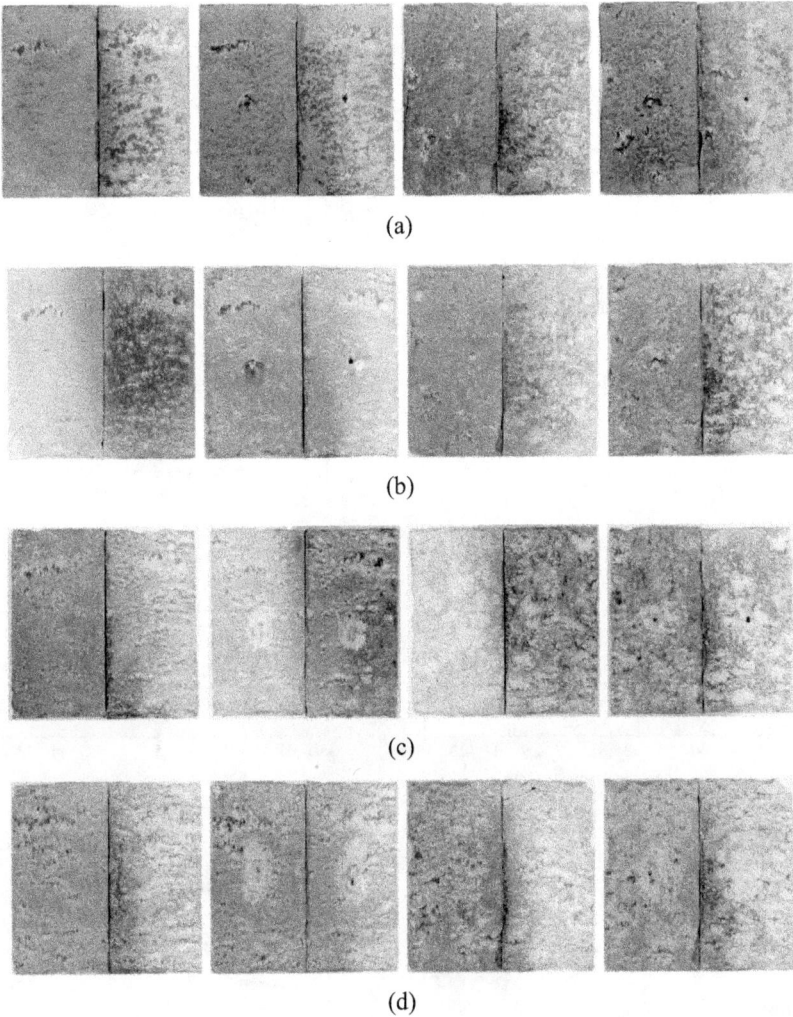

Figure 3.17 Rock joint samples used in the test. (a) Direct shear test; (b) cyclic displacement = 2 mm; (c) cyclic displacement = 4 mm; (d) cyclic displacement = 6 mm.

3.3.2 Cyclic Loading Effect on Shear Performance of Non-Reinforced Jointed Rock

The shear performance of rock joints without rock bolts under cyclic loading conditions is shown in Figure 3.18 by black lines. The shear performance of rock joints in direct shear conditions is shown in Figure 3.18 by blue lines with small circles. Moreover, the shear performance of rock joints after five

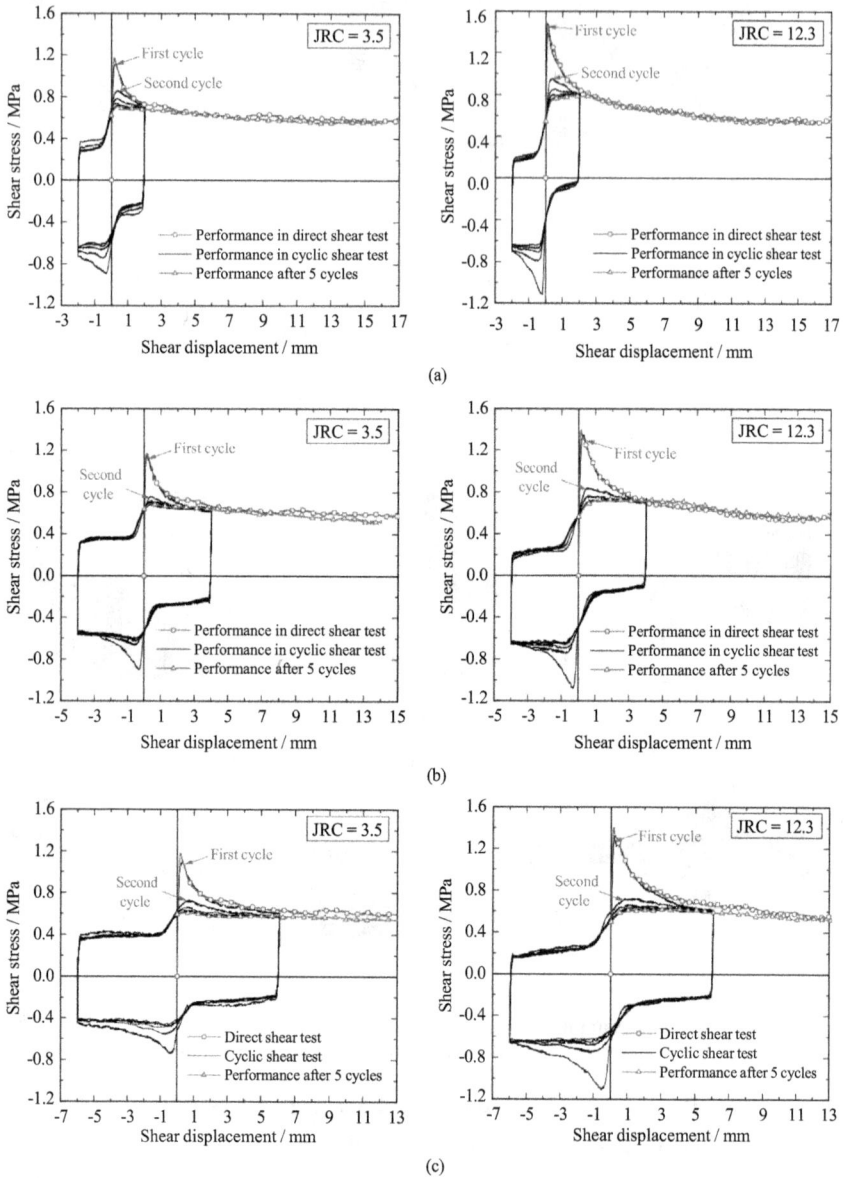

Figure 3.18 Shear behavior of rock joints without rock bolt. (a) Cyclic displacement = 2 mm; (b) cyclic displacement = 4 mm; (c) cyclic displacement = 6 mm.

cycles, namely the termination cycle, is shown in Figure 3.18 by red lines with small triangles. The ordinate axis indicates the shear stress, which is the ratio of shear force-to-total joint area. The axis of abscissa indicates the shear displacement, which is the sliding distance between upper and lower blocks in the horizontal direction. In the first stage of shearing, the shear stress is defined as a positive value. In reverse shear, the shear stress is defined as a negative value. The negative stress/displacement in the results was used to emphasize that the shear directions were different.

In the case of the un-bolted joints, the peak shear strength of the rock joints decreased rapidly with the increasing number of cycles. After five cycles, the peak shear strength was substantially equal to the residual strength. These properties were similar to the results of Lee et al. (2001) and White (2014). When the cyclic displacement was 2 mm as shown in Figure 3.18a, the peak shear strength decreased with the increasing of the number of cycles, indicating that asperity degradation increased gradually. When the cyclic displacement was 4 mm and 6 mm as shown in Figure 3.18b and 3.18c, the shear curves basically would not change after the third cycle, indicating that the asperity degradation had been basically completed in the first two cycles in the case of large cyclic displacements. Comparing the results of the direct shear test and the termination cyclic after five cycles, it was found that the curves were coincident after the cyclic displacement was exceeded.

3.3.3 Cyclic Loading Effect on Shear Performance of Reinforced Jointed Rock

The shear performance of rock joints reinforced by a fully encapsulated rock bolt in cyclic shear tests is shown in Figure 3.19. The shear performance of bolted joints in direct shear conditions and termination cyclic is also shown in Figure 3.19. The sudden drops in the shear stress-displacement curves implied that the rock bolts were broken at these positions. The results of the direct shear tests implied that the rock bolts were broken when the shear displacement reached 10.22 mm and 7.79 mm for the two different JRC conditions. After the failure of rock bolts, the shear resistance dropped to the same level as that for the no-bolt cases. In the condition of the direct shear test, it is easy to find the position where the bolt breaks as shown by the blue lines with small circles in Figure 3.19. However, it is much more complicated in the cyclic shear condition. When the cyclic displacement is small, such as the case of 2 mm, the rock bolt did not break during cyclic shearing. They failed in the termination cycle after five cycles. After the failure of rock bolts, the shear resistance also dropped to the same level as that for the no-bolt cases. When the cyclic displacement is large, such as the case of 6 mm, the rock bolt broke during cyclic shearing. Since the effect of rock bolts is quite small in the second to fifth cycles, it is difficult to determine the exact broken position.

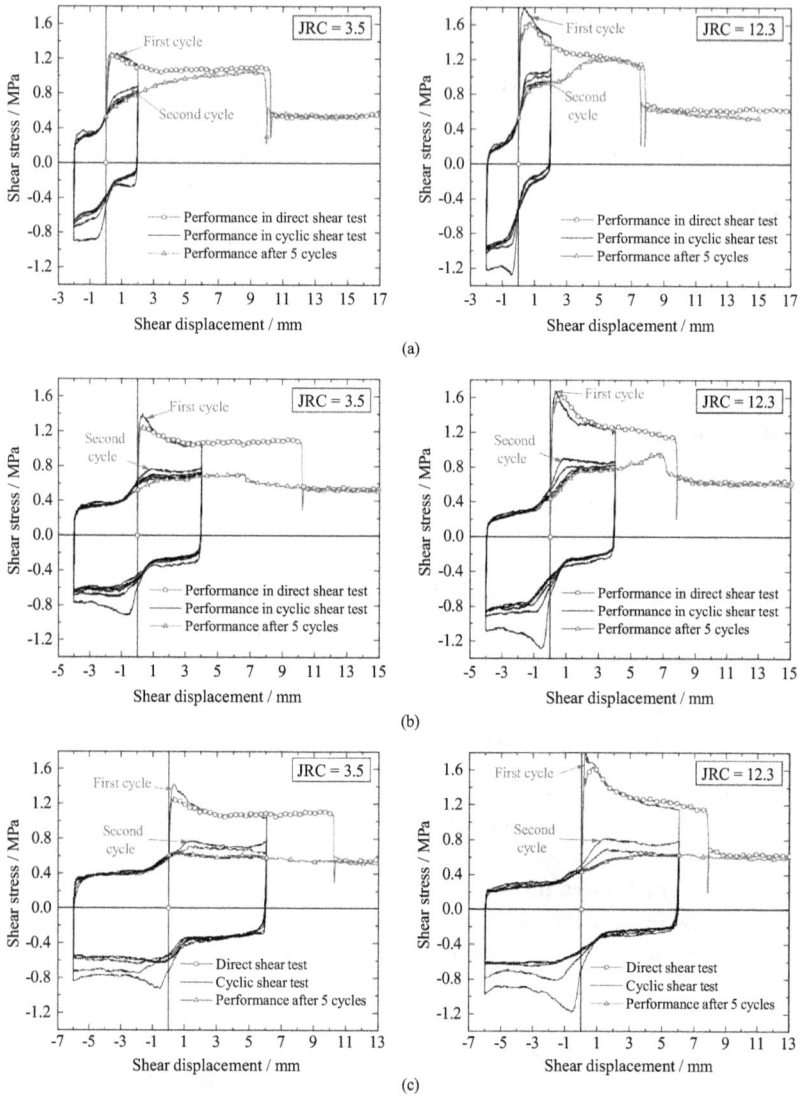

Figure 3.19 Shear behavior of rock joints with fully encapsulated rock bolt. (a) cyclic displacement = 2 mm; (b) cyclic displacement = 4 mm; (c) cyclic displacement = 6 mm.

Compared with the un-bolted joints, the shear strength reduction of bolted joints was much larger under the cyclic loading condition. Therefore, the influence of cyclic loading on the shear resistance of the rock bolt was much larger than on the shear strength of the rock joint itself. When the cyclic displacement was 2 mm as shown in Figure 3.19a, the shear strengths of the bolted rock joints changed greatly with the increasing of the number of cycles, and the rock bolt played a very small role in the second to fifth cycles. According to the results of the termination cyclic after five cycles, it was found that the shear resistance of the rock bolt will gradually recover after the shear displacement has exceeded the cyclic displacement. When the cyclic displacement was 4 mm as shown in Figure 3.19b, the recovery of rock bolt shear resistance was so small that it could be ignored. When the cyclic displacement was 6 mm as shown in Figure 3.19c, no recovery of rock bolt shear resistance was found in the termination cyclic after five cycles, indicating that the rock bolt had completely lost its supporting role after cyclic shear loading.

The evolution of peak shear stress with the increasing of cycles is shown in Figure 3.20. The peak shear strength obtained in direct shear tests is shown by a red cross symbol. In cyclic shear tests, a total of five cycles was conducted. The peak shear strength of the sixth cycle represents the data of

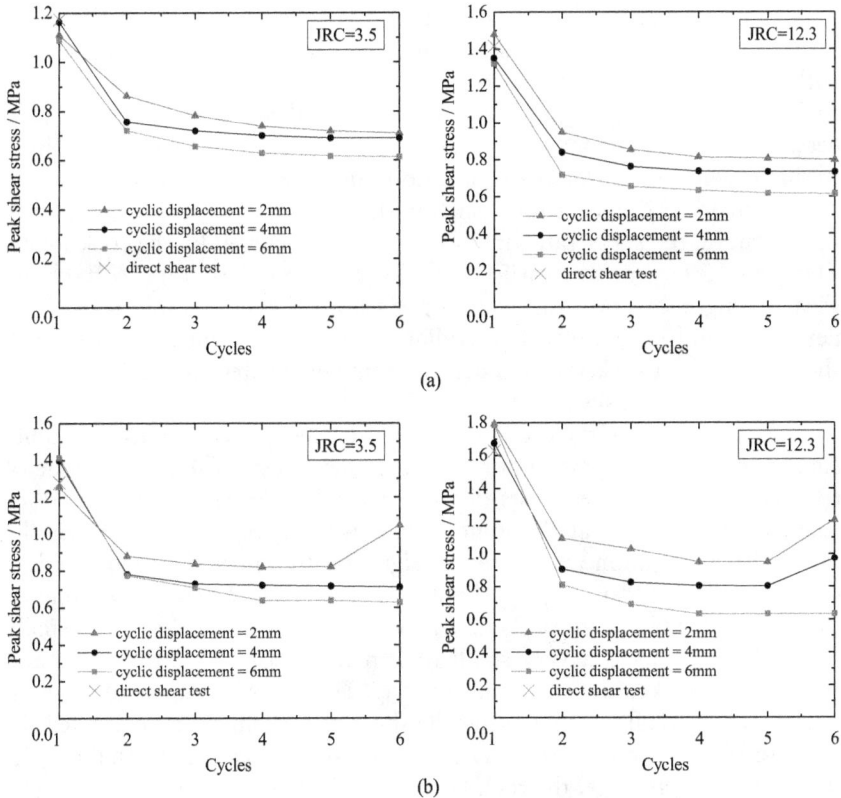

Figure 3.20 Evolution of peak shear stress with the increase of cycles. (a) No-bolt cases; (b) bolted cases.

the termination cyclic after five cycles. The maximum peak shear strength was always observed in the first cycle. For the no-bolt cases, the peak shear strength decreased with the increasing number of cycles. However, for the bolted rock joints, the peak shear strength increased in the termination cyclic in the case of small cyclic displacements.

In the first cycle of the cyclic shear test, the shear performance was quite similar to the performance in direct shear tests, which implied that the test results were repeatable. The difference between the peak shear stress in the direct shear test and in cyclic shear stress is smaller than 15% of its own magnitude. Such errors are normally acceptable in geotechnical tests due to the non-uniformity of rock samples.

3.4 CONCLUSION

This chapter provides a laboratory shear test methodology for the seismic-dependent mechanical behavior of rock mass. Shear rate and cyclic shear loading are two aspects of the seismic-dependent consideration. Jointed rock with/without reinforcement is considered.

The peak shear strength is controlled by the shear rate and joint roughness. The former factor has a nonlinear relationship with the peak shear strength, whereas the latter exhibits good linearity with a high correlation coefficient (40.97, except for results under a shear rate of 24.0 mm/min). Furthermore, the shear rate affects the damage incurred by the rock joints. The joint surface breaks earlier with an increased shear rate. A modified shear strength criterion, which considers both the shear rate and JRC, is developed and is capable of providing an accurate estimate for the peak shear strength of rock joints under different shear rates, as validated by an additional 20 direct shear tests on the Barton joints and rough joints.

In the first cycle of the cyclic shear test, the shear performance was quite similar to its counterpart in normal direct shear tests. The failure mode of rock bolts in cyclic shear tests differed from the normal direct shear tests. The asperity damage at the joint surface was relatively minor under the direct shear condition and increased significantly under the cyclic loading condition. The sudden drops in the shear stress-displacement curves implied that the rock bolts were broken. When the cyclic displacement was small, the rock bolt played a very small role in the second to fifth cycles, and remained unbroken during cyclic shearing. The shear resistance of the rock bolt would gradually recover after the shear displacement has exceeded the cyclic displacement in the termination cyclic after five cycles. When the cyclic displacement was large, the rock bolt broke during cyclic shearing, and the shear resistance could not be recovered in the termination cyclic. It is indicated that the rock bolt had completely lost its supporting role after cyclic shear loading. The shear strength reduction of bolted joints under the cyclic loading condition was much more significant compared to the un-bolted joint. Therefore, the influence of cyclic loading on the shear resistance of the rock bolt was much larger than on the shear strength of the rock joint itself. The shear performance of a rock bolt inserted in a rock joint was strongly influenced by cyclic shear loading. It is worthy to note that normal stress will influence the stress-strain behavior under cyclic shear tests, and a special study on this issue is required.

Chapter 4

Propagation Behavior of Seismic Wave on Rock Fracture

Analytic Method

4.1 METHOD ON SEISMIC PROPAGATION BEHAVIOR OF ROCK MASS

4.1.1 Propagation of Seismic Wave in the Rock Medium

The propagation of seismic waves causes dynamic stress changes in rocks. Generally, the amplitude of the induced stress is smaller than the rock strength, except perhaps near the source. Thereby, the problem of seismic wave propagation is usually analyzed by using the dynamic theory of linear elasticity. As stress waves propagate, the elements within rocks are accelerated as a result of surrounding stresses, then the differential equations of motion representing the stress and motion statuses in the medium can be derived from Newton's second law of motion, as given by

$$
\left.
\begin{aligned}
\partial \sigma_x / \partial x + \partial \tau_{yx} / \partial y + \partial \tau_{zx} / \partial z &= \rho \partial^2 u_x / \partial t^2 \\
\partial \tau_{xy} / \partial x + \partial \sigma_y / \partial y + \partial \tau_{zy} / \partial z &= \rho \partial^2 u_y / \partial t^2 \\
\partial \tau_{xz} / \partial x + \partial \tau_{yz} / \partial y + \partial \sigma_z / \partial z &= \rho \partial^2 u_z / \partial t^2
\end{aligned}
\right\}
\tag{4.1}
$$

where u_x, u_y, and u_z are displacement components of elements in the x, y, and z directions; σ_x, σ_y, σ_z, τ_{xy}, τ_{xz}, and τ_{yz} are stress components; ρ is the material density; and t is time. Based on the theory of elasticity, the stress components can be expressed in the forms of strains, and the above equation can be rewritten as (Achenbach, 1973; Shen, 1998)

$$
\left.
\begin{aligned}
(\lambda + \mu)\partial \varepsilon_v / \partial x + \mu \nabla^2 u_x &= \rho \partial^2 u_x / \partial t^2 \\
(\lambda + \mu)\partial \varepsilon_v / \partial y + \mu \nabla^2 u_y &= \rho \partial^2 u_y / \partial t^2 \\
(\lambda + \mu)\partial \varepsilon_v / \partial z + \mu \nabla^2 u_z &= \rho \partial^2 u_z / \partial t^2
\end{aligned}
\right\}
\tag{4.2}
$$

DOI: 10.1201/9781003401599-4

where λ and μ are Lamé's coefficients; ε_v is the volumetric strain; and ∇^2 is the Laplace operator. Equation (4.2) is the general form of the wave equation, which relates the time and spatial information in the medium and can be used to describe the propagating seismic waves. It can also be represented in the vector form of $(\lambda + \mu)\nabla(\nabla \cdot \boldsymbol{u}) + \mu \nabla^2 \boldsymbol{u} = \rho \ddot{\boldsymbol{u}}$.

To solve this equation, the Helmholtz decomposition (Sternberg, 1960; Jaeger et al., 2007) can be utilized, which decomposes the displacement vector ($\boldsymbol{u}$) into the gradient of a scalar potential (φ) plus the curl of a divergence-free vector potential ($\boldsymbol{\psi}$), as:

$$\boldsymbol{u} = div\varphi + curl\boldsymbol{\psi} = \nabla\varphi + \nabla \times \boldsymbol{\psi} \tag{4.3}$$

The displacement corresponding to φ ($\boldsymbol{u} = \nabla\varphi$) denotes a deformation without rotation, corresponding to the particle motions caused by P-waves. Therefore, the wave equation for the P-wave can expressed in the vector form of $(\lambda + 2\mu)\nabla^2 \boldsymbol{u} = \rho \ddot{\boldsymbol{u}}$, which is equivalent to the following scalar wave equations:

$$V_P^2 \nabla^2 u_x = \partial^2 u_x / \partial t^2, V_P^2 \nabla^2 u_y = \partial^2 u_y / \partial t^2, V_P^2 \nabla^2 u_z = \partial^2 u_z / \partial t^2 \tag{4.4}$$

The other part of the decomposed displacement $\boldsymbol{u} = \nabla \times \boldsymbol{\psi}$ represents a state of pure shear without volumetric strain. It corresponds to the particle motions caused by S-waves. Thereby, the wave equation of the S-wave can be given by $(\lambda + 2\mu)\nabla^2 \boldsymbol{u} = \rho \ddot{\boldsymbol{u}}$, which is equivalent to the three scalar equations of

$$V_S^2 \nabla^2 u_x = \partial^2 u_x / \partial t^2, V_S^2 \nabla^2 u_y = \partial^2 u_y / \partial t^2, V_S^2 \nabla^2 u_z = \partial^2 u_z / \partial t^2 \tag{4.5}$$

The P-wave and S-wave have similar equation forms but with different propagation velocities, as denoted by V_P and V_S, respectively. V_P and V_S depend on the rock properties, and can be given by

$$V_P = \sqrt{(\lambda + 2\mu)/\rho}, V_S = \sqrt{\mu/\rho} \tag{4.6}$$

4.1.2 Method of Seismic Wave Propagation in the Fractured Rock Mass

The propagation of seismic waves in fractured rock masses is affected significantly by the existing fractures. Seismic waves traveling across fractures will be attenuated in the aspects of velocity slowness and energy dissipation. In addition, the wave scattering induced by fractures (e.g., wave reflection, transmission, and diffraction) changes the propagation direction of waves and affects the distribution of the dynamic stress field in the rock mass (Du, 2009). The superposition of multiple reflected and transmitted waves can also cause localized stress concentration around fractures. At present, there are three major methods to describe wave propagation in fractured rock masses, including the effective medium method, the welded interface method and the displacement discontinuity method. They are introduced briefly as follows:

I. Effective medium method

Fractures produce large deformations under external disturbances and hence reduce the elastic moduli of rock masses and the velocities of seismic waves. To date, several effective medium models have been developed based on various assumptions to describe the overall elastic and seismic properties of media containing fractures (e.g., O'Connell & Budiansky, 1974; Anderson et al., 1974; Moreland, 1974; Crampin, 1984; Hudson, 1980, 1981, 1986, 1990a, 1990b, 1994; Schoenberg, 1983; Schoenberg & Douma, 1988; Schoenberg & Muir, 1989; Schoenberg and Sayers, 1995). Here, three representative ones widely used in rock mechanics will be reviewed.

a. O'Connell and Budiansky's effective medium model

The model proposed by O'Connell and Budiansky (1974) can be used to calculate the elastic moduli of solids permeated with an isotropic distribution of flat cracks and to predict the wave velocities in the cracked media. Cracks can be dry or saturated, with sizes much smaller than the wavelength. For solids containing dry cracks, the effective Young's modulus ($\bar{E}$) and the shear modulus ($\bar{G}$) can be expressed by Equations (4.7) and (4.8), corresponding to the circular and elliptical cracks, respectively, where E and G are the moduli of intact material; is the effective Poisson's ratio; $T(b/a,\bar{v})$ is a function related to $\bar{v}$ and the shape of cracks; and e is the crack density parameter.

$$\left.\begin{aligned} \bar{E}/E &= 1 - \left[16\left(1-\bar{v}^2\right)(10-3\bar{v})e\right]/[45(2-\bar{v})] \\ \bar{G}/G &= 1 - [32(1-\bar{v})(5-\bar{v})e]/[45(2-\bar{v})] \end{aligned}\right\} \tag{4.7}$$

$$\left.\begin{aligned} \bar{E}/E &= 1 - \left[16\left(1-\bar{v}^2\right)(3+T(b/a,\bar{v}))e\right]/45 \\ \bar{G}/G &= 1 - [32(1-\bar{v})(1+0.75T(b/a,\bar{v}))e]/45 \end{aligned}\right\} \tag{4.8}$$

The velocities of S- and P-waves in the cracked media ($\bar{V}_S$ and $\bar{V}_P$) can be obtained directly from the effective moduli ($\bar{K}$ is effective bulk modulus), as given by

$$\bar{V}_S / V_S = \sqrt{\bar{G}/G}, \bar{V}_P / V_P = \sqrt{[(1-\bar{v})(1+v)\bar{K}]/[(1+\bar{v})(1-v)K]} \qquad (4.9)$$

b. Hudson's effective medium model

The model proposed by Hudson (1980, 1981, 1986, 1990a, 1990b, 1994) can be used to describe the properties of an elastic solid with a random distribution of penny-shaped cracks, with the assumptions of long wavelength and sparsely distributed cracks (the crack density $e = va^3 \ll 1$, where v is the number density of cracks and a is the crack radius). The effective elastic constants for the cracked material can be expressed by

$$c = c^0 + ec^1 + e^2 c^2 \qquad (4.10)$$

where c^0 is the elastic tensor for uncracked material; c^1 and c^2 are the first-order and second-order corrections accounting for scattering and crack interaction. Hudson (1980, 1981) gave the expressions of effective Lamé's constants ($\bar{\lambda}$ and $\bar{\mu}$) as:

$$(3\bar{\lambda}+2\bar{\mu})/(3\lambda+2\mu) = -e(3\lambda+2\mu)U_{33}/(3\mu), \quad \bar{\mu}/\mu = -2e(3U_{11}+2U_{33})/15 \qquad (4.11)$$

where λ and μ are Lamé's constants of uncracked material; and U_{11} and U_{33} are quantities depending primarily on the conditions imposed on the surface of cracks. Based on this model, the wave velocities for the case of dry cracks can be predicted by

$$\left.\begin{aligned}
(\omega/K)^2 &= V_P^2 \left\{1 - 4e\left[(\lambda+2\mu\cos^2\theta)^2/(\lambda\mu+\mu^2) + (16\mu\sin^2\theta\cos^2\theta)/(3\lambda+4\mu)\right]/3\right\} \\
(\omega/K)^2 &= V_{SV}^2 \left\{1 - 16e(\lambda+2\mu)\left[(\cos^2\theta\sin^2\theta)/(\lambda+\mu) + \cos^2(2\theta)/(3\lambda+4\mu)\right]/3\right\} \\
(\omega/K)^2 &= V_{SH}^2 \left[1 - 16e(\lambda+2\mu)\cos^2\theta/(9\lambda+12\mu)\right]
\end{aligned}\right\} \qquad (4.12)$$

where V_P, V_{SV}, and V_{SH} are velocities of P-, S_V-, and S_H-waves; ω is the angular frequency of wave; K is the wave number; and θ denotes the propagation direction.

c. Schoenberg's effective medium model

The slip-interface model proposed by Schoenberg (1983), Schoenberg and Douma (1988), and Schoenberg and Muir (1989) can simulate the behavior of fractured solids containing large, closely spaced, and aligned joints. This model represents the stiffness matrix of a fracture system by using three

3×3 matrices: $\mathbf{M}$ relating transverse stresses and strains, $\mathbf{N}$ relating normal stresses and strains, and $\mathbf{P}$ relating transverse stresses to normal strains and normal stresses to transverse strains. The stress-strain relation for the ith layer can be given by

$$\mathbf{S}_{1i} = \mathbf{M}_i \mathbf{E}_1 + \mathbf{P}_i \mathbf{E}_{2i}, \ \mathbf{S}_2 = \mathbf{P}_i^T \mathbf{E}_1 + \mathbf{N}_i \mathbf{E}_{2i} \tag{4.13}$$

in which $\mathbf{S}_{1i}$ is the layer-dependent stress vector; $\mathbf{S}_2$ is the layer-independent stress vector; $\mathbf{E}_{2i}$ is the layer-dependent strain vector; and $\mathbf{E}_1$ is the layer-independent strain vector. Over all the constituent layers, the thickness-weighted averages for those vectors and matrixes are denoted by $\mathbf{M}_e$, $\mathbf{N}_e$, $\mathbf{P}_e$, $\langle \mathbf{S}_1 \rangle$, and $\langle \mathbf{E}_2 \rangle$. The stress-strain relation for the fractured system can be written in the following form, which can be used to calculate the effective moduli of fractured systems.

$$\langle \mathbf{S}_1 \rangle = \mathbf{M}_e \mathbf{E}_1 + \mathbf{P}_e \langle \mathbf{E}_2 \rangle, \mathbf{S}_2 = \mathbf{P}_e^T \mathbf{E}_1 + \mathbf{N}_e \langle \mathbf{E}_2 \rangle \tag{4.14}$$

Schoenberg and Douma (1988) proposed the specific forms of those expressions for different fracture systems, and gave the expressions of the wave velocities for the transversely isotropic fracture system, as shown by

$$\left.\begin{aligned}
V_S^2(\theta) &\approx \mu_b \left[1 - \cos^2 \theta E_T \right] / \rho_b \\
V_{QS}^2(\theta) &\approx \mu_b \left[1 - \cos^2 2\theta E_T - \gamma_b \sin^2 2\theta E_N \right] / \rho_b \\
V_{QP}^2(\theta) &\approx (\lambda_b + 2\mu_b) \left[1 - \gamma_b \sin^2 2\theta E_T - \left(1 - 2\gamma_b \sin^2 \theta \right)^2 E_N \right] / \rho_b
\end{aligned}\right\} \tag{4.15}$$

where V_S, V_{QS}, and V_{QP} denote the velocities of the pure shear wave, quasi-shear wave, and quasi-compressional wave, respectively; ρ_b, μ_b, and λ_b are the Poisson's ratio and Lamé's constants for background material; θ is the propagation angle of the wave; and E_N and E_T are dimensionless compliances that give the fracture system compliances relative to the background compliances, normal and tangential to the fracture system, respectively.

2. Welded interface method

The welded interface method treats rock fractures as welded interfaces, and thereby the stress and displacement fields across the interfaces are continuous. This method has been commonly adopted to investigate the problems of seismic wave propagation across stratified geomaterials. When solving the equation of motion in the neighborhood of a welded interface between two elastic media, the dynamic solutions in one medium must be matched with those in the second medium through interface conditions.

When a seismic wave impinges upon a welded interface between two types of rocks, some portion of the energy is reflected back into the first medium, and the remaining portion is transmitted into the second medium. The reflection and transmission of seismic waves at a welded interface have been studied extensively (e.g., Jeffreys, 1926; Muskat & Meres, 1940; Ewing et al., 1957; Brekhovskikh, 1960; Du, 1996). In the welded interface model, the coordinate system can be established along the interface plane (see Figure 4.1). If a P-wave or an S_V-wave (particles undergo in-plane motions) impinges obliquely on an interface, four types of elastic waves will be generated: a reflected P-wave, a reflected S_V-wave, a transmitted P-wave, and a transmitted S_V-wave (Figures 4.1a and 4.1b). While in the case of an obliquely incident S_H-wave (particles move in the out-of-plane direction), only two types of waves will be created: a reflected S_H-wave and a transmitted S_H-wave (Figure 4.1c). Taking the case of incident harmonic P-wave as an example (Figure 4.1a), the expressions of the incident, reflected, and transmitted waves can be written in the following complex forms:

$$
\begin{aligned}
\boldsymbol{u}_{IP} &= \{u_{IP}, v_{IP}\} = I_P \exp\left[iK_{P1}\left(x\sin\alpha_1 - y\cos\alpha_1 - V_{P1}t\right)\right] \times \{\sin\alpha_1, -\cos\alpha_1\} \\
\boldsymbol{u}_{RP} &= \{u_{RP}, v_{RP}\} = R_P \exp\left[iK_{P1}\left(x\sin\alpha_1 + y\cos\alpha_1 - V_{P1}t\right)\right] \times \{\sin\alpha_1, \cos\alpha_1\} \\
\boldsymbol{u}_{RST} &= \{u_{RSV}, v_{RST}\} = R_{SV} \exp\left[iK_{S1}\left(x\sin\beta_1 + y\cos\beta_1 - V_{S1}t\right)\right] \times \{\cos\beta_1, -\sin\beta_1\} \\
\boldsymbol{u}_{TP} &= \{u_{TP}, v_{IP}\} = T_P \exp\left[iK_{P2}\left(x\sin\alpha_2 + y\cos\alpha_2 - V_{P2}t\right)\right] \times \{\sin\alpha_2, -\cos\alpha_2\} \\
\boldsymbol{u}_{TST} &= \{u_{TSV}, v_{TSV}\} = T_{SV} \exp\left[iK_{S2}\left(x\sin\beta_2 - y\cos\beta_2 - V_{S2}t\right)\right] \times \{-\cos\beta_2, -\sin\beta_2\}
\end{aligned} \tag{4.16}
$$

in which $\boldsymbol{u}_i$ is the particle displacement vector composed of the horizontal component u_i and the vertical component v_i ($i = IP, RP, RSV, TP,$ or TSV, representing the incident P-wave, reflected P-wave, reflected S_V-wave, transmitted P-wave, and transmitted S_V-wave, respectively); $I_P, R_P, R_{SV}, T_P,$ and T_{SV} denote the amplitudes of different waves; K_{ij} and V_{ij} ($i = P$ or $S; j = 1$ or 2) are the wave number and velocity for the P- or S-wave in the first (or second) medium, respectively; α_i and β_i ($i = 1$ or 2) are propagation angles of P- and S-wave, respectively, in different media; $x, y,$ and t denote the coordinates and time.

According to the assumptions of the welded interface, the stress and displacement fields at both sides of the interface need to be matched. Therefore, the boundary conditions at the interface ($y = 0$) can be denoted by $\sigma_{y1} = \sigma_{y2}$, $\tau_{xy1} = \tau_{xy2}$, $u_1 = u_2$, and $v_1 = v_2$ (subscripts 1 and 2 denote the medium number). The stress and displacement components can be obtained from the wave equation and the elastic theory, and the boundary conditions at the welded interface can be expressed as

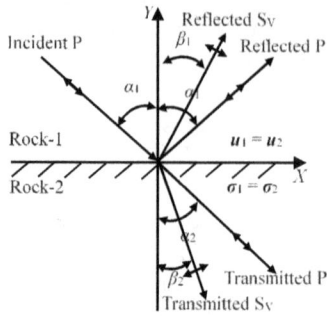

(a)

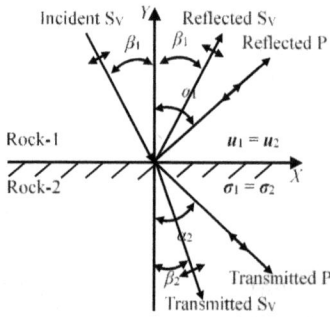

(b)

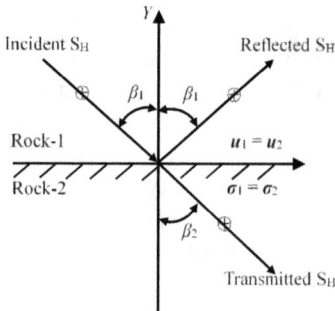

(c)

Figure 4.1 Welded interface model for describing the reflection and transmission of seismic waves at an interface between different rock media. (a) Incidence of P-wave; (b) incidence of S_V-wave; (c) incidence of S_H-wave.

$$
\left.\begin{array}{c}
u_{IP} + u_{RP} + u_{RSV} = u_{TP} + u_{TSV} \ (u_1 = u_2) \\[4pt]
v_{IP} + v_{RP} + v_{RSV} = v_{TP} + v_{TSV} \ (v_1 = v_2) \\[4pt]
\mu_1\left(\dfrac{\partial v_1}{\partial x} + \dfrac{\partial u_1}{\partial y}\right) = u_2\left(\dfrac{v_2}{\partial x} + \dfrac{\partial u_2}{\partial y}\right)(\tau_{xy1} = \tau_{xy2}) \\[6pt]
\lambda_1 \partial u_1/\partial x + (\lambda_1 + 2\mu_1)\partial v_1/\partial y = \lambda_2 \partial u_2/\partial x + (\lambda_2 + 2\mu_2)\partial v_2/\partial y \left(\sigma_{y1} = \sigma_{y2}\right)
\end{array}\right\}
\qquad (4.17)
$$

By substituting the P-wave expressions into the boundary conditions, four algebraic equations for the amplitudes of the four waves generated by the incident wave can be found (Achenbach, 1973; Jaeger et al., 2007), as given by

$$
\begin{bmatrix}
\sin\alpha_1 & \cos\beta_1 & -\sin\alpha_2 & \cos\beta_2 \\[4pt]
\cos\alpha_1 & -\sin\beta_1 & \cos\alpha_2 & \sin\beta_2 \\[4pt]
-\cos 2\beta_1 & \dfrac{Z_{S1}}{Z_{P1}}\sin 2\beta_1 & \dfrac{Z_{P2}}{Z_{P1}}\cos 2\beta_2 & \dfrac{Z_{S2}}{Z_{P1}}\sin 2\beta_2 \\[8pt]
\sin 2\alpha_1 & \dfrac{Z_{P1}}{Z_{S1}}\cos 2\beta_1 & \dfrac{Z_{S2}^2 Z_{P1}}{Z_{S1}^2 Z_{P2}}\sin 2\alpha_2 & -\dfrac{Z_{S2}Z_{P1}}{Z_{S1}^2}\cos 2\beta_2
\end{bmatrix}
\begin{bmatrix} R_P \\ R_{SV} \\ T_P \\ T_{SV} \end{bmatrix}
= I_P
\begin{bmatrix} -\sin\alpha_1 \\ \cos\alpha_1 \\ \cos 2\beta_1 \\ \sin 2\alpha_1 \end{bmatrix}
\qquad (4.18)
$$

where Z_{ij} ($i = P$ or S; $j = 1$ or 2) is the seismic impedance of the P- or S-wave in the first or the second medium. Reflection and transmission coefficients, which are the ratios of amplitudes of reflected and transmitted waves to that of the incident wave, can also be obtained. Similarly, when considering the incidence of an S_V-wave or an S_H-wave, the amplitudes of the reflected and transmitted waves can be derived from

$$
\begin{bmatrix}
\sin\alpha_1 & \cos\beta_1 & -\sin\alpha_2 & \cos\beta_2 \\[4pt]
\cos\alpha_1 & -\sin\beta_1 & \cos\alpha_2 & \sin\beta_2 \\[4pt]
\dfrac{Z_{P1}}{Z_{S1}}\cos 2\beta_1 & -\sin 2\beta_1 & -\dfrac{Z_{P2}}{Z_{S1}}\cos 2\beta_2 & -\dfrac{Z_{S2}}{Z_{S1}}\sin 2\beta_2 \\[8pt]
\dfrac{Z_{S1}}{Z_{P1}}\sin 2\alpha_1 & \cos 2\beta_1 & \dfrac{Z_{S2}^2}{Z_{S1}Z_{P2}}\sin 2\alpha_2 & -\dfrac{Z_{S2}}{Z_{S1}}\cos 2\beta_2
\end{bmatrix}
\begin{bmatrix} R_P \\ R_{SV} \\ T_P \\ T_{SV} \end{bmatrix}
= I_{SV}
\begin{bmatrix} -\cos\beta_1 \\ -\sin\beta_1 \\ \sin 2\beta_1 \\ \cos 2\beta_1 \end{bmatrix}
\qquad (4.19)
$$

$$
R_{SH} = I_{SH}\frac{Z_{S1}\cos\beta_1 - Z_{S2}\cos\beta_2}{Z_{S1}\cos\beta_1 + Z_{S2}\cos\beta_2},\ T_{SH} = I_{SH}\frac{2Z_{S1}\cos\beta_1}{Z_{S1}\cos\beta_1 + Z_{S2}\cos\beta_2}
\qquad (4.20)
$$

The closed-form solutions to those equations can be derived in a straight-forward manner using matrix inversion algorithms. It can be found that the seismic amplitudes for reflected and transmitted waves depend on the impedances of two neighboring rock media and the angles of incidence, reflection, and transmission. The frequency is not involved in those equations, indicating that the reflected and transmitted waves will have identical frequency to the incident wave, but have different amplitudes.

3. Displacement discontinuity method

In the welded interface method, the displacement field around the welded interface is assumed to be continuous. When subjected to external loads, fractures undergo deformations normal to or along their original planes, leading to a discontinuous displacement field. To account for the deformation behavior of fractures and its effects on the dynamic behavior of rock masses, the displacement discontinuity method was developed based on the works by Jones and Whittier (1967), Kendall and Tabor (1971), Schoenberg (1980), Pyrak-Nolte et al. (1988, 1990) and Myer et al. (1990).

The displacement discontinuity method considers that the stress field is continuous across rock fractures, but the displacement field is discontinuous due to the fracture deformation. This method has a significant advantage in that it takes into account the effects of fracture deformation on the propagation of seismic waves. To apply this method to the seismic analysis of fractured rock masses, a proper representation of the deformation behavior of fractures is necessary. To date, several constitutive models describing the strength and deformation characteristics of fractures have been developed, such as the linear Coulomb model, the nonlinear Barton-Bandis model, and the nonlinear continuous-yielding joint model as mentioned before. The Coulomb model is adequate for the simulations of dry fractures under low-amplitude stress waves, where no obvious nonlinear deformation of fractures (such as failure) occurs, and it has been widely used in previous studies on the propagation of seismic waves across single or multiple fractures (Schoenberg, 1980; Pyrak-Nolte et al., 1990; Cai & Zhao, 2000; Zhao et al., 2006a, 2006c; Zhu et al., 2011). The Barton-Bandis model can represent the nonlinear deformational behavior of fractures under high-amplitude stress waves; however, its application in seismic analysis is usually limited due to the difficulties in determining the fracture parameters such as initial stiffness and maximum closure and in solving the complicated nonlinear equation system. Currently, the studies by using the Barton-Bandis model mainly focused on relatively simple problems considering that a seismic wave impinges normally upon fractures (Zhao & Cai, 2001; Zhao et al., 2003, 2006b, 2008a, 2008b).

The displacement discontinuity model is much more complicated than the welded interface model, due to the involvement of fracture deformations. For the simplest one-dimensional problem that considers a normally incident wave propagating across a linearly deformable fracture (the displacement of fracture is proportional to external stress in the neighborhood of the fracture, and the fracture stiffness is k), as shown in Figure 4.2a, an incident P-wave can induce opening and closure of the fracture, and an incident S-wave will cause slippage along the fracture plane. Therefore, the stress and displacement boundary conditions at the fracture can be expressed by

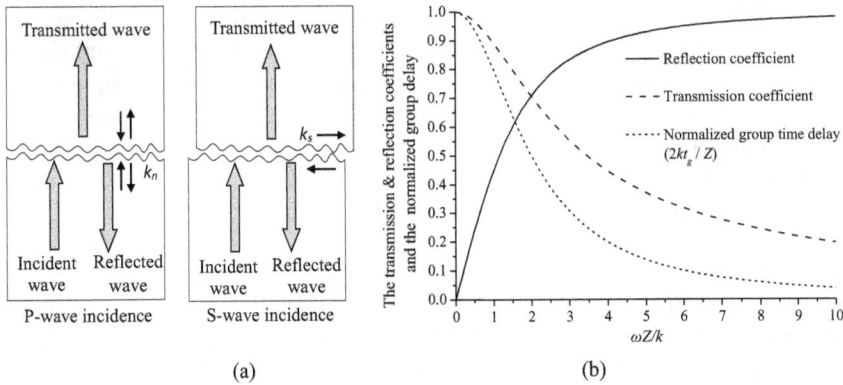

Figure 4.2 Reflection and transmission coefficients and the normalized group time delay of a normally incident wave at a displacement discontinuity. (a) 1-D theoretical model; (b) reflection and transmission coefficients and time delay.

Source: Pyrak-Nolte et al., 1990.

$$\sigma_l = \sigma_u, \, u_l = u_u + u_f \qquad (4.21)$$

where σ_l and u_l denote the stress and displacement at the lower surface of the fracture, caused by the incident and reflected waves; σ_u and u_u represent the stress and displacement along the upper surface of the fracture, caused only by the transmitted wave; and u_f is the fracture deformation as given by $u_f = \sigma/k$. By substituting the wave expressions into the boundary conditions, the reflection and transmission coefficients (R and T) can be obtained, in the form of (Pyrak-Nolte et al., 1990)

$$R = i/[-i + 2(k/Z\omega)], \, T = 2(k/Z\omega)/[-i + 2(k/Z\omega)] \, (\text{for P-wave}) \qquad (4.22)$$

$$R = -i/\left[-i + 2(k/Z\omega)\right], \, T = 2(k/Z\omega)/\left[-i + 2(k/Z\omega)\right] \, (\text{for S-wave}) \qquad (4.23)$$

where Z is the seismic impedance; ω is the angular frequency of the wave; and i is the imaginary unit. The group time delay for the transmitted and reflected waves can be defined by $t_g = d\theta/d\omega$ (θ is the wave phase), with its expression given by

$$t_g = 2(k/Z)/\left[\omega^2 + 4(k/Z)^2\right] \qquad (4.24)$$

These theoretical formulas indicate that the propagation of a seismic wave across a fracture is principally governed by the material impedance, the fracture stiffness, and the wave frequency. For simplification, Pyrak-Nolte et al. (1990) defined a dimensionless frequency ($\omega Z/k$) to describe the seismic response of fracture. The variation of reflection and transmission coefficients and the normalized group time delay with the dimensionless frequency is plotted in Figure 4.2b. With the increase of the dimensionless frequency, the transmission coefficient and the time delay decrease continuously, while the reflection coefficient increases. It reveals that an increasing wave frequency (ω) leads to a decrease in the amplitude of the transmitted wave, exhibiting a low-pass filter effect of fracture. A lower stiffness of fracture (k) can result in a decrease in the transmission coefficient, indicating an attenuation effect of fracture on the wave amplitude. The fracture also has a time delay effect on the wave propagation, as predicted by Equation (4.24).

When a seismic wave impinges upon a fracture at an oblique angle of incidence, the situation becomes much more complicated. An obliquely incident P-wave or S_V-wave will produce four types of reflected and transmitted waves, and an incident S_H-wave can produce two types of S_H-waves. Schoenberg (1980) and Pyrak-Nolte et al. (1990) proposed the expressions of the amplitudes of reflected and transmitted waves for an obliquely incident wave propagating across a linearly deformable fracture, based on the displacement discontinuity models. The seismic waves usually travel upwards in rock masses, especially near the ground surface, and propagate across inclined fractures (Japan Society of Civil Engineers, 1989). The previous model is not convenient to analyze the horizontal and vertical motions of rock masses, which are commonly utilized in earthquake-resistant designs for rock foundations and surface buildings. Therefore, some modifications to the previous model need to be made to provide a more intuitive understanding of the seismic behavior of rock masses.

4.2 PROPAGATION OF P-WAVE ACROSS AN INCLINED ROCK FRACTURE

4.2.1 Problem Formulation Based on the Displacement Discontinuity Method

During earthquakes, the seismic waves propagate from the deep rock formation to the ground surface. Since the deep rock formations usually have higher elastic moduli than the shallower ones, the incidence angles of seismic waves at different rock formations decrease with the upward propagation of seismic waves, according to Snell's law. Therefore, it is reasonable to assume that the seismic wave travels vertically in the rock masses near the ground surface when evaluating the seismic behavior of rock foundations and surface buildings (Japan Society of Civil Engineers, 1989).

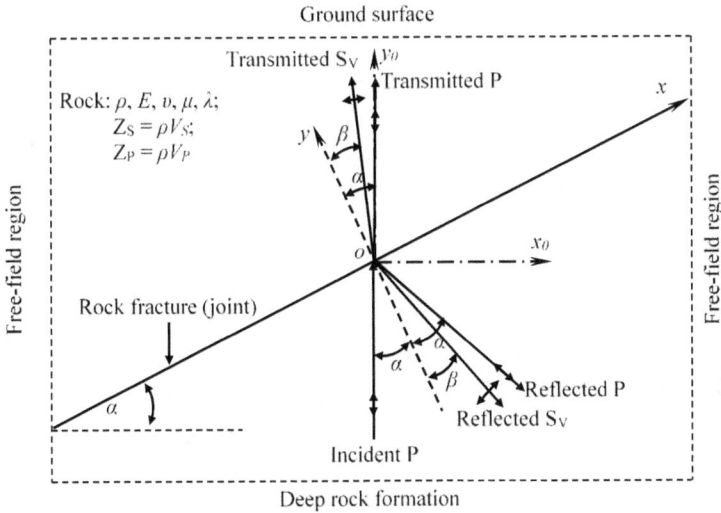

Figure 4.3 The displacement discontinuity model for the propagation of a P-wave across an inclined rock fracture.

Considering that a rock mass contains an inclined joint, which is large in extent size and thin in thickness relative to the wavelength, the analytical model considering an upward-propagating P-wave can be established, as shown in Figure 4.3. The rock in this model is assumed to be elastic, homogeneous, and isotropic, and has identical physico-mechanical properties: ρ (density); E (elastic modulus); υ (Poisson's ratio); and λ and μ (Lamé coefficients). The intrinsic attenuation mechanisms of the material (such as the material damping) as well as the wave reflections at model boundaries are neglected so that the attenuation effect of the joint can be concentrated. When an incident P-wave impinges upon the inclined joint, four types of seismic waves will be generated, including a reflected P-wave, a reflected S_V-wave, a transmitted P-wave, and a transmitted S_V-wave. A local coordinate system (x, y) is firstly defined by placing the origin (o) to the incident point of the seismic wave at the joint, with the x-axis going along the extension direction of the joint, and the y-axis normal to the joint plane. If the dip angle of the rock joint to the horizontal direction is α, the angle of incidence, reflection, and transmission of P-waves (corresponding to the incident, reflected, and transmitted P-waves, respectively) is equal to α, and the reflection and transmission angle of S_V-waves (corresponding to the reflected and transmitted S_V-waves, respectively) is assumed to be β. The relation between the angles α and β can be derived from Snell's law, as given by Equation (4.25), where V_P and V_S represent the propagation velocities of the P-wave and the S_V-wave, respectively.

$$\sin\alpha / \sin\beta = V_p / V_S \qquad (4.25)$$

Without loss of generality, the incident P-wave is assumed to be a simple harmonic wave. In the local coordinate system, the expressions of the incident P-wave, the reflected P-wave, the reflected S_V-wave, the transmitted P-wave, and the transmitted S_V-wave can be written in the displacement forms as:

$$\left.\begin{aligned}
\boldsymbol{u}_{IP} &= \{u_{IP}, v_{IP}\} = I_p exp\left[iK_p\left(xsin\alpha + ycos\alpha - V_p t\right)\right] \times \{sin\alpha, cos\alpha\} \\
\boldsymbol{u}_{RP} &= \{u_{RP}, v_{RP}\} = R_p exp\left[iK_p\left(xsin\alpha - ycos\alpha - V_p t\right)\right] \times \{sin\alpha, -cos\alpha\} \\
\boldsymbol{u}_{RSV} &= \{u_{RSV}, v_{RSV}\} = R_{SV} exp\left[iK_s\left(xsin\beta - ycos\beta - V_s t\right)\right] \times \{cos\beta, sin\beta\} \\
\boldsymbol{u}_{TP} &= \{u_{TP}, v_{TP}\} = T_p exp\left[iK_p\left(xsin\alpha + ycos\alpha - V_p t\right)\right] \times \{sin\alpha, cos\alpha\} \\
\boldsymbol{u}_{TSV} &= \{u_{TSV}, v_{TSV}\} = T_{SV} exp\left[iK_s\left(xsin\beta + ycos\beta - V_s t\right)\right] \times \{cos\beta, -sin\beta\}
\end{aligned}\right\} \qquad (4.26)$$

where $\boldsymbol{u}_i$ ($i = IP, RP, RSV, TP$, or TSV, corresponding to the incident P-wave, reflected P-wave, reflected S_V-wave, transmitted P-wave, and transmitted S_V-wave, respectively) is the particle displacement vector composed of the horizontal component u_i and the vertical component v_i; I_P, R_P, R_{SV}, T_P, and T_{SV} denote the wave amplitudes; K_i and V_i ($i = P$ or S) are the wave number and velocity for the P- or S-wave; α and β are propagation angles of the P- and S-wave, respectively; x, y, and t denote the coordinates and time.

In the cases of low-amplitude waves, the dry joint behaves linearly due to that the incident wave is not strong enough to mobilize the nonlinear deformation of the joint caused by the rock failure on its contact surface. Moreover, the joint deformation is tiny relative to the joint dimension. Therefore, it is reasonable to use the linear Coulomb joint model to describe the mechanical and deformational behavior of the joint. Based on the assumptions of the displacement discontinuity method (continuous stress and discontinuous displacement), the boundary conditions at the joint ($y = 0$) are given by:

$$\sigma_y^- = \sigma_y^+ = \sigma_y, \tau_{xy}^- = \tau_{xy}^+ = \tau_{xy} \text{ (stress conditions)} \qquad (4.27)$$

$$u^+ - u^- = \tau_{xy} / k_s, v^+ - v^- = \sigma_y / k_n \text{ (displacement conditions)} \qquad (4.28)$$

where the superscripts + and – denote the stresses and displacements in the rocks after and before the joint, respectively; k_n and k_s are the normal stiffness and shear stiffness of the joint, respectively; and the components of displacements and stresses in the x and y directions can be expressed by:

$$\left.\begin{array}{l} u^{-} = u_{IP} + u_{RP} + u_{RSV} \\ u^{+} = u_{TP} + u_{TSV} \\ v^{-} = v_{IP} + v_{RP} + v_{RSV} \\ v^{+} = v_{TP} + v_{TSV} \end{array}\right\} \tag{4.29}$$

$$\left.\begin{array}{l} \sigma_{y}^{i} = \lambda \dfrac{\partial u^{i}}{\partial x} + (\lambda + 2\mu)\dfrac{\partial v^{i}}{\partial y} \\[2ex] \tau_{xy}^{i} = \mu\left(\dfrac{\partial v^{i}}{\partial x} + \dfrac{\partial u^{i}}{\partial y}\right) \end{array}\right\} (i = + \text{ or } -) \tag{4.30}$$

By substituting the displacements and stresses into the boundary conditions, the amplitudes of the reflected and transmitted waves (or the reflection and transmission coefficients) can be obtained, in the matrix form of Equation (4.31). During the derivation process, some basic relations were used: $K_P = \omega/V_P$, $K_S = \omega/V_S$, $\mu/V_S = Z_S$, and $\mu/V_P = Z_S^2/Z_P$. It can be found that Equation (4.31) is a simplified form of the expression given by Pyrak-Nolte et al. (1990), with the assumption that the rock medium on both sides of the rock joint has the same physico-mechanical properties. The reflection and transmission coefficients, as defined by the ratios between the amplitudes of reflected and transmitted waves to that of the incident wave, are related to six independent variables, which fall into four types: the joint stiffness (k_n and k_s); the wave impedances of the medium (Z_S and Z_P); the angular frequency of the incident wave (ω); and the dip angle of the joint (α). The propagation angle of S_V-wave (β) can be derived from Equation (4.25), which depends on the angle α and the medium impedances.

$$\begin{bmatrix} Z_P\cos 2\beta & -Z_S\sin 2\beta & -Z_P\cos 2\beta & Z_S\sin 2\beta \\[2ex] \dfrac{Z_S^2}{Z_P}\sin 2\alpha & Z_S\cos 2\beta & \dfrac{Z_S^2}{Z_P}\sin 2\alpha & Z_S\cos 2\beta \\[2ex] k_s\sin\alpha & k_s\cos\beta & \dfrac{i\omega Z_S^2}{Z_P}\sin 2\alpha - k_s\sin\alpha & i\omega Z_S\cos 2\beta - k_s\cos\beta \\[2ex] -k_n\cos\alpha & k_n\sin\beta & i\omega Z_P\cos 2\beta - k_n\cos\alpha & k_n\sin\beta - i\omega Z_S\sin 2\beta \end{bmatrix} \begin{bmatrix} R_P \\ R_{SV} \\ T_P \\ T_{SV} \end{bmatrix} =$$

$$I_P \begin{bmatrix} -Z_P\cos 2\beta \\[2ex] \dfrac{Z_S^2}{Z_P}\sin 2\alpha \\[2ex] -k_s\sin\alpha \\[2ex] -k_n\cos\alpha \end{bmatrix} \tag{4.31}$$

The closed-form solutions of the reflection and transmission coefficients (assuming that $I_P = 1$, without loss of generality) based on the above equation system are given by Equations (4.32) to (4.35). After the reflection and transmission coefficients (R_P, R_{SV}, T_P, and T_{SV}) are solved, the expressions of the reflected and transmitted waves can be obtained. From the above solutions, it can be found that the reflection and transmission coefficients are independent of the coordinate system. They can effectively predict the influences of the rock joint on the propagation of seismic waves, in the form of wave scattering. Those coefficients have complex forms, the magnitudes (or absolute values) of which represent the amplitudes of reflected and transmitted waves (assuming $I_P = 1$), and the phases (or argument) of which represent the change of phase angles ($\Delta\theta$) of different waves relative to that of the incident wave. Then the time delay of different waves caused by the joint can be obtained from the derivative of the changed phase angle concerning the wave angular frequency, in the form of $t_g = d\theta/d\omega$.

$$R_P = -\frac{\left[\begin{array}{l}\omega^2 Z_P^4 Z_S \cos^4 2\beta + 2i\omega k_s Z_P^4 \cos\beta \cos^3 2\beta + 2i\omega k_s Z_P^3 Z_S \sin\alpha \cos^2 2\beta \sin 2\beta \\ -\omega^2 Z_S^5 \sin^2 2\alpha \sin^2 2\beta - i\omega k_n Z_P Z_S^3 \cos\alpha \sin 2\alpha \sin 4\beta - 2i\omega k_s Z_S^4 \sin^2 2\alpha \sin\beta \sin 2\beta\end{array}\right]}{\left[\begin{array}{l}\left(\omega Z_P^2 \cos^2 2\beta + \omega Z_S^2 \sin 2\alpha \sin 2\beta + 2ik_n Z_P \cos\alpha \cos 2\beta + 2ik_n Z_S \sin 2\alpha \sin\beta\right) \\ \times\left(\omega Z_P^2 Z_S \cos^2 2\beta + \omega Z_S^3 \sin 2\alpha \sin 2\beta + 2ik_s Z_P^2 \cos\beta \cos 2\beta + 2ik_s Z_P Z_S \sin 2\beta \sin\alpha\right)\end{array}\right]} \quad (4.32)$$

$$R_{SV} = \frac{2\omega Z_P Z_S \sin 2\alpha \cos 2\beta \left[\begin{array}{l}\omega Z_P^2 Z_S \cos^2 2\beta + \omega Z_S^3 \sin 2\alpha \sin 2\beta + ik_s Z_P^2 \cos\beta \cos 2\beta \\ +ik_n Z_P Z_S \cos\alpha \cos 2\beta + ik_s Z_P Z_S \sin\alpha \sin 2\beta + ik_n Z_S^2 \sin 2\alpha \sin\beta\end{array}\right]}{\left[\begin{array}{l}\left(\omega Z_P^2 Z_S \cos^2 2\beta + \omega Z_S^3 \sin 2\alpha \sin 2\beta + 2ik_s Z_P^2 \cos\beta \cos 2\beta + 2ik_s Z_P Z_S \sin\alpha \sin 2\beta\right) \\ \times\left(\omega Z_P^2 \cos^2 2\beta + \omega Z_S^2 \sin 2\alpha \sin 2\beta + 2ik_n Z_P \cos\alpha \cos 2\beta + 2ik_n Z_S \sin 2\alpha \sin\beta\right)\end{array}\right]} \quad (4.33)$$

$$T_P = -\frac{\left[\begin{array}{l}2k_n \omega Z_P^2 Z_S \cos\alpha \cos^3 2\beta + 2k_n \omega Z_P^2 Z_S^2 \sin 2\alpha \sin\beta \cos^2 2\beta + k_s \omega Z_P^2 Z_S^2 \sin 2\alpha \cos\beta \sin 4\beta \\ +2\omega k_s Z_P Z_S^3 \sin\alpha \sin 2\alpha \sin^2 2\beta + ik_n k_s Z_P^2 Z_S^2 \sin 2\alpha \sin 4\beta + iZ_P^2 Z_S k_n k_s \sin 2\alpha \sin 4\beta \\ +4ik_n k_s Z_P^3 \cos\alpha \cos\beta \cos^2 2\beta + 4ik_n k_s Z_P Z_S^2 \sin\alpha \sin 2\alpha \sin\beta \sin 2\beta\end{array}\right]}{\left[\begin{array}{l}\left(\omega Z_P^2 Z_S \cos^2 2\beta + \omega Z_S^3 \sin 2\alpha \sin 2\beta + 2ik_s Z_P^2 \cos\beta \cos 2\beta + 2ik_s Z_P Z_S \sin\alpha \sin 2\beta\right) \\ \times\left(2k_n Z_P \cos\alpha \cos 2\beta + 2k_n Z_S \sin 2\alpha \sin\beta - i\omega Z_P^2 \cos^2 2\beta - i\omega Z_S^2 \sin 2\alpha \sin 2\beta\right)\end{array}\right]} \quad (4.34)$$

$$T_{SV} = \frac{\left[\begin{array}{l}-2\omega Z_P Z_S \sin 2\alpha \cos 2\beta \left[\begin{array}{l}ik_n Z_S^2 \sin 2\alpha \sin\beta - ik_s Z_P^2 \cos 2\beta \cos\beta \\ +ik_n Z_P Z_S \cos\alpha \cos 2\beta - ik_s Z_P Z_S \sin\alpha \sin 2\beta\end{array}\right]\end{array}\right]}{\left[\begin{array}{l}\left(\omega Z_P^2 Z_S \cos^2 2\beta + \omega Z_S^3 \sin 2\alpha \sin 2\beta + 2ik_s Z_P^2 \cos\beta \cos 2\beta + 2ik_s Z_P Z_S \sin 2\beta \sin\alpha\right) \\ \times\left(\omega Z_P^2 \cos^2 2\beta + \omega Z_S^2 \sin 2\alpha \sin 2\beta + 2ik_n Z_P \cos\alpha \cos 2\beta + 2ik_n Z_S \sin 2\alpha \sin\beta\right)\end{array}\right]} \quad (4.35)$$

The reflected and transmitted waves play an important role in the seismic response of the rock mass and subsequently affect the stability and safety of rock foundations and surface buildings and structures. To estimate the stability of the surface structures, the vertical and horizontal motions of the rock mass need to be investigated, which are generally utilized in geotechnical earthquake engineering. Therefore, it is necessary to make some modifications to the expressions of the reflected and transmitted waves in the local coordinate system. A global coordinate system (x_0, y_0) is firstly established, with its origin (o_0) located at the same point as the pre-defined local coordinate system and its x_0-axis and y_0-axis going along the horizontal and vertical directions, respectively (as shown in Figure 4.3). The relationship of the coordinates between the global and local systems can be given by:

$$x_0 = x \cos \alpha - y \sin \alpha, \, y_0 = x \sin \alpha + y \cos \alpha \qquad (4.36)$$

$$x = x_0 \cos \alpha + y_0 \sin \alpha, \, y = -x_0 \sin \alpha + y_0 \cos \alpha \qquad (4.37)$$

Then the expressions of all seismic waves in the local system can be transformed into their global system forms by performing the rotation of coordinate axes, as given by:

$$\left. \begin{aligned}
u_{IP0} &= \{u_{IPD}, v_{PP0}\} = I_P \exp[iK_P(y_0 - V_P t)] \times \{0,1\} \\
u_{RP0} &= \{u_{RP0}, v_{RP0}\} = R_P \exp[iK_P(x_0 \sin 2\alpha - y_0 \cos 2\alpha - V_P t) \times \{\sin 2\alpha, -\cos 2\alpha\} \\
u_{RSV0} &= \{u_{RSV0}, v_{RSV0}\} = R_{SV} \exp[iK_S(x_0 \sin(\alpha + \beta) - y_0 \cos(\alpha + \beta) - V_s t] \times \{\cos(\alpha + \beta), \sin(\alpha + \beta)\} \\
u_{IP0} &= \{u_{IP0}, v_{IP0}\} = T_P \exp[iK_P(y_0 - V_P t)] \times \{0,1\} \\
u_{ISV0} &= \{u_{ISV0}, v_{ISV0}\} = T_{SV} \exp[iK_S(-x_0 \sin(\alpha - \beta) + y_0 \cos(\alpha - \beta) - V_s t] \times \{\cos(\alpha - \beta), \sin(\alpha - \beta)\}
\end{aligned} \right\} \qquad (4.38)$$

The horizontal and vertical motions (u^- and v^-) of the rock particles below the rock joint, which represent the seismic response of the rock mass under the joint, can be derived from the superposition of the incident P-wave, reflected P-wave, and reflected S_V-wave, as given by:

$$u^- = u_{IP0} + u_{RP0} + u_{RSV0} = R_P \sin 2\alpha \exp\left[iK_P(x_0 \sin 2\alpha - y_0 \cos 2\alpha - V_P t)\right] +$$
$$R_{SV} \cos(\alpha + \beta) \exp\left[iK_S(x_0 \sin(\alpha + \beta) - y_0 \cos(\alpha + \beta) - V_s t)\right] \qquad (4.39)$$

$$v^- = v_{IP0} + v_{RP0} + v_{RSV0} = I_P \exp\left[iK_P(y_0 - V_P t)\right] - R_P \cos 2\alpha \exp\left[iK_P(x_0 \sin 2\alpha -\right.$$
$$y_0 \cos 2\alpha - V_P t)\Big] + R_{SV} \sin(\alpha + \beta) \exp\left[iK_S(x_0 \sin(\alpha + \beta) - y_0 \cos(\alpha + \beta) - V_s t)\right] \qquad (4.40)$$

Similarly, the horizontal and vertical motions (u^+ and v^+) of the rock particles above the rock joint can be obtained from the superposition of the transmitted P-wave and the transmitted S_V-wave, in the forms of:

$$u^+ = u_{TP0} + u_{TSV0} = T_{SV} \cos(\alpha - \beta) \exp\left[iK_S(-x_0 \sin(\alpha - \beta) + y_0 \cos(\alpha - \beta) - V_s t)\right] \qquad (4.41)$$

$$v^+ = v_{TP0} + v_{TSV0} = T_P \exp\left[iK_P\left(y_0 - V_P t\right)\right] + T_{SV} \sin(\alpha - \beta)\exp$$
$$\left[iK_S\left(-x_0 \sin(\alpha - \beta) + y_0 \cos(\alpha - \beta) - V_S t\right)\right] \tag{4.42}$$

Equations (4.39) to (4.42) give the complex forms of the seismic motions (horizontal and vertical motions) at both sides of the rock joint. Only the real components of those wave expressions have physical significance; therefore, the seismic motions at the original point of the global coordinate system ($x_0 = y_0 = 0$; Figure 4.3) can be rewritten in the following forms (without loss of generality, the amplitude of an incident wave is assumed to be 1):

$$\left.\begin{aligned}
u^+ &= |T_{SV}|\cos(\alpha - \beta)\cos\left[-\omega t + \Delta\left(T_{S_V}\right)\right]\\
v^+ &= |T_P|\cos\left[-\omega t + \Delta\left(T_P\right)\right] + |T_{SV}|\sin(\alpha - \beta)\cos\left[-\omega t + \Delta\left(T_{SV}\right)\right]
\end{aligned}\right\} \tag{4.43}$$

$$\left.\begin{aligned}
u^- &= |R_P|\sin 2\alpha \cos\left[-\omega t + \Delta\left(R_P\right)\right] + |R_{SV}|\cos(\alpha + \beta)\cos\left[-\omega t + \Delta\left(R_{SV}\right)\right]\\
v^- &= \cos(-\omega t) - |R_P|\cos 2\alpha \cos\left[-\omega t + \Delta\left(R_P\right)\right] + |R_{SV}|\sin(\alpha + \beta)\cos\left[-\omega t + \Delta\left(R_{SV}\right)\right]
\end{aligned}\right\} \tag{4.44}$$

where $|R_P|$, $|R_{SV}|$, $|T_P|$, and $|T_{SV}|$ represent the magnitudes of the complex reflection and transmission coefficients; and $\Delta(R_P)$, $\Delta(R_{SV})$, $\Delta(T_P)$, and $\Delta(T_{SV})$ represent the phase angles of those coefficients.

The following properties of rock medium are used in the following analysis of Section 4.2: the rock density (ρ) is 2600 kg/m^3; the P-wave velocity (V_P) is 4900 m/s; and the S-wave velocity (V_S) is 2700 m/s. Those rock properties correspond to the typical granite (Jaeger et al., 2007). Some other properties of this medium can be derived from the relationship between wave velocities and elastic moduli (e.g., Young's modulus E is 48.60 GPa; the Poisson's ratio v is 0.282). The dip angle of the rock joint (α), equaling the incidence angle of the P-wave, ranges from 0° to 89° (when $\alpha = 90°$, surface waves may be generated along the joint plane, and this case will not be considered). Generally, the normal stiffness (k_n) of a rock joint is much larger than the shear stiffness (k_s), with the ratio $R = k_n/k_s \geq 1$, which depends seriously on the level of applied normal stress (Bandis et al., 1983; Oda et al., 1992). In this study, a relatively weak joint is considered to represent the effects of the joint on wave propagation, and the ratio R is assumed to be 10 without loss of generality ($k_n = 10k_s$). k_n ranges from 500 MPa/m to 3000 MPa/m, with an increment of 500 MPa/m, and the corresponding k_s is from 50 MPa/m to 300 MPa/m with an increment of 50 MPa/m. To investigate the effects of the wave frequency, f ranges from 20 Hz to 60 Hz with an increment of 10 Hz.

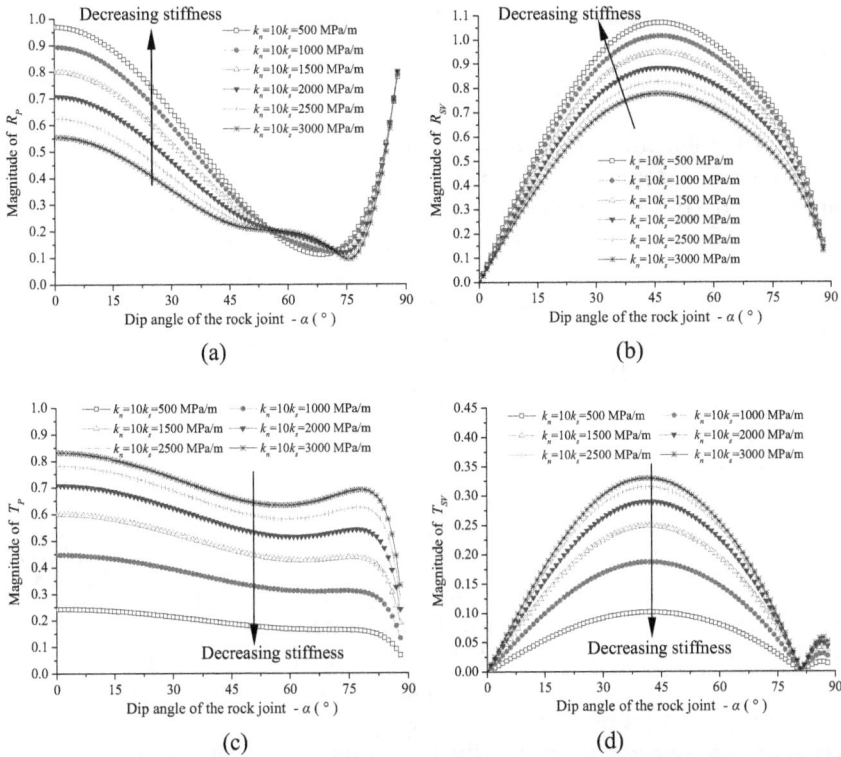

Figure 4.4 Variation of the reflection and transmission coefficients with the stiffness and dip angle of the rock joint (P-wave incidence and f = 50 Hz). (a) Reflection coefficient of P-wave; (b) reflection coefficient of S_V-wave; (c) transmission coefficient of P-wave; (d) transmission coefficient of S_V-wave.

4.2.2 Effect of Joint Properties on the Wave Propagation

Figure 4.4 shows the variation tendency of the reflection and transmission coefficients (amplitudes) with the properties of the rock joint (e.g., f = 50 Hz). It can be found that the magnitudes of the reflection and transmission coefficients, which represent the wave amplitudes, are influenced significantly by the dip angle and stiffness of the joint. As shown in Figure 4.4a, when the joint dip angle (α) is 0° (the incident P-wave impinges normally upon the joint), the reflection coefficient of the P-wave ($|R_P|$) has an initial value, which is identical to the result predicted by using the 1-D displacement discontinuity model (Pyrak-Nolte et al., 1990; Cai & Zhao, 2000). The $|R_P|$ firstly decreases with the increase of α, and reaches its minimum value when

α is around 72° (from 69° to 75° depending on the joint stiffness). After that, the $|R_p|$ increases quickly and reaches a large value close to 1 as the α approaches 90°. The analytical result reveals that the reflected P-wave has a large amplitude when the joint dip angle is in the ranges of 0° < α < 15° and 85° < α < 90°, and has a small amplitude in the dip angle range of 50° < α < 80°. The joint stiffness also has a significant influence on the value of R_p. With the decrease of the joint stiffness (k_n and k_s), the R_p increases continuously in most dip angle ranges, indicating that the weak joint is more likely to produce reflected P-waves with large amplitudes. On the contrary, the strong joint can reduce wave reflection and promote wave transmission across the joint. It should be noted that in a small dip angle range (56° < α < 72°), the influence of joint stiffness on the R_p is not obvious.

The distribution of the reflection coefficient of S_V-wave ($|R_{SV}|$) with varying dip angles of rock joint exhibits different characteristics from the distribution of R_p, as shown in Figure 4.4b. The $|R_{SV}|$ firstly increases with the increase of the joint dip angle, and reaches its maximum value when α = 47°, indicating that the reflected S_V-wave has the largest amplitude. After that, the $|R_{SV}|$ subsequently decreases as α increases to 90°. When α approaches 0° or 90°, the $|R_{SV}|$ has its minimum value approaching 0, which implies that if the rock joint is normal or parallel to the incident P-wave, no reflected S_V-wave will be generated. The joint stiffness has a similar influence on the $|R_{SV}|$ value to the case of $|R_p|$. As the joint stiffness decreases, the $|R_{SV}|$ increases continuously, revealing that a weak joint can promote the reflection of the S_V-wave, and conversely a strong joint will reduce the S_V-wave reflection. When the joint stiffness is low enough (e.g., k_n = 10k_s = 500 MPa/m), the maximum displacement amplitude of the reflected S_V-wave at α = 47° is even larger than that of the incident wave ($|R_{SV}|$ > 1). However, the energy conservation is still preserved due to that the wave energy is proportional to the integration of the square of particle velocity (the wave impedance is a constant).

Figure 4.4c shows the variation of the transmission coefficient of P-wave ($|T_p|$) with the dip angle and stiffness of the joint. As α increases, the $|T_p|$ firstly decreases from its initial value to a critical value when α is around 62° (from 58° to 67°, varying with the joint stiffness). Then the $|T_p|$ increases gradually until the α reaches around 76° (from 74° to 78°). After that, the $|T_p|$ decreases quickly with the increase of α. From the result, it can be derived that when the joint dip angle is small, the transmission of the P-wave easily occurs and the transmitted P-wave has a large amplitude. Additionally, when the joint dip angle approaches 90°, the incident P-wave travels nearly parallel to the joint, and will not transmit across the joint, thereby the transmission coefficient approaches 0. The joint stiffness has an opposite effect on the value of $|T_p|$, compared with the cases of $|R_p|$ and $|R_{SV}|$. With the decrease of joint stiffness, the $|T_p|$ decreases continuously, indicating that the weak joint

can reduce the transmission of P-waves. On the contrary, a strong discontinuity will enhance the P-wave transmission.

The variation of the transmission coefficient of S_V-wave ($|T_{SV}|$) with the
joint dip angle and joint stiffness is shown in Figure 4.4d. The T_{SV} curve con-
sists of two parabolic parts. With the increase of α, the $|T_{SV}|$ firstly increases
from 0 to its maximum value at about $\alpha = 42°$, and then decreases to 0 when
the α reaches 81°. After that, the $|T_{SV}|$ curve turns into the second parabolic
part with the turning point located at about $\alpha = 87°$. It can be found that
when the joint dip angle approaches 0° or 90° (the joint is normal or parallel to the incident P-wave), the transmitted S_V-wave will not be generated,
and the related coefficient $|T_{SV}|$ approaches 0. With the decrease of the joint
stiffness, the $|T_{SV}|$ decreases continuously, indicating that the weak joint with
low stiffness can restrain the transmission of the S_V-wave.

The analytical results shown in Figure 4.4 demonstrate that the coefficients of reflection and transmission, which represent the amplitudes of the
reflected and transmitted waves, exhibit different distribution characteristics with the change of the joint dip angle. The joint stiffness also affects
significantly the magnitudes of the reflection and transmission coefficients.
The weak rock joint with low stiffness can increase the amplitudes of the
reflected P-wave and S_V-wave, but decrease the amplitudes of the transmitted
P-wave and S_V-wave. On the contrary, the strong rock joint with high stiffness has the opposite effect on the amplitudes of the reflected and transmitted waves. Additionally, the transmission coefficients of P-wave and S_V-wave
are smaller than 1, exhibiting an attenuation effect of the rock joint on the
amplitudes of transmitted waves.

To investigate the time delay effect of the rock joint on wave propagation,
parametric studies were performed, mainly focusing on the variation of the
time delay of waves (time gap caused by the presence of the joint) with the
dip angle and stiffness of the joint, and the results are shown in Figure 4.5.
The time delay of the reflected P-wave (t_{RP}, Figure 4.5a) firstly decreases
with the increase of α until it reaches the minimum value when the α is
equal to 66°. After that, the t_{RP} increases quickly with the joint dip angle
and finally reaches its maximum value as α approaches 90°. It can be found
that when the joint dip angle approaches 0° or 90°, the joint can cause a
long time delay. While the joint with a dip angle near 66° will induce a short
time delay. On the other hand, joint stiffness has an obvious influence on the
time delay of the reflected P-wave. With the increase of stiffness, the time
delay of the reflected P-wave decreases continuously, implying that the strong
joint can induce a shorter time delay than the weak joint. Figure 4.5b shows
the change of the time delay of reflected S_V-wave (t_{RS}) with the varying joint
properties. At the beginning, the t_{RS} (absolute value) increases slowly with
the joint dip angle from its initial value to the maximum at about $\alpha = 77°$,
and then t_{RS} decreases quickly to its maximum value as the α approaches 90°.

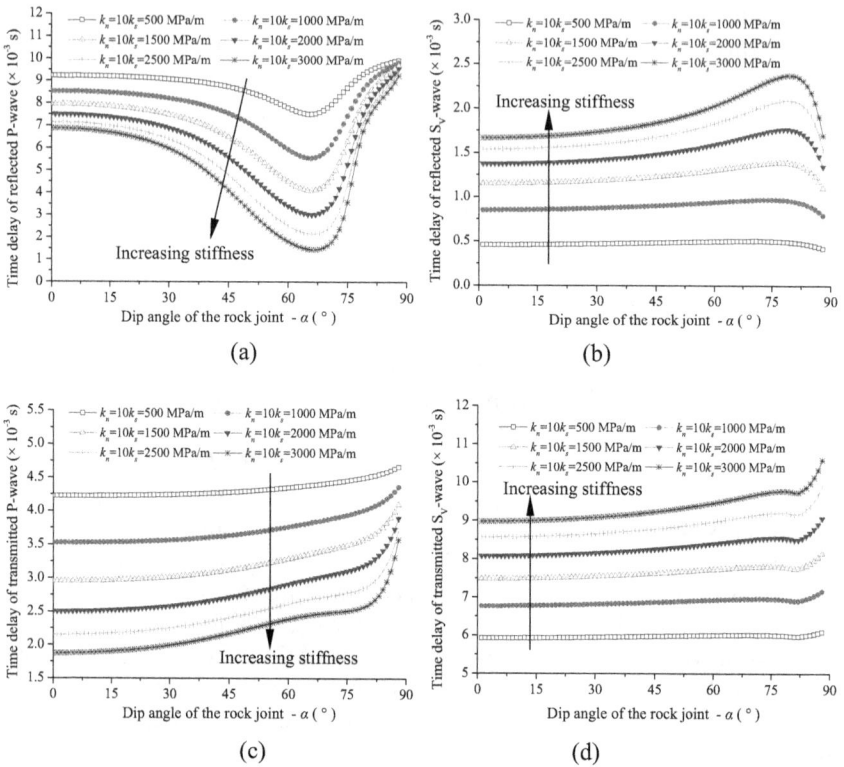

Figure 4.5 Variation of the time delay of reflected and transmitted waves with the stiffness and dip angle of the rock joint (P-wave incidence and $f = 50$ Hz). (a) Time delay of reflected P-wave; (b) time delay of reflected S_V-wave; (c) time delay of transmitted P-wave; (d) time delay of transmitted S_V-wave.

The influence of the joint stiffness on the t_{RS} is different from that on the t_{RP}: an increasing joint stiffness can lead to the increase of the time delay of the reflected S_V-wave, and vice versa, indicating that the strong joint with high stiffness can prolong the time gap between the incident P-wave and the reflected S_V-wave.

The time delay of the transmitted P-wave and transmitted S_V-wave caused by the rock joint is shown in Figure 4.5c and Figure 4.5d, respectively. It can be found that the time delay of transmitted waves (as denoted by t_{TP} and t_{SV}) increases gradually with the increases of the joint dip angle, and reaches its maximum value as the α approaches 90°. Additionally, joint stiffness has an obvious effect on the time delay of transmitted waves. For the transmitted P-wave (Figure 4.5c), the increasing joint stiffness will lead to a decrease of the time delay (t_{TP}), revealing that the strong joint can help reduce the time gap between the incident and the transmitted P-waves. While for the transmitted S_V-wave (Figure 4.5d), the time delay caused by the joint (t_{SV}) increases significantly with the increase of the joint stiffness, indicating that a strong joint with high stiffness can cause a long time delay for the transmitted S_V-wave.

From the analytical results, it can be found that the time delay effect of rock joints on wave propagation is affected significantly by the joint dip angle and the joint stiffness. A strong rock joint with high stiffness can reduce the time delay of the reflected and transmitted P-waves but prolong the time delay of the reflected and transmitted S_V-waves. On the contrary, a weak joint with low stiffness has the opposite effect on the time delay of different waves.

4.2.3 Effect of Wave Frequency on the Wave Propagation

The frequency of the incident P-wave enters into the expressions of reflection and transmission coefficients, and consequently affects the amplitudes of reflected and transmitted waves as well as the time delay. To investigate the influences of wave frequency on wave propagation, parametric studies were carried out by changing the frequency of the incident P-wave (from 20 Hz to 60 Hz with an increment of 10 Hz). In the parametric studies, the magnitudes of the reflection and transmission coefficients as well as the time delay caused by the joint were calculated and discussed.

Figure 4.6 shows the variation of the reflection and transmission coefficients (magnitudes) with the joint dip angle (α) and the wave frequency (f), assuming that the rock joint has the stiffness of $k_n = 10k_s = 1000$ MPa/m. It can be found that the magnitudes of the reflection and transmission coefficients, which reflect the amplitudes of reflected and transmitted waves, have a similar variation tendency with the change of the joint dip angle to the cases in Figure 4.4, despite the difference in the wave frequency. Since the influence of joint dip angle on the reflection and transmission coefficients has been presented before, here only the influence of the joint stiffness is discussed. As shown in Figure 4.6a, the reflection coefficient of the P-wave ($|R_P|$) increases continuously with the increasing wave frequency, indicating that a high-frequency seismic wave can promote the reflection of the P-wave with a large amplitude. A similar effect of wave frequency on the wave amplitude can also be found in the case of the reflected S_V-wave, as shown in Figure 4.6b. With the increase of wave frequency, the reflection coefficient of the S_V-wave ($|R_{SV}|$), which represents the amplitude of the reflected S_V-wave, also increases gradually, and vice versa. However, for the cases of transmitted waves (including the transmitted P-wave and the transmitted S_V-wave), the effects of wave frequency on the magnitudes of transmission coefficients ($|T_P|$ and $|T_{SV}|$) are different, as shown in Figure 4.6c and 4.6d. The analytical results demonstrate that the values of

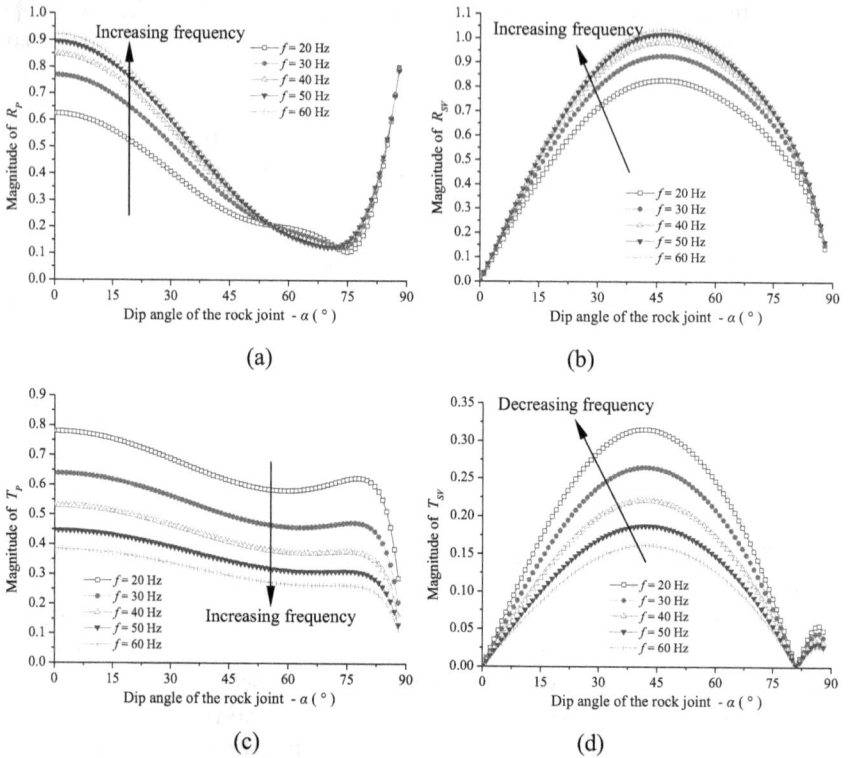

Figure 4.6 Variation of the reflection and transmission coefficients with the joint dip angle and the wave frequency (P-wave incidence and $k_n = 10k_s = 1000$ MPa/m). (a) Reflection coefficient of P-wave; (b) reflection coefficient of S_V-wave; (c) transmission coefficient of P-wave; (d) transmission coefficient of S_V-wave.

$|T_P|$ and $|T_{SV}|$ decrease with the increase of the wave frequency, indicating that the high-frequency seismic wave will produce transmitted P- and S_V-waves with small amplitudes.

From the analytical results, it can be derived that when a P-wave impinges upon an inclined rock joint, the amplitudes of the reflected waves (the reflected P- and S_V-wave) increase with the increase of wave frequency. However, the amplitudes of the transmitted waves (the transmitted P- and S_V-wave) decrease with the increasing wave frequency. The results indicate that the rock joint behaves like a low-pass filter, which can filter the high-frequency content of the wave signal by reducing the amplitude.

To investigate the effect of the wave frequency on the time delay, parametric studies were carried out. Figure 4.7 shows the time delay of the reflected

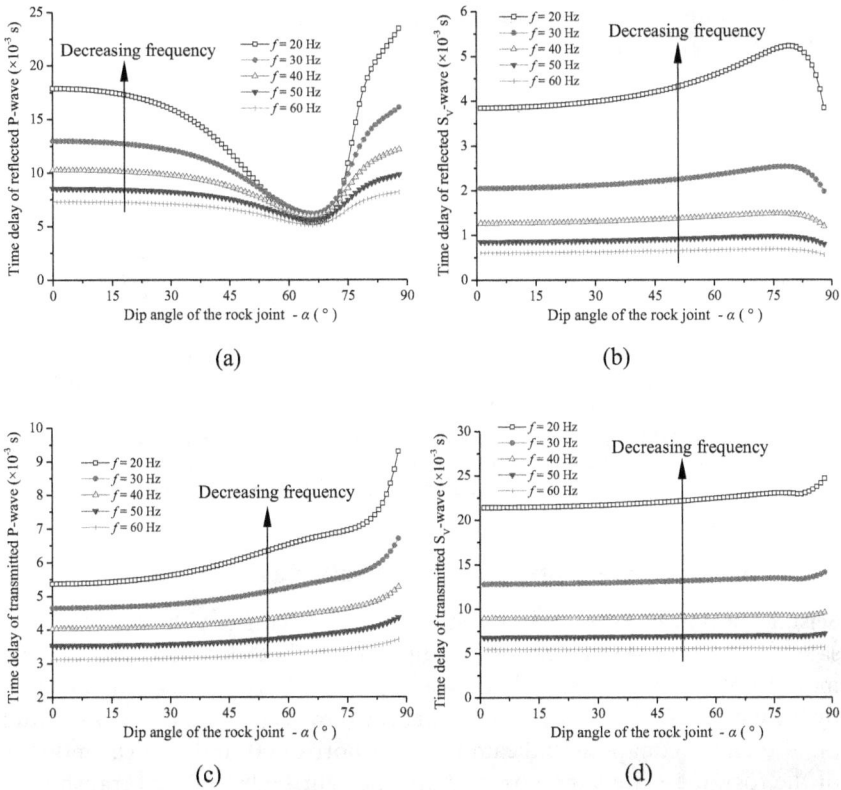

Figure 4.7 Variation of the time delay of reflected and transmitted waves with the joint dip angle and the wave frequency (P-wave incidence and $k_n = 10k_s = 1000$ MPa/m). (a) Time delay of reflected P-wave; (b) time delay of reflected S_V-wave; (c) time delay of transmitted P-wave; (d) time delay of transmitted S_V-wave.

and transmitted waves, which changes with the frequency of the incident P-wave and the dip angle of the rock joint. The variation of time delay with the joint dip angle has already been described earlier, therefore, this part principally focuses on the influence of wave frequency on the time delay. The analytical results indicate that the wave frequency has a similar effect on the time delay of all the reflected and transmitted waves (reflected P- and S_V-waves, transmitted P- and S_V-waves): with the persistent decrease of the wave frequency, the time delay increases continuously, revealing that a low-frequency incident wave which travels across a rock joint will be delayed by a longer time gap than the high-frequency seismic wave.

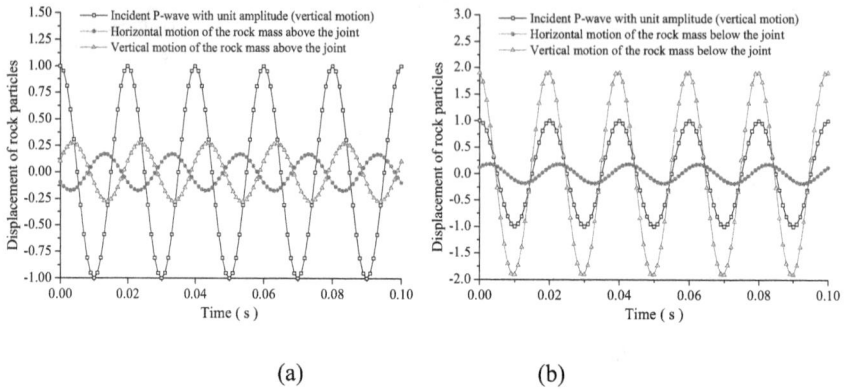

Figure 4.8 Seismic motions of the rock masses above and below the rock joint caused by an incident P-wave with unit amplitude ($\alpha = 45°, f = 50$ Hz, and $k_n = 10k_s = 1000$ MPa/m). (a) Seismic motions above the joint; (b) seismic motions below the joint.

4.2.4 Seismic Response of the Rock Mass

Seismic responses of the fractured rock masses concern the stability and safety of rock-based structures and buildings. The horizontal and vertical motions of the rock mass at both sides of the rock joint, which are aroused by the incidence of a P-wave, can be analyzed by using Equations (4.43) and (4.44). Those equations indicate that the horizontal and vertical motions of the rock mass are superposed by the incident, reflected, and transmitted waves. To illuminate the seismic motions below and above the rock joint, a specific case is firstly given in Figure 4.8 by assuming that the joint dip angle (α) is 45°, the joint stiffness is $k_n = 10k_s = 1000$ MPa/m, and the frequency of the incident P-wave is 50 Hz. Since the reflected and transmitted waves have the same frequency as the incident wave, the superposed motions of the rock particle in the horizontal and vertical directions also have an identical frequency and period to the incident wave. However, due to the time delay effect of the rock joint, the reflected and transmitted waves are delayed to some extent at the joint, and thereby the phase angles of the vertical and horizontal waves are different from that of the incident wave. On the other hand, the superposition of waves controls the amplitudes of horizontal and vertical motions. In the rock mass above the discontinuity, the amplitudes of the horizontal and vertical motions are 0.172 and 0.275, respectively, which are smaller than that of the incident P-wave ($I_P = 1$), due to the attenuation effect of the rock joint on the wave amplitude. In the rock mass below the joint, the amplitudes of the horizontal and vertical motions are 0.180 and 1.90, respectively. The vertical motion below the joint has a much large amplitude than the incident wave, due to the superposition of the incident and reflected waves.

In the geotechnical earthquake engineering, the maximum amplitude of the seismic motion (or seismic wave) is one of the most important factors to assess the seismic behavior of rock masses as well as the stability of rock-based structures and buildings. Moreover, the maximum amplitude of the seismic motion is also affected significantly by the properties of rock joints and the properties of seismic waves. To obtain an insight into the seismic response of the fractured rock mass, parametric studies were carried out, mainly focusing on the influences of the joint dip angle, the joint stiffness, and the wave frequency on the maximum amplitudes of the horizontal and vertical motions of rock masses at both sides of the rock joint.

Figure 4.9 shows the variation of the maximum amplitudes of horizontal and vertical motions at both sides of the rock joint with varying dip angles and stiffness (assuming that the incident P-wave has a frequency of 50 Hz and has unit amplitude). In the rock mass above the joint, the seismic motions of rock particles are caused by the transmitted P-wave and S_V-wave. As shown

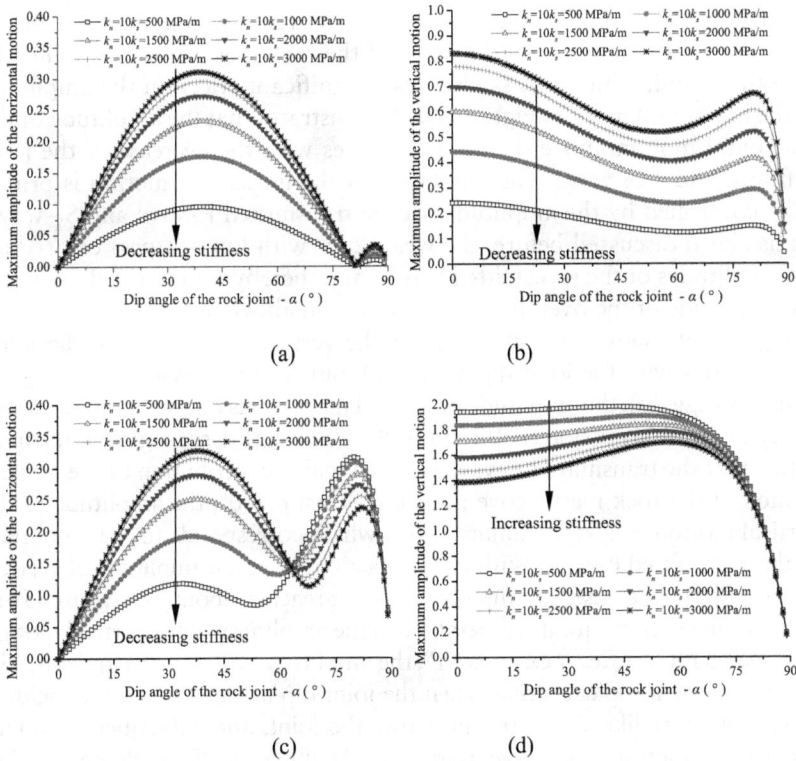

Figure 4.9 Maximum amplitudes of the horizontal and vertical motions at both sides of the rock joint with varying dip angles and stiffness (P-wave incidence and f = 50 Hz). (a) Horizontal amplitude above the joint; (b) vertical amplitude above the joint; (c) horizontal amplitude below the joint; (d) vertical amplitude below the joint.

in Figure 4.9a, the amplitude of the horizontal motion above the joint is affected significantly by the joint dip angle (α), and the variation curve is composed of two parabolic parts. When the incident P-wave impinges normally upon the joint ($\alpha = 0°$), the amplitude of the horizontal motion is 0. In this case, only the reflected and transmitted P-waves that travel in the vertical direction are generated, and no S_V-wave is produced. Those P-waves induce particle motions only in the vertical direction, and thereby the amplitude of the horizontal motion is 0. With the increase of α, the amplitude of horizontal motion firstly increases from 0 to its maximum value at about $\alpha = 40°$, and then decreases to 0 when α reaches 81°. After that, the curve turns into the second parabolic part with the turning point at about $\alpha = 86°$. The amplitude of the horizontal motion is small when α ranges from 81° to 89°. In this case, the incident P-wave propagates approximately parallel to the existing joint, and the transmitted S_V-wave has a small amplitude (see Figure 4.4d), leading to a small amplitude of the overall horizontal motion. It can be found that the variation curve of the horizontal motion amplitude above the joint with the joint dip angle has a similar shape to the $|T_{SV}|$ curve (see Figure 4.4d), indicating that the transmitted S_V-wave makes a major contribution to the horizontal motion of the rock mass above the joint. On the other hand, joint stiffness also has a significant effect on the amplitude of the horizontal motion. Figure 4.9a demonstrates that the amplitude of the horizontal motion above the joint decreases with the decrease of the joint stiffness, and vice versa. The amplitude of the horizontal motion is principally controlled by the amplitudes of the transmitted P-wave and S_V-wave. As has been discussed before, the weak joint with low stiffness can reduce the amplitudes of the transmitted waves, and thereby lead to the decrease of the amplitude of the overall horizontal motion above the joint.

Figure 4.9b shows the amplitude of the vertical motion above the joint, which varies with the joint dip angle and joint stiffness. With the change of joint dip angle (α), the amplitude of vertical motion has a similar variation tendency to the transmission coefficient of the P-wave ($|T_P|$, see Figure 4.4c), indicating that the transmitted P-wave makes a major contribution to the vertical motion of the rock mass above the joint. When $\alpha = 0°$, the amplitude of the vertical motion has its maximum value, which corresponds to the amplitude of the transmitted P-wave. With the increase of the α, the amplitude of vertical motion firstly decreases continuously until α reaches about 58° (from 55° to 62°, varying with the joint stiffness). Then the amplitude of the vertical motion increases with α. After α exceeds 81°, the amplitude will decrease quickly. The analytical result indicates that when the joint dip angle is small, the incident P-wave is more likely to transmit across the joint, and subsequently induce the vertical motion with large amplitude. As the joint dip angle approaches 90°, the P-wave that travels nearly parallel to the joint is difficult to transmit across the joint, and thereby the vertical motion above the joint has a small amplitude. The joint stiffness has a similar effect on the amplitude of vertical motion to that on the amplitude of horizontal motion. With the decrease of the joint stiffness, the amplitude of the vertical motion above the joint decreases gradually because the weak joint with low stiffness can reduce the amplitudes of the transmitted P- and S_V-waves (see Figure 4.4c and 4.4d).

The seismic behavior of the rock mass below the rock joint, which is controlled by the incident P-wave, the reflected P-wave, and the reflected S_V-wave, exhibits different characteristics compared with the seismic behavior of the rock mass above the joint. Figure 4.9c shows the amplitude of the horizontal motion below the rock joint. The variation curve of the amplitude in the horizontal direction has a complicated shape with the change of the joint dip angle (α). When $\alpha = 0°$, the amplitude of horizontal motion is 0. In this case, only reflected and transmitted P-waves are generated when the incident P-wave travels across the normally distributed joint. Those P-waves propagate in the vertical direction and cannot induce the horizontal motion of the rock mass. With the increase of α, the amplitude of horizontal motion firstly increases until it reaches the first critical value (denoted by A_{C1}) when α is around 36° (the corresponding α increases from 34° to 38° with the increase of joint stiffness). After that, the amplitude of horizontal motion decreases continuously until α reaches about 61° (the corresponding α increases from 54° to 68° with the increase of joint stiffness). Then the horizontal amplitude starts to increase and reaches the second critical value (denoted by A_{C2}) when α is about 81° (from 80° to 82°). Then the horizontal amplitude will decrease quickly and approach 0. Figure 4.9c indicates that the minimum amplitude of the horizontal motion appears when the joint dip angle approaches 0° and 90°. When the incident P-wave propagates normal or parallel to the joint, no S_V-wave will be generated, and thereby the amplitude of the horizontal motion approaches 0. The joint dip angle influences significantly the amplitude of the horizontal motion in the rock mass below the joint. If $\alpha < 63°$, the amplitude of the horizontal motion decreases continuously with the decrease of joint stiffness. However, in the dip angle range of $\alpha \geq 63°$, the amplitude of the horizontal motion increases as the joint stiffness decreases. The effect of the joint stiffness on the amplitude of the horizontal motion is complicated, due to the superposition of reflected waves, which also depends on the joint dip angle. Additionally, the joint stiffness also affects the dip angle where the maximum amplitude of horizontal motion appears. In the stiffness range of this study, the maximum amplitude appears at about $\alpha = 81°$, when the joint stiffness is low ($k_n = 10$, $k_s = 500$, 100 and 1500 MPa/m). If the joint stiffness is high ($k_n = 10$, $k_s = 2000$, 2500 and 3000 MPa/m), the maximum amplitude appears at about $\alpha = 38°$.

Figure 4.9d shows the variation in the amplitude of vertical motion in the rock mass below the joint. The amplitude of vertical motion at $\alpha = 0°$ is larger than 1 due to the superposition of the incident P-wave and the reflected P-wave. With the increase of the joint dip angle, the amplitude of vertical motion increases slowly from its initial value to its maximum value. The maximum amplitude appears when the joint dip angle reaches a critical value, which ranges from 42° to 57° with the increase of joint stiffness. After that, the amplitude of vertical motion starts to decrease quickly as the α approaches 90°. The joint stiffness also affects the amplitude of the vertical motion. As the joint stiffness decreases, the amplitude of vertical motion increases gradually, due to that a weak joint can enhance the reflection of seismic waves and enlarge the amplitudes of the reflected P- and S_V-waves.

The analytical results shown in Figure 4.9 demonstrate that the amplitudes of the seismic motions in the fractured rock mass are affected by the joint dip angle and the joint stiffness to a great extent. In another aspect, the properties of the seismic wave, especially the wave frequency, also influence the propagation of seismic waves. To examine the effects of wave frequency on the seismic motions of the fractured rock mass, parametric studies were carried out, with the results shown in Figure 4.10 (assuming $k_n = 10k_s = 1000$

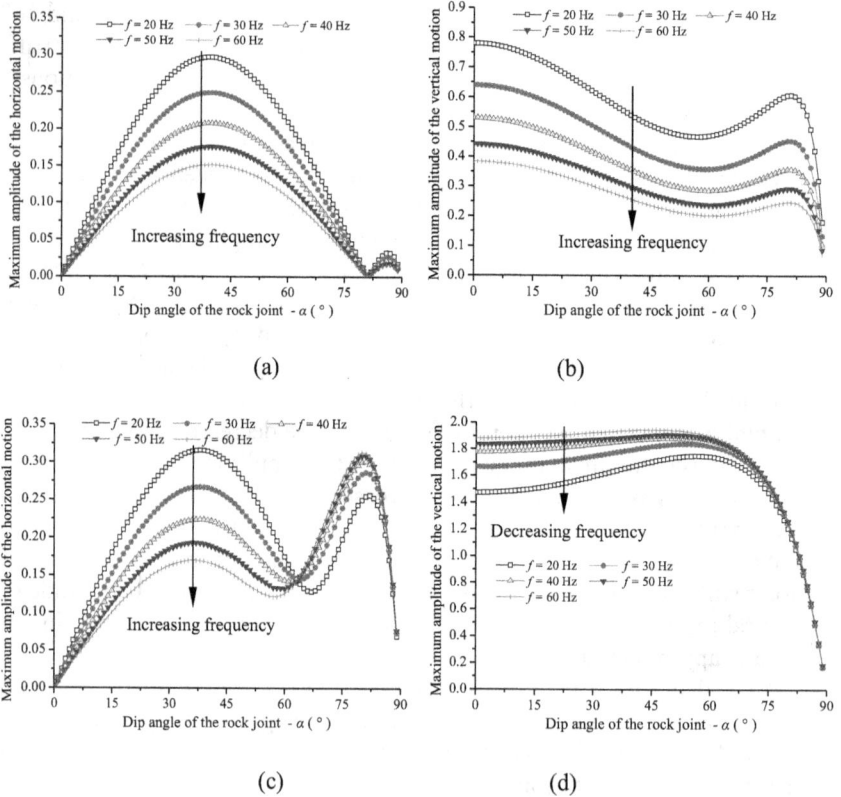

Figure 4.10 Maximum amplitudes of the horizontal and vertical motions at both sides of the rock joint with varying joint dip angles and wave frequencies (P-wave incidence and $k_n = 10k_s = 1000$ MPa/m). (a) Horizontal amplitude above the joint; (b) vertical amplitude above the joint; (c) horizontal amplitude below the joint; (d) vertical amplitude below the joint.

MPa/m). With the change of joint dip angle, the amplitudes of the horizontal and vertical motions at both sides of the rock joint have a similar variation tendency to the cases in Figure 4.9, despite the difference in the wave frequency. Because the influence of joint dip angle on the amplitudes of seismic motions has been discussed before, here only the influence of the wave frequency is focused on. In the rock mass above the joint, the seismic motions of rock particles are induced by the transmitted waves. As can be clearly confirmed in Figure 4.10a and 4.10b, the amplitudes of the seismic motions in both horizontal and vertical directions increase continuously with the decrease of wave frequency, and vice versa. The low-frequency seismic wave can produce transmitted P- and S_V-waves with large amplitudes when it travels across the rock joint, thereby the overall seismic motions above the joint have larger amplitudes than the cases that consider the incidence of high-frequency waves. In the rock mass below the joint, the seismic motions are caused by the incident and reflected waves. As shown in Figure 4.10c, the effect of wave frequency on the amplitude of horizontal motion below the joint exhibits a two-stage characteristic: in the dip angle range of $\alpha < 63°$, the amplitude of horizontal motion increases with the decrease of wave frequency; while in the dip angle range of $\alpha \geq 63°$, the overall horizontal amplitude decreases as the wave frequency decreases. Therefore, $\alpha = 63°$ can be regarded as a critical dip angle to estimate the effect of wave frequency on the overall amplitude of horizontal motion below the rock joint. Figure 4.10d gives the amplitude of the vertical motion below the joint. It can be found that a decreasing wave frequency can lead to a decrease of the overall vertical amplitude. This is because a low-frequency incident wave produces reflected P- and S_V-waves with small amplitudes (as shown in Figure 4.6a and 4.6b).

4.3 PROPAGATION OF SV-WAVE ACROSS AN INCLINED ROCK FRACTURE

4.3.1 Problem Formulation Based on the Displacement Discontinuity Method

Relative to P-waves, incident S-waves induce different types of particle motions, and thereby lead to different seismic behavior of fractured rock masses. There are two basic types of S-waves, including the S_H-wave and the S_V-wave. Compared with the case of S_V-wave, the situation considering the incidence of S_H-wave is much simpler, due to that the S_H-wave does not generate P- or S_V-waves when it impinges upon a fracture. To examine the effects of rock fracture on the propagation of S_H-wave as well as the seismic behavior of the fractured rock mass, an analytical model based on the displacement discontinuity method was built, as shown in Figure 4.11. This unbounded model contains an inclined planar joint, which behaves

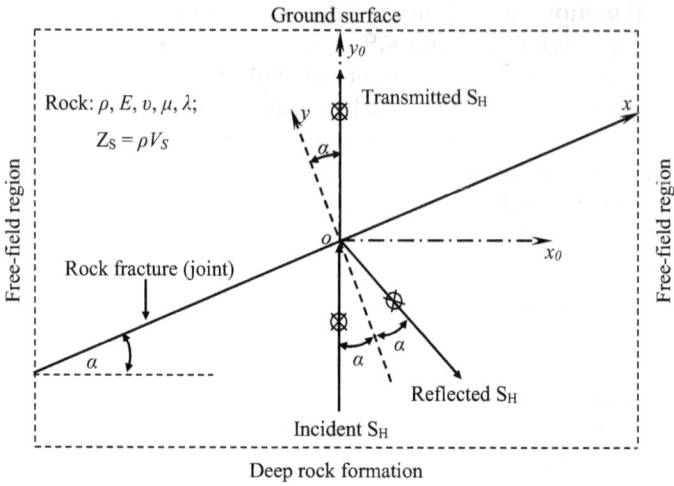

Figure 4.11 The displacement discontinuity model for the propagation of an S_H-wave across an inclined rock fracture.

linearly. The rock is assumed to be elastic, homogeneous, and isotropic, and the wave reflections at the model boundaries are neglected. When the incident S_H-wave impinges upon the joint, a reflected S_H-wave and a transmitted S_H-wave are produced.

A local coordinate system (x, y) is built, with its x-axis going along the joint plane. According to Snell's law, the angle of incidence, the angle of reflection, and the angle of transmission are identical to the joint dip angle (α). The incident, reflected, and transmitted S_H-waves will cause particle motions in the out-of-plane direction (in the z direction). Then the expressions of all seismic waves can be defined as

$$\left.\begin{aligned}
w_{ISH} &= I_{SH} \exp\left[iK_s\left(x\sin\alpha + y\cos\alpha - V_s t\right)\right] \\
w_{RSH} &= R_{SH} \exp\left[iK_s\left(x\sin\alpha - y\cos\alpha - V_s t\right)\right] \\
w_{TSH} &= T_{SH} \exp\left[iK_s\left(x\sin\alpha + y\cos\alpha - V_s t\right)\right]
\end{aligned}\right\} \tag{4.45}$$

where w_i $(I = ISH, RSH,$ and $TSH)$ represents the out-of-plane particle displacements caused by incident, reflected, and transmitted S_H-waves; and I_{SH}, R_{SH}, and T_{SH} denote the wave amplitudes. According to the assumptions of displacement discontinuity, the boundary conditions along the fracture plane $(y = 0)$ can be written as

$$w^+ - w^- = \tau_{yz}^- / k_s, \tau_{yz}^- = \tau_{yz}^+ \tag{4.46}$$

$$w^+ = w_{TSH}, w^- = w_{ISH} + w_{RSH}, \tau_{yz}^+ = \mu \partial w^+ / \partial y, \tau_{yz}^- = \mu \partial w^- / \partial y \tag{4.47}$$

After substituting the wave expressions into the boundary conditions, the amplitudes of the reflected and transmitted S_H-waves can be obtained (Pyrak-Nolte et al., 1990):

$$R_{SH} = \frac{-i I_{SH} Z_s \omega \cos \alpha}{2k_s - i Z_s \omega \cos \alpha}, T_{SH} = \frac{2 I_{SH} k_s}{2k_s - i Z_s \omega \cos \alpha} \tag{4.48}$$

To examine the overall seismic behavior of the fractured rock mass, a global coordinate system (x_0, y_0) is built, with its x_0-axis and y_0-axis going along the horizontal and vertical directions, respectively (see Figure 4.11). Then the wave expressions in the local coordinate system can be transformed into their corresponding forms in the global coordinate system, as given by

$$\left. \begin{array}{l} w_{ISH} = I_{SH} \exp\left[iK_s \left(y_0 - V_s t \right) \right] \\ w_{RSH} = R_{SH} \exp\left[iK_s \left(x_0 \sin 2\alpha - y_0 \cos 2\alpha - V_s t \right) \right] \\ w_{TSH} = T_{SH} \exp\left[iK_s \left(y_0 - V_s t \right) \right] \end{array} \right\} \tag{4.49}$$

The seismic motions of the rock mass caused by the S_H-waves are in the out-of-plane direction. In the region below the rock joint, the seismic motion (w^-) is induced by the incident and reflected waves. While in the region above the joint, the corresponding seismic motion (w^+) is only caused by the transmitted wave. Therefore, the seismic motions at both sides of the joint can be written as

$$\begin{array}{l} w^- = w_{ISH} + w_{RSH} = I_{SH} \exp\left[iK_s \left(y_0 - V_s t \right) \right] + \\ R_{SH} \exp\left[iK_s \left(x_0 \sin 2\alpha - y_0 \cos 2\alpha - V_s t \right) \right] \end{array} \tag{4.50}$$

$$w^+ = w_{TSH} = T_{SH} \exp\left[iK_s \left(y_0 - V_s t \right) \right] \tag{4.51}$$

Since only the real components of those wave expressions have physical significance, therefore, the seismic motions at the original point of the global coordinate system $(x_0 = y_0 = 0)$ can be given in the following forms (without loss of generality, the amplitude of an incident wave is assumed to be 1):

$$w^- = \cos(-\omega t) + |R_{SH}| \cos\left[-\omega t + \Delta\left(R_{SH}\right)\right] \tag{4.52}$$

$$w^+ = |T_{SH}| \cos\left[-\omega t + \Delta\left(T_{SH}\right)\right] \tag{4.53}$$

The same rock properties in the case of P-wave incidence were used in the following analysis of Section 4.3: the rock density is 2600 kg/m³, and the S_H-wave velocity (V_S) is 2700 m/s. The joint dip angle (α), which equals to the angle of incidence, the angle of reflection, and the angle of transmission, ranges from 0° to 90°. Likewise, the shear stiffness of the joint (k_s) ranges from 50 MPa/m to 300 MPa/m with an increment of 50 MPa/m to examine the effects of joint stiffness. Additionally, the wave frequency (f) ranges from 20 Hz to 60 Hz with an increment of 10 Hz to investigate the effects of wave frequency.

4.3.2 Effect of Joint Properties on the Wave Propagation

The magnitudes of the reflection and transmission coefficients represent the amplitudes of the reflected and transmitted waves, which are important parameters to estimate the attenuation effect of rock joints on wave propagation. When considering the incidence of an S_H-wave, the variation of the reflection and transmission coefficients (magnitudes) with the joint dip angle and the joint stiffness (e.g., $f = 50$ Hz) is shown in Figure 4.12. It can be found that the variation tendency of those coefficients with the joint dip angle is different from that in the case of P-wave incidence. With the increase of the joint dip angle (α), the reflection coefficient of S_H-wave ($|R_{SH}|$) decreases continuously from its initial value to about 0 (see Figure 4.12a). The initial value corresponds to the amplitude of the reflected S_H-wave in the

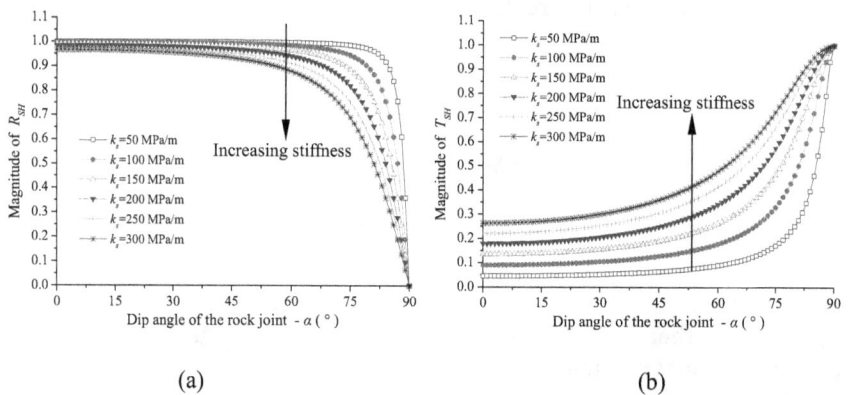

(a)   (b)

Figure 4.12 Variation of the reflection and transmission coefficients with the dip angle and stiffness of the rock joint (S_H-wave incidence and $f = 50$ Hz). (a) Reflection coefficient of S_H-wave; (b) transmission coefficient of S_H-wave.

case of normal incidence. As α approaches 90°, the value of $|R_{SH}|$ approaches 0, indicating that the incident wave travels parallel to the joint, and no wave reflection will occur at the joint. In the range of small joint dip angle (e.g., 0° < α < 45°), the $|R_{SH}|$ decreases slowly with the increase of α; while in the range of large dip angle (e.g., $\alpha \geq$ 45°), the decrease of $|R_{SH}|$ becomes much quicker. On the other hand, the joint stiffness (k_s) affects significantly the value of $|R_{SH}|$. Increasing joint stiffness can lead to the continuous decrease of $|R_{SH}|$ at any dip angle, and vice versa. It indicates that the strong rock joint can restrain the wave reflection and reduce the amplitude of the reflected S_H-wave, similar to the case of P-wave incidence.

As shown in Figure 4.12b, the transmission coefficient of S_H-wave ($|T_{SH}|$) exhibits different variation tendencies with the change of joint dip angle and joint stiffness. With the increase of α, the magnitude of $|T_{SH}|$ increases from its initial value (corresponding to the wave amplitude of normal incidence) to the maximum value located at α = 90°. When α = 90°, the value of $|T_{SH}|$ approaches 1, implying that the incident S_H-wave travels parallel to the joint and the wave amplitudes at both sides of the joint is equal to that of the incident wave. The joint stiffness has an opposite effect on the value of $|T_{SH}|$. An increasing joint stiffness leads to the increase of $|T_{SH}|$, indicating that the strong joint can promote wave transmission and increase the amplitude of the transmitted S_H-wave.

Parametric studies were also carried out to investigate the effects of joint properties on the time delay of reflected and transmitted S_H-waves, with the results shown in Figure 4.13. With the increase of joint dip angle, the time delay of the reflected S_H-wave (t_{RSH}, Figure 4.13a), which is caused by the presence of the rock joint, will increase continuously to its maximum value at α = 90°. However, the effect of joint dip angle on the time delay of transmitted S_H-wave is different (t_{TSH}, Figure 4.13b). The t_{TSH} decreases gradually with the increase of the joint dip angle and reaches its minimum when α = 90°. On the other hand, the obvious effect of joint stiffness on the time delay can be confirmed. For the reflected S_H-wave, an increasing joint stiffness leads to an increase of the time delay; while for the transmitted S_H-wave, an increasing joint stiffness causes a decrease of the time delay.

4.3.3 Effect of Wave Frequency on the Wave Propagation

The expressions of the reflection and transmission coefficients shown in Equation (4.48) indicate that the seismic response of the joint is frequency dependent. The wave frequency f (or the angular frequency ω) affects the amplitudes and the time delay of reflected and transmitted S_H-waves, which subsequently control the overall behavior of the fractured rock mass. To examine the effects of wave frequency on wave propagation, parametric studies were carried out, with their results shown in Figures 4.14 and 4.15. Figure 4.14 shows the variation of the reflection and transmission coefficients (magnitudes) with

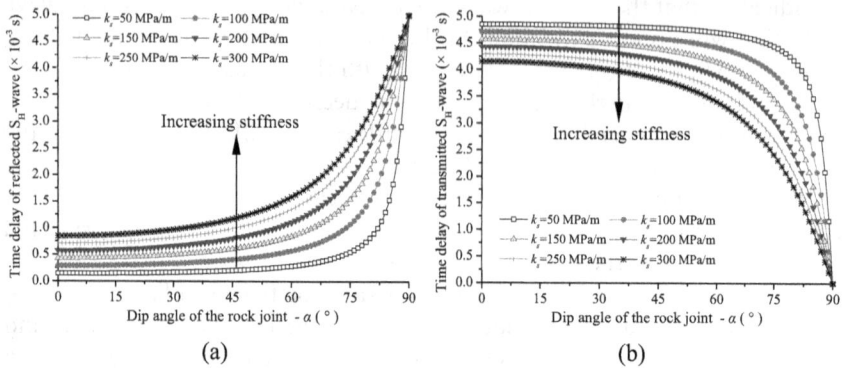

Figure 4.13 Variation of the time delay of reflected and transmitted waves with the dip angle and stiffness of the rock joint (S_H-wave incidence and $f = 50$ Hz). (a) Time delay of reflected S_H-wave; (b) time delay of transmitted S_H-wave.

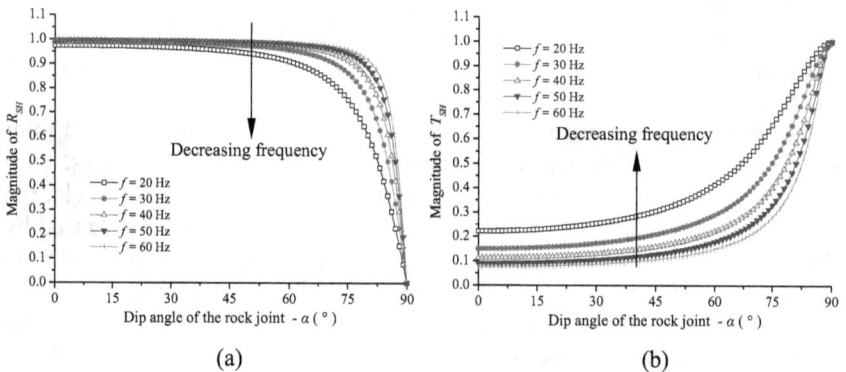

Figure 4.14 Variation of the reflection and transmission coefficients with the joint dip angle and the wave frequency (S_H-wave incidence and $k_s = 100$ MPa/m). (a) Reflection coefficient of S_H-wave; (b) transmission coefficient of S_H-wave.

the wave frequency. It can be found that with the decrease of the wave frequency from 60 Hz to 20 Hz, the magnitude of the reflection coefficient ($|R_{SH}|$) decreases continuously, while the magnitude of the transmission coefficient ($|T_{SH}|$) increases gradually. Those results indicate the low-pass filter effect of the rock joint. Relative to the high-frequency seismic wave (S_H-wave), the low-frequency wave is more likely to propagate across the joint. Thereby, the transmitted wave is attenuated to a smaller extent and has larger amplitude. Correspondingly, the reflected wave exhibits a smaller amplitude.

Figure 4.15 shows the time delay of the reflected and transmitted S_H-waves (t_{RSH} and t_{TSH}), which varies with the wave frequency. It can be found

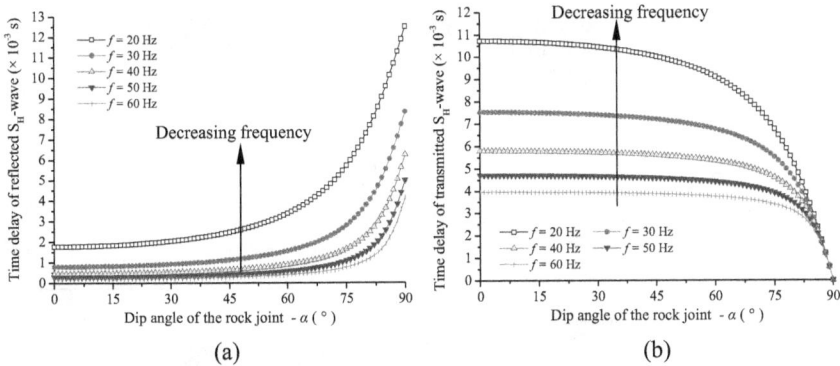

Figure 4.15 Variation of the time delay of reflected and transmitted waves with the joint dip angle and the wave frequency (S_H-wave incidence and k_s = 100 MPa/m). (a) Time delay of reflected S_H-wave; (b) time delay of transmitted S_H-wave.

that the wave frequency has accordant effects on the values of t_{RSH} and t_{TSH}. With the decrease of wave frequency, the t_{RSH} and t_{TSH} increase continuously at any joint dip angle, indicating that a low-frequency incident wave can cause longer time gaps between the incident wave and the reflected and transmitted waves.

4.3.4 Seismic Response of the Rock Mass

Due to the wave scattering occurring at the rock joint, the seismic motions of rocks below and above the joint exhibit different characteristics. When an S_H-wave propagates across a rock joint, the seismic motion of rock particles in the rock mass is in the out-of-plane direction. In the region above the joint, the seismic motion is caused by the transmitted S_H-wave; while in the region below the joint, the seismic motion is controlled by the superposition of incident and reflected S_H-waves.

The seismic motions at both sides of the joint can be estimated by using Equations (4.50) and (4.51). Firstly, parametric studies were carried out to investigate the effects of joint properties on the seismic motions of rock mass, with the results shown in Figure 4.16 (e.g., f = 50 Hz). In the region above the joint, the seismic motion is contributed only by the transmitted wave; thereby, the amplitude of the out-of-plane motion shown in Figure 4.16a is identical to the magnitude of the transmitted wave (see Figure 4.12b). Due to the attenuation effect of the rock joint, the amplitude of the seismic motion above the joint is smaller than that of the incident S_H-wave. With the increase of joint dip angle (α), the amplitude of seismic motion increases gradually from an initial value at α = 0° to the maximum

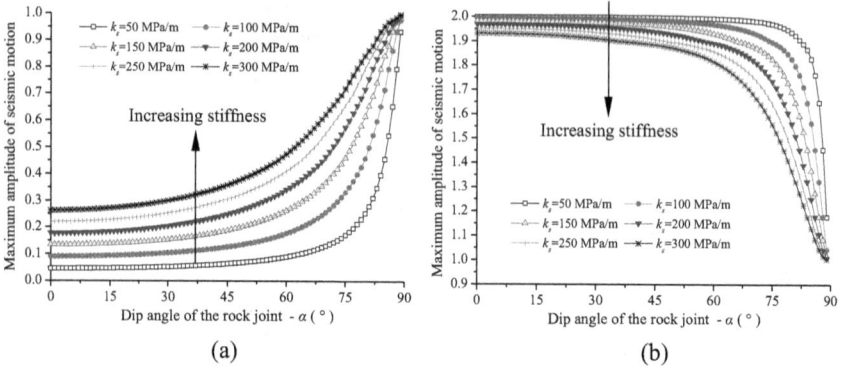

Figure 4.16 Maximum amplitudes of the seismic motions at both sides of the rock joint with varying dip angles and stiffness (S_H-wave incidence, and f = 50 Hz). (a) Maximum amplitude above the joint; (b) maximum amplitude below the joint.

value at α = 90°. The maximum amplitude appearing at α = 90° equals 1, because the transmitted wave has the same amplitude as the incident wave when the seismic wave propagates parallel to the joint. With the increase of joint stiffness, the amplitude of seismic motion above the joint increases gradually, implying that a strong joint with high stiffness can help to reduce the attenuation effect of the joint on wave propagation. Figure 4.16b shows the amplitude of seismic motion below the rock joint. The seismic amplitude is larger than 1 at most dip angles, due to the superposition of the incident and reflected waves. The seismic amplitude decreases from its initial value to the minimum value as α reaches 90°. When α = 90°, the amplitude of the reflected wave approaches 0 (see Figure 4.12a), and the overall seismic motion is only contributed by the incident wave, with the overall amplitude approaching 1. The joint stiffness has a different effect on the seismic amplitude below the joint. With the increase of joint stiffness, the amplitude of seismic motion below the joint will decrease continuously, due to that the strong joint can effectively reduce the amplitude of the reflected wave.

Besides the joint properties, wave frequency is another important parameter to affect the overall seismic behavior of fractured rock masses. To examine the effect of wave frequency on the seismic motions above and below the rock joint, parametric studies were carried out, with the results shown in Figure 4.17 (e.g., k_s = 100 MPa/m). For the seismic motion above the joint (Figure 4.17a), the amplitude increases gradually with the decrease of wave frequency. This is due because low-frequency waves can transmit easily

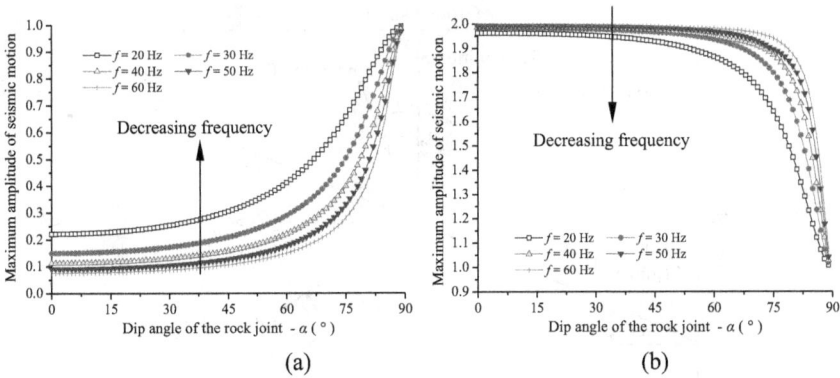

Figure 4.17 Maximum amplitudes of seismic motions at both sides of the rock joint with varying joint dip angles and wave frequencies (S_H-wave incidence and k_s = 100 MPa/m). (a) Maximum amplitude above the joint; (b) maximum amplitude below the joint.

across the joint with its amplitude being attenuated to a smaller extent. For the seismic motion below the joint (Figure 4.17b), the seismic amplitude decreases with the decrease of wave frequency. As has been discussed before, the amplitude of the reflected wave will decrease as the wave frequency decreases, which can help reduce the amplitude of the overall seismic motion below the joint.

4.4 PROPAGATION OF SV-WAVE ACROSS AN INCLINED ROCK FRACTURE

To investigate the seismic responses of rock joints to an incident S_V-wave, a displacement discontinuity model was established, as shown in Figure 4.18. Similarly, this unbounded model contains an inclined linearly deformable joint, and a local coordinate system (x, y) is defined with its x-axis going along the joint plane. An S_V-wave impinging upon the joint will produce four new seismic waves: a reflected P-wave, a reflected S_V-wave, a transmitted P-wave, and a transmitted S_V-wave. If the joint dip angle is α, the angle of incidence, reflection, and transmission for S_V-waves is equal to α. The reflection and transmission angle of P-waves (β) can be derived from Snell's law. In the local system, expressions of the incident, reflected, and transmitted waves can be given by Equation (4.54), in which the symbols have the same meanings as those in Equation (4.26).

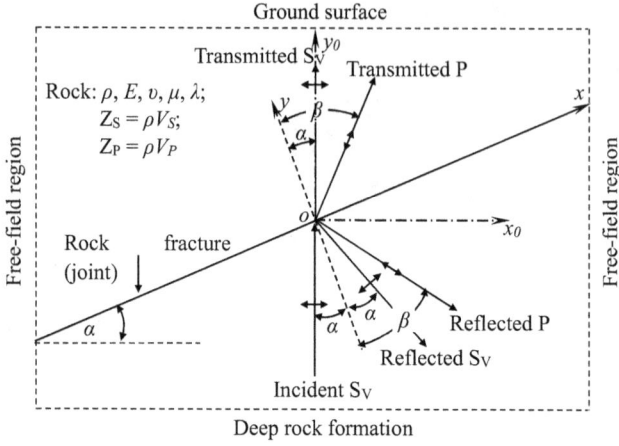

Figure 4.18 The displacement discontinuity model for the propagation of an S_V-wave across an inclined rock fracture.

$$
\left.
\begin{aligned}
\boldsymbol{u}_{ISV} &= \left\{u_{ISV}, v_{ISV}\right\} = I_{SV}\,exp[iK_S\left(xsin\alpha + ycos\alpha - V_St\right)] \times \left\{cos\alpha, -sin\alpha\right\} \\
\boldsymbol{u}_{RP} &= \left\{u_{RP}, v_{RP}\right\} = R_P\,exp[iK_P\left(xsin\beta - ycos\beta - V_pt\right)] \times \left\{sin\beta, -cos\beta\right\} \\
\boldsymbol{u}_{RSV} &= \left\{u_{RSV}, v_{RSV}\right\} = R_{SV}\,exp[iK_S\left(xsin\alpha - ycos\alpha - V_St\right)] \times \left\{cos\alpha, sin\alpha\right\} \\
\boldsymbol{u}_{TP} &= \left\{u_{TP}, v_{IP}\right\} = T_P\,exp[iK_P\left(xsin\beta + ycos\beta - V_pt\right)] \times \left\{sin\beta, cos\beta\right\} \\
\boldsymbol{u}_{TSV} &= \left\{u_{TSV}, v_{TSV}\right\} = T_{SV}\,exp[iK_S\left(xsin\alpha + ycos\alpha - V_St\right)] \times \left\{cos\alpha, -sin\alpha\right\}
\end{aligned}
\right\} \quad (4.54)
$$

For the linearly deformable joint, the displacement and stress boundary conditions along the joint ($y = 0$) have been given by Equations (4.27) and (4.28), in which the stress and displacement components at both sides of the joint can be expressed by

$$
\left.
\begin{aligned}
u^- &= u_{ISV} + u_{RP} + u_{RSV} \\
u^+ &= u_{TP} + u_{TSV} \\
v^- &= v_{ISV} + v_{RP} + v_{RSV} \\
v^+ &= v_{TP} + v_{TSV}
\end{aligned}
\right\} \quad (4.55)
$$

$$
\left.
\begin{aligned}
\sigma_y^i &= \lambda\frac{\partial u^i}{\partial x} + (\lambda + 2\mu)\frac{\partial v^i}{\partial y} \\
\tau_{xy}^i &= \mu\left(\frac{\partial v^i}{\partial x} + \frac{\partial u^i}{\partial y}\right)
\end{aligned}
\right\} (i = + \text{ or} -) \quad (4.56)
$$

After solving the boundary conditions, the amplitudes of reflected and transmitted waves (denoted by R_P, R_{SV}, T_P, and T_{SV}, respectively) can be expressed in the matrix form of Equation (4.57), which is a simplified form of the expression given by Pyrak-Nolte et al. (1990). Additionally, the reflection and transmission coefficients can be derived conveniently by assuming that the amplitude of the incident wave is 1 ($I_{SV} = 1$). The closed-form solutions of the reflection and transmission coefficients are expressed in Equations (4.58) to (4.61). After the coefficients of reflection and transmission are obtained, the expressions of all the reflected and transmitted waves can be subsequently obtained.

$$
\begin{bmatrix}
Z_P \cos 2\alpha & -Z_S \sin 2\alpha & -Z_P \cos 2\alpha & Z_S \sin 2\alpha \\
\dfrac{Z_S^2}{Z_P}\sin 2\beta & Z_S \cos 2\alpha & \dfrac{Z_S^2}{Z_P}\sin 2\beta & Z_S \cos 2\alpha \\
-k_n \cos\beta & k_n \sin\alpha & iZ_P\omega\cos 2\alpha - k_n\cos\beta & k_n\sin\alpha - Z_S\omega\sin 2\alpha \\
k_s \sin\beta & k_s\cos\alpha & i\dfrac{Z_S^2}{Z_v}\omega\sin 2\beta - k_s\sin\beta & iZ_S\omega\cos 2\alpha - k_s\cos\alpha
\end{bmatrix}
\begin{bmatrix} R_P \\ R_{SV} \\ T_P \\ T_{SV} \end{bmatrix}
= I_S
\begin{bmatrix} Z_S\sin 2\alpha \\ Z_S\cos 2\alpha \\ k_n\sin\alpha \\ -k_s\cos\alpha \end{bmatrix}
\quad (4.57)
$$

$$
R_P = \frac{-Z_P Z_S \omega \sin 4\alpha \begin{bmatrix} Z_P^2 Z_S \omega \cos^2 2\alpha + Z_S^3 \omega \sin 2\alpha \sin 2\beta + 2iZ_P^2 k_s \cos^3\alpha - iZ_P^2 k_s \cos\alpha + \\ iZ_S^2 k_n \sin 2\beta \sin\alpha + iZ_P Z_S k_s \cos 2\alpha \cos\beta + iZ_P Z_S k_s \sin 2\alpha \sin\beta \end{bmatrix}}{\begin{bmatrix} \left(2Z_P^2 k_s \cos\alpha - 4Z_P^2 k_s \cos^3\alpha - 2Z_P Z_S k_s \sin 2\alpha \sin\beta + iZ_P^2 Z_S \omega \cos^2 2\alpha + iZ_S^3 \omega \sin 2\alpha \sin 2\beta\right) \\ \times\left(-2Z_P k_n \cos 2\alpha \cos\beta - 2Z_S k_n \sin 2\beta \sin\alpha + iZ_S^2 \omega \sin 2\alpha \sin 2\beta + iZ_P^2 \omega \cos^2 2\alpha\right) \end{bmatrix}}
\quad (4.58)
$$

$$
R_{SV} = \frac{-Z_S \omega \begin{bmatrix} \omega Z_P^4 \cos^4 2\alpha + 2iZ_P^3 k_n \cos^3 2\alpha \cos\beta + 2iZ_P^2 Z_S k_n \cos^2 2\alpha \sin 2\beta \sin\alpha \\ -Z_S^4 \omega \sin^2 2\alpha \sin^2 2\beta - 2iZ_P Z_S^2 k_s \sin^2 2\alpha \sin 2\beta \sin\beta - iZ_P^2 Z_S k_s \sin 4\alpha \sin 2\beta \cos\alpha \end{bmatrix}}{\begin{bmatrix} \left(2Z_P^2 k_s \cos\alpha - 4Z_P^2 k_s \cos^3\alpha - 2Z_P Z_S k_s \sin 2\alpha \sin\beta + iZ_P^2 Z_S \omega \cos^2 2\alpha + iZ_S^3 \omega \sin 2\alpha \sin 2\beta\right) \\ \times\left(-2Z_P k_n \cos 2\alpha \cos\beta - 2Z_S k_n \sin 2\beta \sin\alpha + iZ_S^2 \omega \sin 2\alpha \sin 2\beta + iZ_P^2 \omega \cos^2 2\alpha\right) \end{bmatrix}}
\quad (4.59)
$$

$$
T_P = \frac{Z_P Z_S \omega \sin 4\alpha \begin{bmatrix} iZ_P^2 k_s \cos\alpha - 2iZ_P^2 k_s \cos^3\alpha + iZ_S^2 k_n \sin 2\beta \sin\alpha + iZ_P Z_S k_n \cos 2\alpha \cos\beta \\ -iZ_P Z_S k_n \sin 2\alpha \sin\beta \end{bmatrix}}{\begin{bmatrix} \left(2Z_P^2 k_s \cos\alpha - 4Z_P^2 k_s \cos^3\alpha - 2Z_P Z_S k_s \sin 2\alpha \sin\beta + iZ_P^2 Z_S \omega \cos^2 2\alpha + iZ_S^3 \omega \sin 2\alpha \sin 2\beta\right) \\ \times\left(-2Z_P k_n \cos 2\alpha \cos\beta - 2Z_S k_n \sin 2\beta \sin\alpha + iZ_S^2 \omega \sin 2\alpha \sin 2\beta + iZ_P^2 \omega \cos^2 2\alpha\right) \end{bmatrix}}
\quad (4.60)
$$

$$
T_{SV} = \frac{\begin{bmatrix} 2Z_P^2 Z_S k_s k_n \sin 4\alpha \sin 2\beta + 4Z_P^3 k_s k_n \cos^2 2\alpha \cos\alpha \cos\beta - 2iZ_P Z_S^4 k_s \omega \cos^3 2\alpha \cos\alpha \\ +4Z_P Z_S^2 k_s k_n \sin 2\alpha \sin\alpha \sin 2\beta \sin\beta - 2iZ_S^4 k_n \omega \sin 2\alpha \sin\alpha \sin^2 2\beta - \\ iZ_P^3 Z_S k_s \omega \sin 4\alpha \cos 2\alpha \sin\beta - iZ_P Z_S^3 k_n \omega \sin 4\alpha \sin 2\beta \cos\beta \end{bmatrix}}{\begin{bmatrix} \left(2Z_P^2 k_s \cos\alpha - 4Z_P^2 k_s \cos^3\alpha - 2Z_P Z_S k_s \sin 2\alpha \sin\beta + iZ_P^2 Z_S \omega \cos^2 2\alpha + iZ_S^3 \omega \sin 2\alpha \sin 2\beta\right) \\ \times\left(-2Z_P k_n \cos 2\alpha \cos\beta - 2Z_S k_n \sin 2\beta \sin\alpha + iz_S^2 \omega \sin 2\alpha \sin 2\beta + iZ_P^2 \omega \cos^2 2\alpha\right) \end{bmatrix}}
\quad (4.61)
$$

To investigate the seismic motions of the rock mass, a global coordinate system (x_0, y_0) is built, with its x_0-axis and y_0-axis going along the horizontal and vertical directions, respectively (see Figure 4.18). Then the expressions of all seismic waves in the local system can be transformed into their global system forms by performing the rotation of coordinate axes, as given by

$$\left. \begin{aligned} \boldsymbol{u}_{ISV0} &= \{u_{ISV0}, v_{SVV0}\} = I_{SV} \exp\left[iK_S\left(y_0 - V_S t\right)\right] \times \{1, 0\} \\ \boldsymbol{u}_{RF0} &= \{u_{RP0}, v_{RP00}\} = R_P \exp\left[iK_P\left(x_0 \sin(\alpha + \beta) - y_0 \cos(\alpha + \beta) - V_P t\right]\times\{\sin(\alpha + \beta), -\cos(\alpha + \beta)\} \\ \boldsymbol{u}_{RSV0} &= \{u_{RSV0}, v_{RSV0}\} = R_{SV} \exp\left[iK_S\left(x_0 \sin 2\alpha - y_0 \cos 2\alpha - V_S t\right)\right]\times\{\cos 2\alpha, \sin 2\alpha\} \\ \boldsymbol{u}_{TP0} &= \{u_{IP0}, v_{IPO}\} = T_P \exp\left[iK_P\left(-x_0 \sin(\alpha - \beta) + y_0 \cos(\alpha - \beta) - V_P t\right]\times\{-\sin(\alpha - \beta), \cos(\alpha - \beta)\} \\ \boldsymbol{u}_{ISV0} &= \{u_{ISV0}, v_{ISV0}\} = T_{SV} \exp\left[iK_S\left(y_0 - V_S t\right)\right]\times\{1, 0\} \end{aligned} \right\} \quad (4.62)$$

In the region below the rock joint, the particle motions are induced by the incident S_V-wave, the reflected S_V-wave, and the reflected P-wave, and thereby the seismic motions in the horizontal and vertical directions (u^- and v^-) can be given by

$$u^- = u_{ISV0} + u_{RSV0} + u_{RP0} = I_{SV} \exp\left[iK_S\left(y_0 - V_S t\right)\right] + R_{SV} \cos 2\alpha \exp\left[iK_S\left(x_0 \sin 2\alpha - y_0 \cos 2\alpha - V_S t\right)\right] + R_P \sin(\alpha + \beta)\exp\left[iK_P\left(x_0 \sin(\alpha + \beta) - y_0 \cos(\alpha + \beta) - V_P t\right)\right] \quad (4.63)$$

$$v^- = v_{ISV0} + v_{RP0} + v_{RSV0} = -R_P \cos(\alpha + \beta)\exp\left[iK_P\left(x_0 \sin(\alpha + \beta) - y_0 \cos(\alpha + \beta) - V_P t\right)\right] + R_{SV} \sin 2\alpha \exp\left[iK_S\left(x_0 \sin 2\alpha - y_0 \cos 2\alpha - V_S t\right)\right] \quad (4.64)$$

Additionally, the particle motions in the region above the rock joint (u^+ and v^+) are induced by the transmitted P-wave and S_V-wave, and can be written as

$$u^+ = u_{TP0} + u_{TSV0} = -T_P \sin(\alpha - \beta)\exp\left[iK_P\left(-x_0 \sin(\alpha - \beta) + y_0 \cos(\alpha - \beta) - V_P t\right)\right] + T_{SV} \exp\left[iK_S\left(y_0 - V_S t\right)\right] \quad (4.65)$$

$$v^+ = v_{TP0} + v_{TSV0} = T_P \cos(\alpha - \beta)\exp\left[iK_P\left(-x_0 \sin(\alpha - \beta) + y_0 \cos(\alpha - \beta) - V_P t\right)\right] \quad (4.66)$$

Since only the real components of those expressions have physical significance, the seismic motions at the original point of the global coordinate system ($x_0 = y_0 = 0$) can be given in the following real forms (I_{SV} is assumed to be 1):

$$u^- = u_{ISV0} + u_{RSV0} + u_{RP0} = \cos(-\omega t) + |R_{SV}|\cos 2\alpha \cos\left[-\omega t + \Delta\left(R_{SV}\right)\right] + |R_P|\sin(\alpha + \beta)\cos\left[-\omega t + \Delta\left(R_P\right)\right] \quad (4.67)$$

$$v^- = v_{ISV0} + v_{RP0} + v_{RSV0} = -|R_P|\cos(\alpha + \beta)\cos\left[-\omega t + \Delta\left(R_P\right)\right] + |R_{SV}|\sin 2\alpha \cos\left[-\omega t + \Delta\left(R_{SV}\right)\right] \quad (4.68)$$

$$u^+ = u_{TP0} + u_{TSV0} = -|T_P|\sin(\alpha - \beta)\cos\left[-\omega t + \Delta\left(T_P\right)\right] + |T_{SV}|\cos\left[-\omega t + \Delta\left(T_{SV}\right)\right] \quad (4.69)$$

$$v^+ = v_{TP0} + v_{TSV0} = |T_P|\cos(\alpha - \beta)\cos\left[-\omega t + \Delta\left(T_P\right)\right] \quad (4.70)$$

However, the S_V-wave has not been considered in the parametric studies, due to the limit of the critical incidence angle of S_V-wave, which is defined as

$\alpha_c = \sin^{-1}(V_S/V_P)$ (V_S and V_P represent the velocities of S_V-wave and P-wave, respectively). If the incidence angle of S_V-wave exceeds the critical angle ($\alpha > \alpha_c$; e.g., $\alpha_c = 33.4°$ for the rock properties in this study), the emergence angle of newly generated P-waves is no longer real-valued, which is difficult to be analyzed by using the analytical method mentioned in this chapter.

4.5 CONCLUSION

This chapter provides an analytical study of the effects of rock fractures on the propagation of seismic waves by using the displacement discontinuity method. Analytical models considering that different types of seismic waves (P-wave, SH-wave, and SV-wave) propagate across an inclined rock fracture (joint) were formulated. To obtain an insight into the seismic responses of the rock joint and the fractured rock mass, a series of parametric studies were also carried out based on the displacement discontinuity models that take into account the incidences of a P-wave and an S_H-wave, respectively.

For the case of P-wave incidence, the magnitudes of the reflection and transmission coefficients, which represent the amplitudes of the corresponding reflected and transmitted waves, exhibit different distribution characteristics with the change of the joint dip angle. The results of parametric studies indicate that the joint stiffness affects significantly the magnitudes of the reflection and transmission coefficients. The weak rock joint with low stiffness increases the amplitudes of the reflected P-wave and S_V-wave, but decreases the amplitudes of the transmitted P-wave and S_V-wave. Relatively, the strong joint with high stiffness has the opposite effect on the magnitudes of reflection and transmission coefficients. The results demonstrate that the transmission coefficients of P-wave and S_V-wave are smaller than 1, exhibiting an attenuation effect of the rock joint on the amplitudes of transmitted waves. On the other hand, the frequency of the incident P-wave also has an obvious effect on the wave amplitudes. With the increase of wave frequency, the amplitudes of the reflected waves increase gradually, while the amplitudes of the transmitted waves will decrease. These results reveal that the seismic response of rock joints is frequency dependent, and the joint behaves like a low-pass filter, which can filter the high-frequency content of the wave signal by reducing its amplitude. In the parametric studies, the time delay of reflected and transmitted waves caused by the presence of rock joint, which varies with the joint dip angle, the joint stiffness and the wave frequency, was also analyzed.

The overall seismic motions of the fractured rock mass, which are induced by the propagation of P-wave, were investigated. The seismic behavior of rock masses exhibits different characteristics in the regions above and below the joint. In the region above the joint, the seismic motions are governed by the transmitted waves, while in the region below the joint, the seismic motions are controlled by the incident and reflected waves. The overall horizontal and vertical motions of the rock masses at both sides of the joint have complicated variation tendency with the joint dip angle, due to the superposition of different componential waves. The joint stiffness affects significantly the amplitudes of the reflected and transmitted waves and subsequently affects the overall seismic motions. With the decrease of joint stiffness, the amplitudes of the horizontal and vertical motions in the region above the joint decrease continuously, because the weak joint has obvious attenuation effects on the transmitted waves. Oppositely, the amplitude of vertical motion in the region below the joint increases with the decrease of joint stiffness, because the weak joint can enhance the reflection of seismic waves. The joint stiffness has a relatively complicated effect on the horizontal amplitude below the joint. The analytical results indicate that in the joint dip angle range of $\alpha < 63°$, a decreasing joint stiffness leads to a decrease of the horizontal amplitude below the joint, while if $\alpha > 63°$, the decreasing joint stiffness causes the increase of the horizontal amplitude. On the other hand, the effects of wave frequency on the seismic motions were also analyzed. Due to the low-pass filter effect of the joint, the increasing P-wave frequency can lead to a decrease of the horizontal and vertical amplitudes above the joint, but can increase the vertical amplitude below the joint. Similarly, for the horizontal amplitude below the joint, the effect of wave frequency can be divided into two types, which depend on the critical dip angle of the joint ($\alpha = 63°$).

When considering the propagation of S_H-wave across a rock joint, the situation becomes simpler compared with the case of P-wave incidence, because the S_H-wave only generates a reflected S_H-wave and a transmitted S_H-wave. The corresponding parametric studies indicate that the variation tendency of the reflection and transmission coefficients with the joint dip angle is quite different from that in the case of P-wave incidence. With the increase of the joint dip angle, the magnitude of the reflection coefficient decreases continuously to its minimum at $\alpha = 90°$ ($|R_{SH}|_{min} = 0$), while the magnitude of the transmission coefficient will increase gradually to its maximum located at $\alpha = 90°$ ($|T_{SH}|_{max} = 1$). The joint stiffness influences the magnitudes of the reflection and transmission coefficients obviously. Increasing joint stiffness can lead to the continuous decrease of the reflection coefficient, and lead to the increase of the transmission coefficient, indicating that a strong joint with high stiffness can restrain the reflection of S_H-wave and can promote the transmission of S_H-wave. On the other hand, the frequency of the incident S_H-wave also has a significant effect on the magnitudes of the reflection and transmission coefficients. With the decrease of the wave frequency, the magnitude of the reflection coefficient decreases, while the magnitude of the transmission coefficient increases gradually, as a result of the low-pass filter effect of the rock joint. It can be found that the effects of the joint stiffness and the wave frequency on the propagation of S_H-wave across rock joints are approximately similar to the case of P-wave incidence.

The incident S_H-wave causes out-of-plane motions of the fractured rock mass. In the region above the joint, the seismic motion is induced only by the transmitted S_H-wave, and thereby it has the same variation tendency with the transmission coefficient as the change of joint dip angle. Due to the attenuation effect of the joint, the amplitude of the overall motion above the joint is smaller than that of the incident wave. The seismic motion in the region below the joint is superposed by the incident and reflected wave, thereby it has larger amplitude than the incident wave. With the increase of the joint stiffness, the amplitude of the seismic motion above the joint increases gradually, while the amplitude of the seismic motion below the joint will decrease. The results indicate that the strong joint with high stiffness, which promotes the transmission of S_H-wave, can enlarge the seismic motions above the joint, similar to the case of P-wave incidence. On the other hand, a decreasing wave frequency can lead to an increase of the seismic motion amplitude in the region above the joint, and lead to a decrease of motion amplitude in the region below the joint, which can be explained by the low-pass filter effect of the joint.

In this chapter, the effects of joint dip angle, joint stiffness, and wave frequency on the propagation of P-wave and S_H-wave across a rock joint were investigated based on the displacement discontinuity models. Additionally, the overall seismic motions at both sides of the rock joint were also examined. The research results can provide a fundamental understanding of the seismic behavior of fractured rock mass as well as the effects of rock joints on the propagation of seismic waves. However, another type of seismic wave, the S_V-wave, has not been considered in the parametric studies, due to the limit of the critical incidence angle of S_V-wave, which is defined as $\alpha_c = \sin^{-1}(V_S/V_P)$ (V_S and V_P represent the velocities of S_V-wave and P-wave, respectively). If the incidence angle of S_V-wave exceeds the critical angle ($\alpha > \alpha_c$; e.g., $\alpha_c = 33.4°$ for the rock properties in this study), the emergence angle of newly generated P-waves is no longer real-valued, which is difficult to analyze by using the analytical method mentioned in this chapter. Additionally, the displacement discontinuity models in this study do not consider the wave reflections at the model boundaries, and it does not apply to engineering problems with complicated boundary conditions. Therefore, further researches on this topic are necessary in the future.

Chapter 5

Effect of Rock Fracture on Seismic Wave Propagation

Experimental Method

5.1 PRINCIPLES OF DYNAMIC CENTRIFUGE MODEL TEST

Small-scale models are a relatively inexpensive and convenient way to study the mechanical behavior of prototypes with much larger dimensions. However, for most prototypes encountered in engineering practices (e.g., foundations and underground structures), the stresses due to self-weight usually play an important role, which can cause nonlinear deformations of the geotechnical materials such as soil and rock. To make a reliable estimate of the behavior of a prototype, it is necessary to reproduce the same stress status in the model as that in the prototype, which can be accomplished by spinning the model in a geotechnical centrifuge.

The centrifuge applies an increased gravitational acceleration to the reduced-scale model to produce identical self-weight stresses in the model and prototype. The same stress status enhances the similitude between the model and prototype and makes it possible to obtain accurate data to help solve complex engineering problems. In centrifuge modeling, the model has to be scaled to satisfy certain requirements, which are usually called scaling laws or scaling relationships. The scaling laws are used to convert the observed model behavior to prototype behavior (Joseph et al., 1988). Generally, there are two ways to derive the scaling laws of centrifuge tests: the first one consists of using the governing equation of the phenomenon to be modeled (Roscoe, 1968; Dong et al., 2000), and the other way is to employ the method of dimensional analysis (Hoek, 1965).

In dynamic centrifuge model tests, the same geotechnical material (soil, rock, or its related similar material) with the prototype is typically selected to build the small-scale model; therefore, the material density is identical in the model and the prototype ($\rho_m = \rho_p$, where the subscripts m and p represent the properties of model and prototype, respectively). If the sizes of the model (length, height, and thickness) are reduced by a scaling factor of N compared with the prototype dimensions ($L_m = L_p/N$), and simultaneously

DOI: 10.1201/9781003401599-5

the centrifugal acceleration is increased by the same factor N relative to the gravitational acceleration ($a_m = N \times g$), then the stress at any point in the model is the same as it would be at the corresponding point of the prototype ($\sigma_m = \sigma_p$). Therefore, the scaling factors of length, acceleration, and stress can be expressed as $L^* = L_m/L_p = 1/N$, $a^* = a_m/a_p = N$, and $\sigma^* = \sigma_m/\sigma_p = 1$, where the asterisk on a quantity denotes the scaling factor for that quantity. Since the acceleration has the dimension of LT^{-2}, the scaling factor of time t in dynamic problems can be given by $t^* = (L^*/a^*)^{1/2} = 1/N$. Similarly, the stress has the dimension of $ML^{-1}T^{-2}$, thereby the scaling factor of mass m can be derived by $m^* = L^* \times t^{*2} = 1/N^3$. After the scaling factors of mass, time, and length are obtained, the scaling factors for the other parameters, such as force, velocity, and frequency, can also be derived subsequently by employing the method of dimensional analysis, and the results are shown in Table 5.1, which can also be found in the references given by Kutter (1983) and Taylor (1995).

Table 5.1 Scaling Laws of the Dynamic Centrifuge Model Test

Parameter	Index (Unit)	Dimension	Scaling Factor Model/Prototype
Length	L (m)	L	$1/N$
Volume	V (m^3)	L^3	$1/N^3$
Displacement	d (m)	L	$1/N$
Velocity	V (m/s)	LT^{-1}	1
Acceleration	a, g (m/s^2)	LT^{-2}	N
Density	ρ (kg/m^3)	ML^{-3}	1
Mass	M (kg)	M	$1/N^3$
Elastic modulus	E (MPa)	ML^{-1}T^{-2}	1
Poisson's ratio	υ	–	1
Stress	σ (MPa)	ML^{-1}T^2	1
Strain	ε	–	1
Strength	σ_s (MPa)	ML^{-1}T^{-2}	1
Force	F (N)	MLT^{-2}	$1/N^2$
Internal frictional angle	ϕ (degree)	–	1
Cohesion	c (MPa)	ML^{-1}T^{-2}	1
Time (dynamic)	t (s)	T	$1/N$
Frequency	f (Hz)	T^{-1}	N
Damping ratio	ξ	–	1
Joint stiffness	k_n, k_s (MPa/m)	ML^{-2}T^{-2}	N

5.2 EXPERIMENTAL TECHNOLOGIES

5.2.1 Medium-Sized Centrifuge Apparatus

In this study, a medium-sized geotechnical centrifuge apparatus (type: MIS-257–1–15), produced by Marui Co., Ltd. (Osaka, Japan), was utilized to investigate the seismic behavior of fractured rock foundations as well as the effects of rock fractures on seismic motion. Figure 5.1 shows the photo and internal structure of the medium-sized centrifuge. This apparatus belongs to the type of beam centrifuge, which applies the centrifugal acceleration (or force) to the small-scale model through the high-speed rotation of the centrifuge arm. The fundamental hardware configuration of this centrifuge apparatus is mainly made up of three parts, including the rotating system, the model container, and the data acquisition and processing system.

The rotating system, mainly consisting of a spindle, two rotating arms, a rotation drive motor, and the conduction devices, is the most important part of the centrifuge. The spindle is located in the center of the centrifuge, and its high-speed rotation drives the centrifuge arms to generate centrifugal forces. The rotation drive motor installed at the bottom of the apparatus provides the rotation power for the centrifuge spindle and arms through the conduction devices such as the V-belt pulley. The output power of the drive motor can be adjusted by using a motor controller. Additionally, a small tilt motor is also installed in the interior of the spindle to adjust the inclination of the rotating arms for balance, with an adjustment range of 0° to ±15°. The nominal radius of the rotating system is 1500 mm. The centrifugal acceleration acting on the model is proportional to the nominal radius and the rotational speed of the system. When studying quasi-static problems (e.g., the foundation settlement due to self-weight), the maximum loading capability of this centrifuge can reach about 200 g (g represents the gravitational acceleration) at a maximum rotational speed of 345 cycles per minute, while for dynamic problems, the maximum loading capability reduces to about 50 g, due to the bearing capability constraint of the dynamic system.

There are two model containers attached to the ends of centrifuge arms through hinged supports. At the very beginning of tests, the model container is approximately perpendicular to the rotating arm, due to the effect of gravitational force. With the increase in rotational speed, the container rises gradually as a result of the increased centrifugal force in the horizontal direction, until its axis coincides with that of the rotating arm. After that, the centrifugal force, which is much larger than the gravitational force, controls the mechanical and deformational behavior of the model. The dimension of the original containers of the centrifuge is 500 mm × 300 mm × 150 mm (width × height × thickness). According to the scaling law, this container can be used to model a prototype with the largest dimension of 100 m × 60 m × 30 m

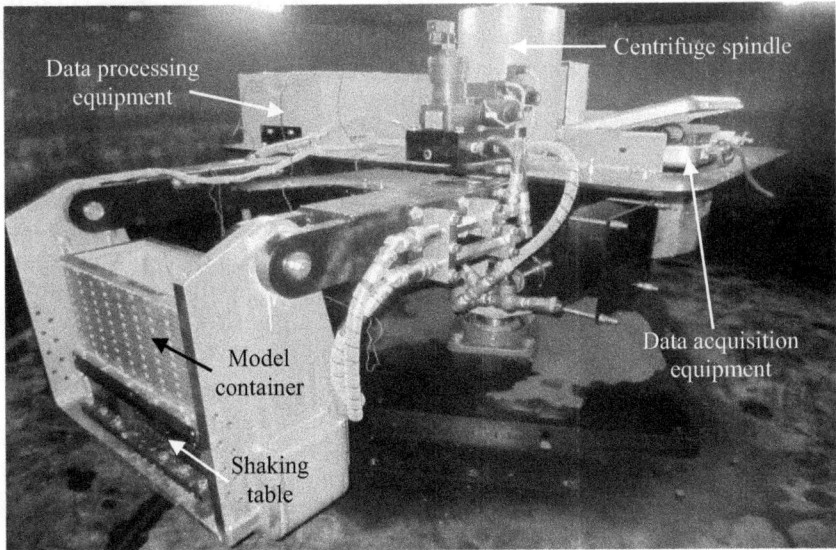

(a)

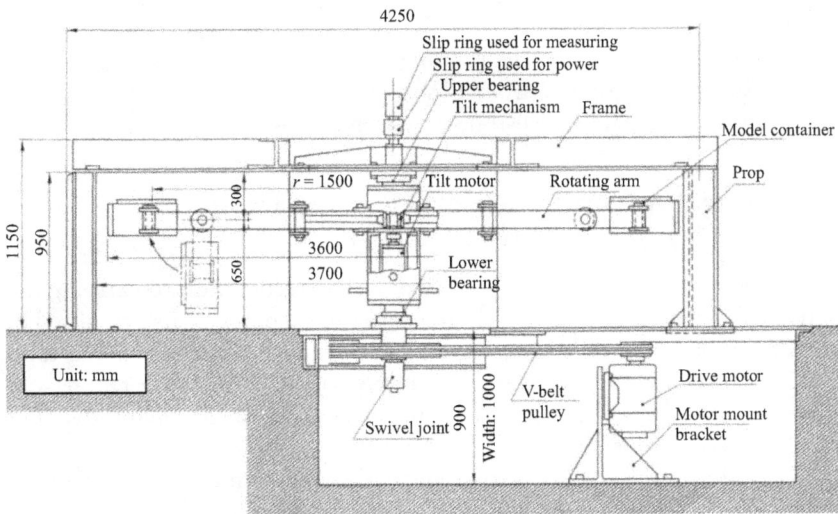

(b)

Figure 5.1 Medium-sized centrifuge apparatus and its structure and sizes. (a) Photo of the centrifuge apparatus; (b) schematic view of the centrifuge apparatus.

under static conditions, and a prototype with the largest dimension of 25 m × 15 m × 7.5 m under dynamic loading conditions. To apply seismic loads to models and to represent reliable boundary conditions during dynamic centrifuge tests, a vibration apparatus that consists of a shaking table and a laminar model container was developed, and utilized in centrifuge tests instead of the original model container. Details of this newly developed vibration apparatus will be described in the following section.

The data acquisition and processing system of this centrifuge consists of various measuring devices (e.g., the displacement, acceleration, and soil pressure sensors) and the data collecting and processing equipment. A displacement sensor with a measuring range of ±30 mm is installed at the bottom of the model container to measure the displacement variation during the tests and to make sure that the vibration of the container starts from its center position. A high-accuracy acceleration sensor with a measuring range of 500 m/s^2 is fixed at the bottom of the model container to measure the input seismic acceleration. In addition, other acceleration sensors can be installed to measure the seismic response of models. All the displacement and acceleration data measured by sensors are collected and processed by using the data collecting and processing equipment during tests, which convert analog signals to digital signals and send them to the operating platform.

The operation of the centrifuge is conducted by using an operating platform located outside of the centrifuge frame. A comprehensive software system is integrated into the operating platform, with the basic functions of (1) controlling and monitoring the running status of the centrifuge; (2) collecting and storing the test data from all sensors; and (3) processing the test data and plotting their related real-time variation curves.

5.2.2 Development of a Vibration Apparatus for Dynamic Centrifuge Tests

A proper vibration apparatus is an important precondition to carrying out dynamic centrifuge tests and to make sure the accuracy of test results. The vibration apparatus should contain two essential parts: the shaking table that applies various seismic motions (simple harmonic waves or real seismic waves) to models; and the model container that can provide reliable dynamic boundary conditions for models.

The shaking table attempts to replicate the true nature of the earthquake input, which induces realistic inertia forces in the model and thereby generates response stresses and displacements. With the wide application of dynamic centrifuge tests in recent decades, the technology of shaking tables has been improved greatly (Ko, 1994; Yu & Chen, 2005; Hou, 2006; Chen & Yu, 2006; Severn, 2011). According to the excitation modes, shaking tables mainly fall into five types: the mechanical shaking table, the explosive-type shaking table, the piezoelectric shaking table, the electromagnetic shaking table, and the electrohydraulic shaking table. The mechanical shaking table was utilized in the earliest studies, which employed different techniques, such as the toggle and spring system (Bolton & Steedman, 1982; Ortiz et al., 1983), the 'bumpy road' technique (Kutter, 1983), and the cam and bar system (Kimura et al., 1988a), to produce seismic motions. This type of shaking table can only generate simple harmonic waves within a small range of frequency. The explosive-type shaking table (Zelikson et al., 1983) provides a possible method to produce seismic signals by using explosives. It can excite large models; however, it is difficult to control accurately the characteristics of seismic waves generated via explosion, and thereby the tests have poor reproducibility. The piezoelectric shaking table (Arulanandan et al., 1982) consists of several piezoelectric ceramic elements, which produce deformations when they are placed in an electric field. The piezoelectric shaking table can excite random vibrations, and the model mass is usually small due to the constraint of the deformation capability of piezoelectric elements. The electromagnetic shaking table (Fujii, 1991) contains two coils that carry alternating currents. The alternating current cause attractive and repulsive forces between coils, which excite seismic motions. This type of shaking table can generate large-amplitude seismic waves with various frequencies; however, its application is usually limited in centrifuge tests due to its large size and weight. Compared with the above types of shaking tables, the electrohydraulic shaking table, which uses high-pressure oil to excite the vibration of models, has significant performance advantages in the aspects that it can (1) simulate accurately different kinds of seismic waves with various frequencies and amplitudes; (2) automatically control the vibration process by using a servo valve; and (3) instantly return and record the information about seismic motions. In recent years, the electrohydraulic shaking table has been widely utilized in dynamic centrifuge tests (Kimura et al., 1988b; Zhang et al., 2002; Zhang et al., 2004).

In the dynamic centrifuge tests, the model is enclosed within finite boundaries provided by the container walls. Another important consideration is the presence of artificial boundaries on the model behavior. Due to the stiffness difference between the model and container walls, waves that travel outwards may be reflected back into the model (Coe et al., 1985; Weissman & Prevost, 1989; Chen et al., 2010a). The reflected waves can distort the stress and strain fields in the model, leading to that the test results can not represent accurately the actual situation. When studying the seismic behavior of rock foundations, the proper container needs to make sure the free motion of the model and to permit the seismic wave to travel freely out of the model through the lateral boundaries. Currently, three types of containers, including the rigid-wall container, the equivalent shear beam container, and the laminar container, are commonly used in dynamic centrifuge tests (Lee et al., 2012). The rigid-wall container is usually made up of hard materials such as steel. It has a much higher stiffness than the test material; thereby, wave reflections are inevitably generated at container walls (Zeng, 1998). Additionally, the rigid walls can also restrict the model deformation. The equivalent shear beam container is made up of rectangular frames separated by rubber layers (Zeng & Schofield, 1996; Teymur & Madabhushi, 2003; Coelho et al., 2003). To perform a reliable simulation, the container stiffness should match to the model stiffness. However, it is usually difficult to make sure the stiffness matches when using the equivalent shear beam container, due to the change of model stiffness during dynamic tests. To simulate the reliable boundary conditions of models, the laminar containers with a variety of shapes and dimensions have been designed by Whitman and Lambe (1986), Hushmand et al. (1988), Turan et al. (2009), and Chen et al. (2010a) for use in 1-g shaking table tests or N-g centrifuge tests. The laminar containers are built from stacked rings or rectangular frames separated by bearings that are used to reduce the frictional forces among rings or frames. This type of container can ensure the free movements of the model and the container frames and can provide reliable boundary conditions for models. Therefore, it has become the most popular type of container and has been utilized increasingly in dynamic centrifuge tests.

To perform dynamic centrifuge tests in this study, a vibration apparatus composed of a shaking table and a model container was firstly developed, with its photo shown in Figure 5.2 and its details shown in Figure 5.3. The electrohydraulic servo-controlled shaking table system and the corresponding laminar model container were selected during the design process, as a result of their significant advantages mentioned above. This vibration apparatus can accurately simulate different types of horizontal shear waves

Figure 5.2 Photo of the vibration apparatus for dynamic centrifuge tests.

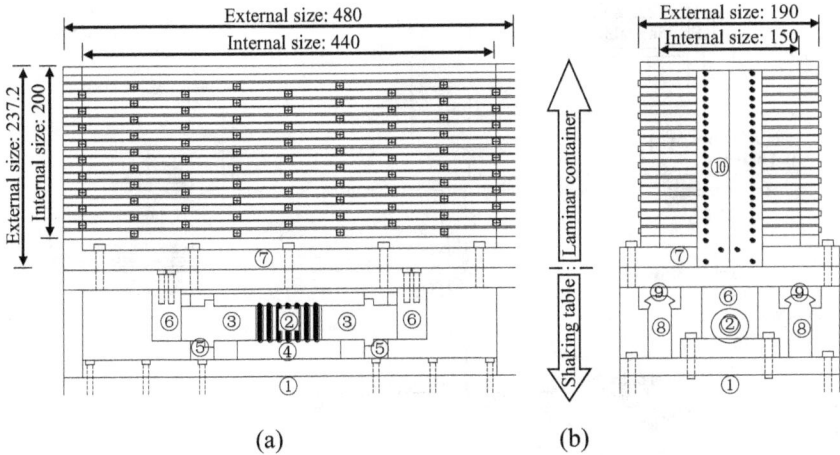

Figure 5.3 Details of the vibration apparatus (① Bedplate of shaking table; ② piston actuator; ③ actuator bush; ④ actuator cylinder; ⑤ cylinder flange; ⑥ piston bearing plate; ⑦ bedplate of laminar container; ⑧ rail spacer; ⑨ rail; ⑩ rubber guide plate). (a) Front view (unit: mm); (b) side view (unit: mm).

such as simple harmonic waves and real seismic waves, under a maximum centrifugal acceleration equaling 50 times of the gravitational acceleration. Therefore, it is capable of providing seismic input for the studies of various seismic problems.

This vibration apparatus is fixed on the bucket of the centrifuge and takes the place of the original model container to apply seismic waves and to represent reliable boundary conditions for models. A piston actuator (② in Figure 5.3) is mounted in the interior of the shaking table. During tests, the motion of the actuator stem caused by the high-pressure oil fluid drives the movement of the bedplate of the model container to excite shear seismic motions. To ensure the stability of horizontal motion, two parallel steel rails (⑨ in Figure 5.3), which are fixed to the head plate of the shaking table, as well as the corresponding rail spacers (⑧ in Figure 5.3), are set along the movement direction. The rails with low friction and high stiffness can effectively avoid the swing of the model container in other directions and thereby can improve the test accuracy. The laminar container, which is fixed together with the head plate of the shaking table through high-strength bolts, can slide freely along the rails during tests. The container, built from 20 stacked rectangular frames, has an internal dimension of 440 mm × 200 mm × 150 mm (width × height × thickness). All the rectangular frames with a uniform height of 8 mm are made from lightweight aluminum alloy to reduce the total weight of the container. Several ball bearings are installed between

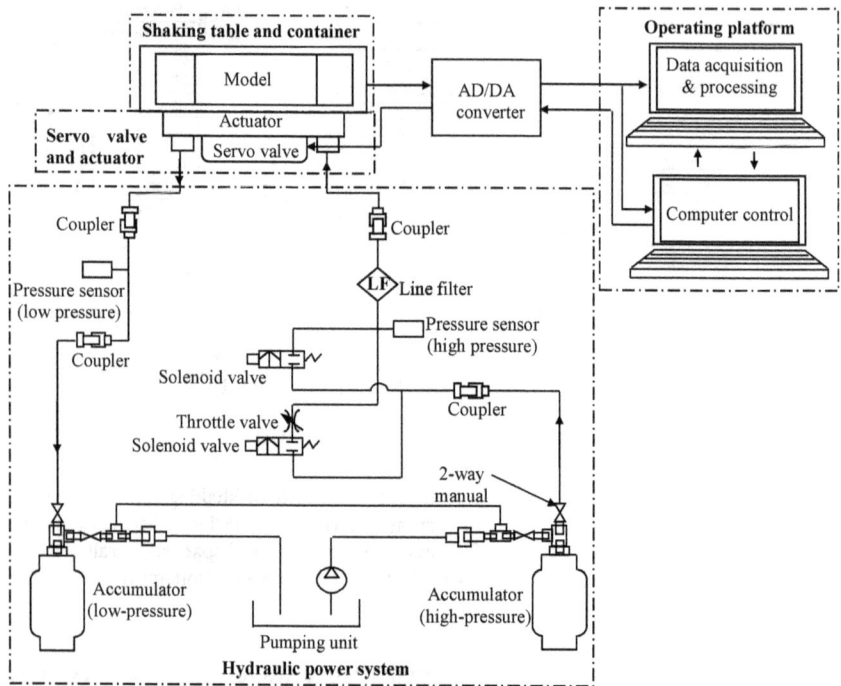

Figure 5.4 Electrohydraulic servo control system for the vibration apparatus.

adjacent frames to permit the free movement of frames in the horizontal direction with minimal frictional resistance. Additionally, two rubber guide plates (⑩ in Figure 5.3) are fixed on the two end walls of the container to avoid lateral swing during dynamic centrifuge tests.

Figure 5.4 shows the schematic diagram for representing the electrohydraulic servo control system of the vibration apparatus. The major components of this control system can be divided into three parts, including the hydraulic power system, the servo valve and actuator, and the operating platform. A brief description of the composition and functions of each component is provided as follows.

The hydraulic power system supplies the vibration apparatus with power to excite the seismic motions of models. It mainly consists of two accumulators, a pumping unit, two pressure sensors, oil supply pipes, couplers, and some control valves such as solenoid valves, throttle valves, and two-way manual valves. Two accumulators (a high-pressure accumulator and a low-pressure accumulator) with a volume capacity of 20 L and a bearing capacity of 25 MPa are mounted on both sides of the shaking table. Before carrying out dynamic tests, the accumulator at the right side (see Figure 5.4) is firstly pressurized by using the pumping. During tests, the high-pressure oil fluid from the right-side accumulator drives the movement of the actuator stem in the shaking table to excite seismic motions. Two pressure sensors connected to the circulation loop are used to monitor the oil pressures in accumulators. Additionally, the flow rate and flow direction in the loop can be regulated by using the different control valves.

The servo valve and the actuator are critical parts concerning the performance of the shaking table. During tests, the servo valve regulates the fluid provided by the hydraulic power system into the appropriate side of the actuator's piston to cause the actuator stem to move to the desired position, according to the electrical signals coming from the operating platform (computer). The movement of the actuator stem then drives the seismic motions of the container and the model. The actuator in this shaking table can be accurately controlled within the stroke displacement of ±2.5 mm (the maximum stroke displacement is ±3.5 mm). The actuator can excite seismic waves in the frequency range from 0 Hz to 50 Hz under a centrifugal acceleration of 50 g.

The operating platform of this vibration apparatus is mainly made up of two personal computers and an AD/DA converter. One of those computers is used to control the working status of the vibration apparatus by sending signals to the servo valve. The corresponding real-time control system software is installed in this computer, with the basic functions of (1) inputting seismic waves with specific waveforms (e.g., sinusoidal wave, triangle wave, sawtooth wave, square wave, and other arbitrary waveforms), frequencies, amplitudes, and duration; and (2) monitoring and regulating the vibration status of the shaking table according to the data feedback from sensors. The other computer is used to collect, process, and store the real-time dynamic data from displacement and acceleration sensors. Those two computers are linked together for real-time data sharing. In addition, an AD/DA converter is installed in the control system to fulfill the interconversion between the digital signals from computers and the analog signals from the servo valve and measuring sensors.

5.3 MODEL PREPARATION AND LOADING CONDITIONS

In the dynamic centrifuge model tests, a type of rock-like gypsum material was developed to simulate soft-medium rocks, such as sandstone and limestone, which are usually met in engineering practices. This rock-like material is a mixture of gypsum, water, and retardant with a mixing ratio by weight of 1:0.2:0.005 (see Figure 5.5a). The gypsum manufactured by the Tokyo Tokuyama Dental Corp. has a compressive strength of 54 MPa. The reason for adding retardant is to delay the hardening time of the gypsum material, to provide sufficient time for the model preparation. To measure

Figure 5.5 Experimental material and mold for manufacturing rock-like models. (a) Experimental material; (b) model mold.

Table 5.2 Physico-Mechanical Properties of the Rock-Like Gypsum Material

Parameter	Index	Unit	Value
Density	ρ	g/cm^3	2.066
Modulus of elasticity	E	GPa	28.7
Poisson's ratio	v	–	0.23
Compressive strength	σ_c	MPa	47.4
Tensile strength	σ_t	MPa	2.5
Cohesion	c	MPa	5.3
Internal friction angle	ϕ	Degree	63.3
Propagation velocity of P-wave	V_P	m/s	4013
Propagation velocity of S-wave	V_S	m/s	2376

the physico-mechanical properties of this rock-like material, a series of laboratory tests, including unconfined and confined compression tests and the Brazil test, were carried out, with the results listed in Table 5.2. It can be found that this rock-like material has similar physico-mechanical properties to typical soft-medium rocks, with good brittleness due to the high ratio of compressive strength to tensile strength.

A rigid acrylic mold was designed for casting different types of fractured models. This mold was made up of five pieces of acrylic plates with a thickness of 20 mm, which were fixed together by using stationary fixtures (see Figure 5.5b). After the mold was assembled, aluminum plates with a thickness of 0.3 mm were fixed in the interior of the mold to create rock fractures (joints), which penetrated through the model in its thickness direction. The mold and aluminum plates were pre-greased to help mold the model. During the casting process, the mold was vibrated to reduce the number of small cavities in the models and to improve the homogeneity of the material.

The casted model within the mold was placed in the room-temperature environment (about 25°C) for 24 hours until it reached the maximum strength. After that, the aluminum plate was taken out of the model to form the frictional penetrating joint. Finally, the model released from the mold can be used for dynamic centrifuge tests.

Two sets of rock-like models, which are characterized by the different distribution of rock joints, were processed to study the seismic behavior of fractured rock foundations, as shown in Figure 5.6. All the test models have a rectangular shape with the same dimension of $L \times H \times T$ (length × height × thickness) = 400 mm × 180 mm × 100 mm. Set-1 models with single joints passing through their geometrical centers were tested to investigate the effect of joint dip angle on the seismic behavior of the rock foundation.

Six types of models containing single joints were processed, and the joint dip angle α, as defined by the angle from the joint plane to the horizontal plane, was set as 0°, 15°, 24°, 45°, 60°, and 75°, respectively (see Figure 5.6a). Set-2 models were processed to investigate the effect of joint intensity on the seismic behavior of fractured rock foundations. This set of models contains different numbers of inclined parallel joints, with the joint dip angle α of 24° and the joint spacing s of 80 mm. The joint intensity, which corresponds to

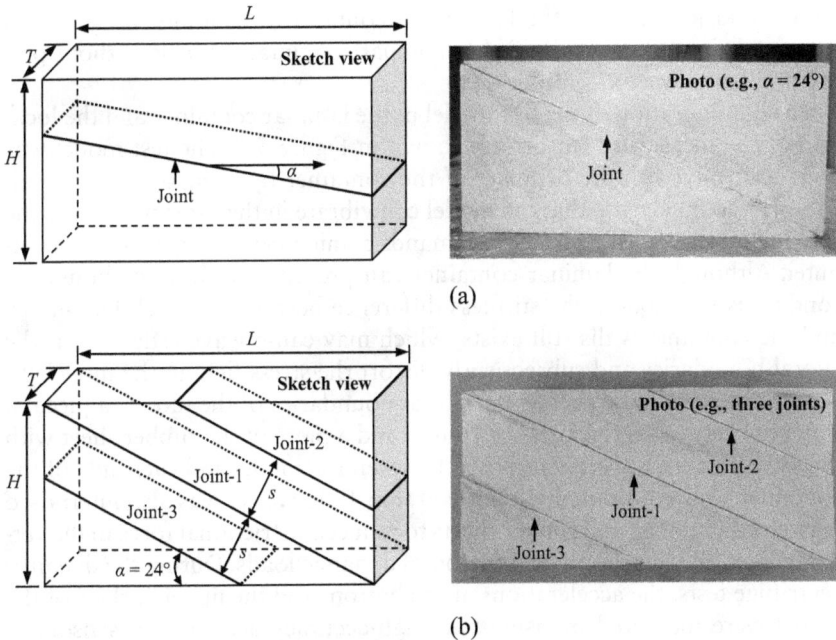

(a)

(b)

Figure 5.6 Shape and dimension of test models and the distribution of pre-existing joints. (a) Set-1 models containing single joints; (b) Set-2 models containing multiple joints.

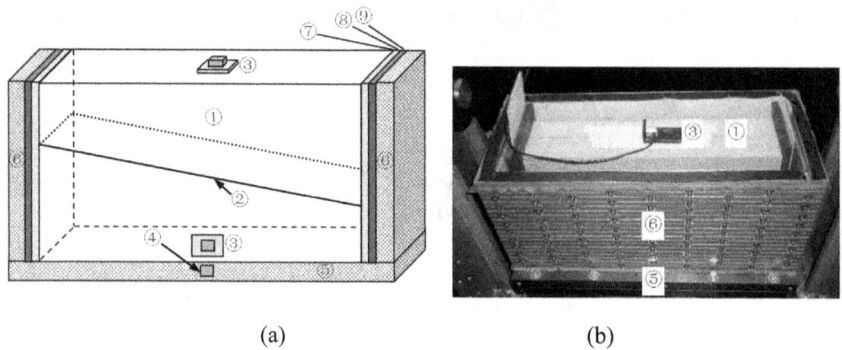

(a)   (b)

Figure 5.7 Configuration of test model in the container and the location of measuring sensors (①: model; ②: joint; ③: acceleration sensor; ④: displacement sensor; ⑤: shaking table; ⑥: container; ⑦: sponge sheet; ⑧:Teflon sheet; ⑨ rubber sheet). (a) Sketch view; (b) photo.

the joint number, ranges from 1 to 3. The distribution of pre-existing joints in Set-2 models is shown in Figure 5.6b: a model contains a single joint, as indicated by joint-1; a model contains two parallel joints, as indicated by joint-1 and joint-2; and the last model contains three joints, as indicated by joint-1, joint-2, and joint-3. For comparison, intact models without pre-existing joints were also processed.

The configuration of the test model in the laminar container and the location of the measuring sensors are shown in Figure 5.7. The test model was bonded firmly with the bedplate of the container by using a high-strength adhesive, to make sure that the model can vibrate in the same manner as the shaking table, according to the commands coming from the controlling computer. Although the laminar container can provide the dynamic boundary conditions for models, the stiffness difference between the rock-like model and the container walls still exists, which may cause wave reflection at the lateral boundaries and subsequently distort the stress field in the model. To absorb the reflected waves at the lateral boundaries of the model, a piece of sponge sheet with a thickness of 10 mm and a piece of the rubber sheet with the same thickness were interposed between the model and each side of the container wall. Additionally, a piece of thin Teflon sheet was also interposed between the sponge and rubber sheets to reduce the frictional force in the vertical direction when applying horizontal dynamic loads. During the dynamic centrifuge tests, the accelerations at the bottom and the upper surface of the model were measured by using two high-accuracy acceleration sensors to estimate the seismic responses of the fractured model. The acceleration-time course measured with the upper sensor represents the response seismic wave on the ground surface, and the acceleration-time course measured with the

Table 5.3 Dynamic Loading Conditions during the 50-*g* Centrifuge Tests

Model	Waveform	Wave Amplitude	Wave Frequency
Set-1 models (α: 0°, 15°, 24°, 45°, 60°, and 75°)	Horizontal shear wave (sine wave)	1.5 mm 2.0 mm	20 Hz 40 Hz
Set-2 models (joint number: 1, 2, and 3)		2.5 mm	50 Hz

lower sensor represents the input seismic wave. In addition, a displacement sensor was fixed on the bedplate of the model container to monitor the container deformation during tests.

In the dynamic centrifuge tests, the centrifugal acceleration for all models was set as 50 *g*; hence, the scaling factor for the acceleration is 50. According to the scaling laws, the fractured rock-like model can simulate a prototype rock foundation with the dimension of length × height × thickness = 20 m × 9 m × 5 m and with the same geometrical distribution of joints. The horizontal shear wave (S_V-wave), which brings greater damage to foundations and surface structures relative to the compressional wave, was adopted during tests. Although real seismic waves observed during earthquakes usually have complex waveforms, they can be treated as a superposed form of a series of simple harmonic waves with different frequencies, amplitudes, and phase angles. In this study, the shear sine wave with the displacement form of $u = A\sin(2\pi ft)$ (A is the displacement amplitude and f is the wave frequency), as the most fundamental seismic wave, was chosen as the input wave. The seismic wave was input from the bottom of the model by using the shaking table and traveled upwards to the upper surface of the model. To investigate the effects of the wave characteristics on the seismic behavior of rock foundations, the wave amplitude was set as 1.5 mm, 2.0 mm, and 2.5 mm, respectively, and the wave frequency was set as 20 Hz, 40 Hz, and 50 Hz respectively. The loading conditions for dynamic centrifuge tests are listed in Table 5.3.

5.4 COMPARISON OF TEST RESULTS BETWEEN MODELS WITH AND WITHOUT JOINTS

During the dynamic centrifuge tests, the artificial rock-like model without pre-existing joint was firstly tested to examine the seismic responses of intact rock foundations to horizontal shear waves. Additionally, by comparing the test results of the intact model with those of the fractured model, the effects of the rock joint on the dynamic behavior of the rock foundation were also illuminated.

The acceleration waves measured synchronously at the bottom and upper surface of the intact model are shown in Figure 5.8 (e.g., the amplitudes of

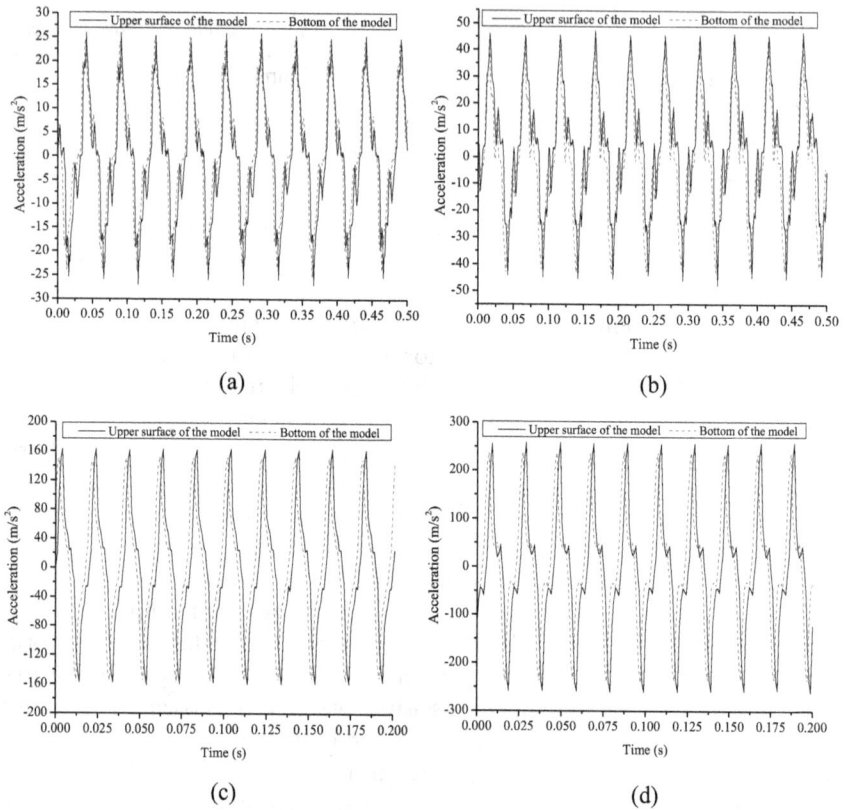

Figure 5.8 Acceleration waves measured at the bottom and upper surface of the intact model. (a) $A = 1.5$ mm and $f = 20$ Hz; (b) $A = 2.5$ mm and $f = 20$ Hz; (c) $A = 1.5$ mm and $f = 50$ Hz; (d) $A = 2.5$ mm and $f = 50$ Hz.

the input waves A are 1.5 mm and 2.5 mm, respectively; and the frequencies of input waves f are 20 Hz and 50 Hz, respectively). It should be noted that only a part of the response acceleration wave, which contains ten periods in the stable stage of the loading process, was plotted in the figure to make the waveform clearer. It can be found from Figure 5.8 that the frequency and the amplitude of the seismic wave measured at the model bottom are approximately identical to those of the designed input wave (the acceleration amplitude of the designed input wave can be determined by $a_{max} = 4\pi^2 A f^2$), thus verifying the validity of the newly developed vibration apparatus. The acceleration wave at the upper surface of the model has approximately the same frequency (or period) as the wave measured at the model bottom, revealing that the seismic wave traveling in the intact model will not change its frequency.

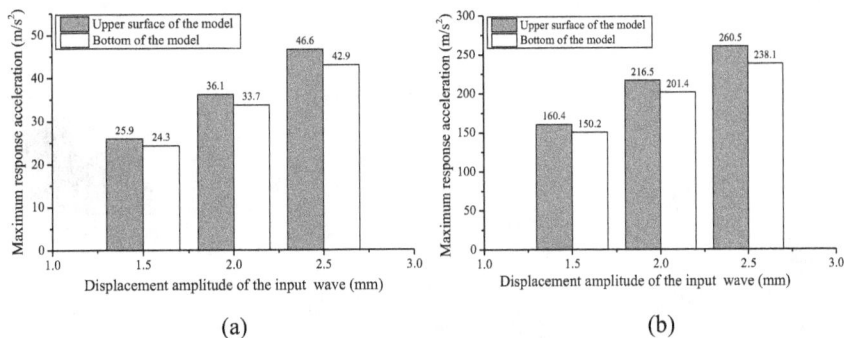

Figure 5.9 Average amplitudes of acceleration waves measured at the bottom and upper surface of the intact model. (a) f = 20 Hz; (b) f = 50 Hz.

In the engineering analysis, the amplitude of the response seismic wave is the most important concern to the safety and stability of surface buildings and structures. Figure 5.9 shows the average amplitudes of waves in the stable stage of the loading process, which were measured at the bottom and upper surface of the intact model. It can be found that the wave amplitude at the model's upper surface is larger than that at the model's bottom, demonstrating that the intact rock foundation has an amplification effect on the wave amplitude during the wave propagation. The amplification effect may be due to the deformation of the model under dynamic loads. Figure 5.9 also indicates that the amplification effect is affected to some extent by the frequency and amplitude of the input displacement wave. With the increase of the wave frequency and amplitude, the dynamic stresses in the model increase, and consequently lead to the increase of the model deformation; thereby, the amplification effect of the intact model becomes more obvious.

To examine the effects of rock joint on the overall dynamic response of the model, a comparison of seismic waves obtained from the intact model without pre-existing joints and the fractured model containing a single joint with the dip angle of α = 0° was carried out, as shown in Figure 5.10 (e.g., the amplitude and frequency of the input displacement wave are 1.5 mm and 50 Hz, respectively).

Figure 5.10a gives the acceleration waves measured at the bottom and the upper surface of the fractured model. It can be found that the response wave at the model top exhibits some different characteristics compared with the wave at the model bottom (representing the input wave), due to the presence of a joint. Firstly, the average amplitude of the response acceleration wave at the model top is smaller than that of the input wave, indicating that the joint has an attenuation effect on the wave amplitude. Secondly, the phase difference between the response wave and the input wave becomes

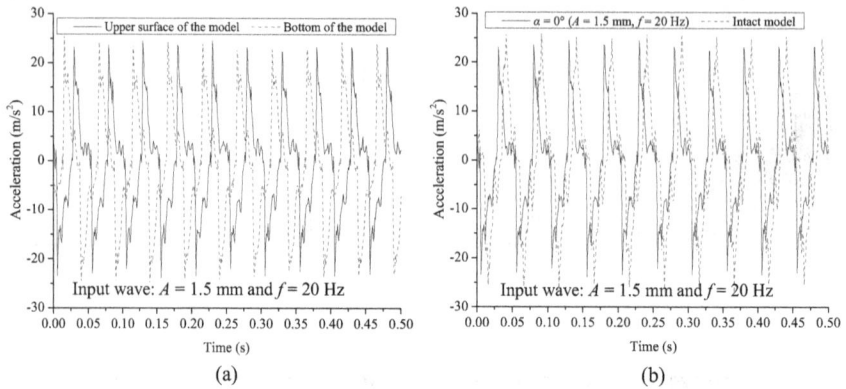

Figure 5.10 Comparison of seismic waves measured from the intact model and the fractured model containing a single joint with the dip angle of $\alpha = 0°$. (a) Acceleration waves at the bottom and the upper surface of the fractured model;
(b) response waves at the upper surfaces of the intact and fractured models.

more obvious than the case of the intact model (see Figure 5.8c), revealing
that the joint presents a time delay effect on the seismic wave. The attenuation and time delay effects of joint observed in this experimental study
correspond well with the results of the qualitative experiment reported by
Pyrak-Nolte (1996). During the dynamic loading process, the joint in the
fractured model produces large deformations such as the sliding deformation along its original plane as well as the opening and closing deformations
normal to its original plane (note that in the case of horizontal joint with the
input of shear wave, only sliding deformation of the joint occurs). Those joint
deformations are usually accompanied by energy loss and thereby reduce the
amplitude of the response waves. Additionally, the deformation response of
the joint is usually slower than that of the rock particles, and consequently,
the joint can delay the seismic signals. Figure 5.10a also indicates that the
presence of a joint affects the waveform of the transmitted wave across the
joint. Relatively to the waveform of the input wave at the model's bottom,
the waveform of the response wave at the model's top is distorted to some
extent (the response wave becomes coarser), as a result of the deformation
of the joint. Since the input wave is approximately a simple harmonic wave
with a constant frequency, the effect of the joint on the dominant frequency
of the transmitted wave is not obvious, while for real seismic waves with
complicated waveforms, the rock joint may filter some high-frequency con-
tents of the wave components, leading to a decrease of the dominant frequency of the wave signals.

Figure 5.10b shows the response acceleration waves measured at the upper
surfaces of both the intact model and the fractured model (note that the time

in these two wave curves does not correspond with each other, due to that those waves were recorded from different models). It can be found that the difference in the wave amplitude becomes more obvious between the response waves from the intact model and the fractured model, because the intact model has an amplification effect on the wave amplitude, while the fractured model has an attenuation effect on the wave amplitude.

5.5 SEISMIC RESPONSE OF TEST MODELS CONTAINING SINGLE JOINTS

5.5.1 Effect of Joint Dip Angle on the Seismic Response of Models to Shear Waves

Fractures such as joints existing in the rock mass and rock structures have various geometrical distributions. The joint dip angle, as one of the most important geometrical parameters, has a significant influence on the overall mechanical and deformation behavior of fractured rock masses. To investigate the influence of the joint dip angle on the dynamic behavior of rock foundations, dynamic centrifuge tests on the Set-1 models containing single joints with varying dip angles were carried out, and the variation of the amplitude of response acceleration wave at the model top with the change of joint dip angle was discussed.

During the dynamic centrifuge tests, the acceleration wave measured at the model bottom represents the input seismic wave. Although the model was bonded firmly to the bedplate of the model container by using a high-strength adhesive, the amplitude of the acceleration wave observed at the model bottom is still slightly different from that of the designed input wave (the acceleration amplitude of the designed input wave is in the form of $a = -A(2\pi f)^2 \sin(2\pi ft)$, where A is the displacement amplitude of the designed input wave; f is the frequency of the designed input wave; and t is the time). This slight difference is inevitable due to the accuracy of the shaking table as well as the bonding conditions at the model bottom. To avoid the influence of the difference in input waves on the test results of fractured models, the modification of the test results needs to be performed, by using the following formula

$$a_{modified} = \frac{a_{designed}}{a_{bottom}} \times a_{top} \qquad (5.1)$$

where $a_{modified}$ denotes the modified wave amplitude at the model top; $a_{designed}$ denotes the wave amplitude of the designed input wave; a_{bottom} is the wave amplitude of the measured wave at the model bottom; and a_{top} is the wave amplitude of the measured wave at the model top. It should be noted that a_{bottom} and a_{top} represent the average amplitudes of ten wave periods selected from the stable stage of the loading process.

Figure 5.11 shows the relationship between the joint dip angles and the amplitudes of the response waves measured at the upper surfaces of Set-1

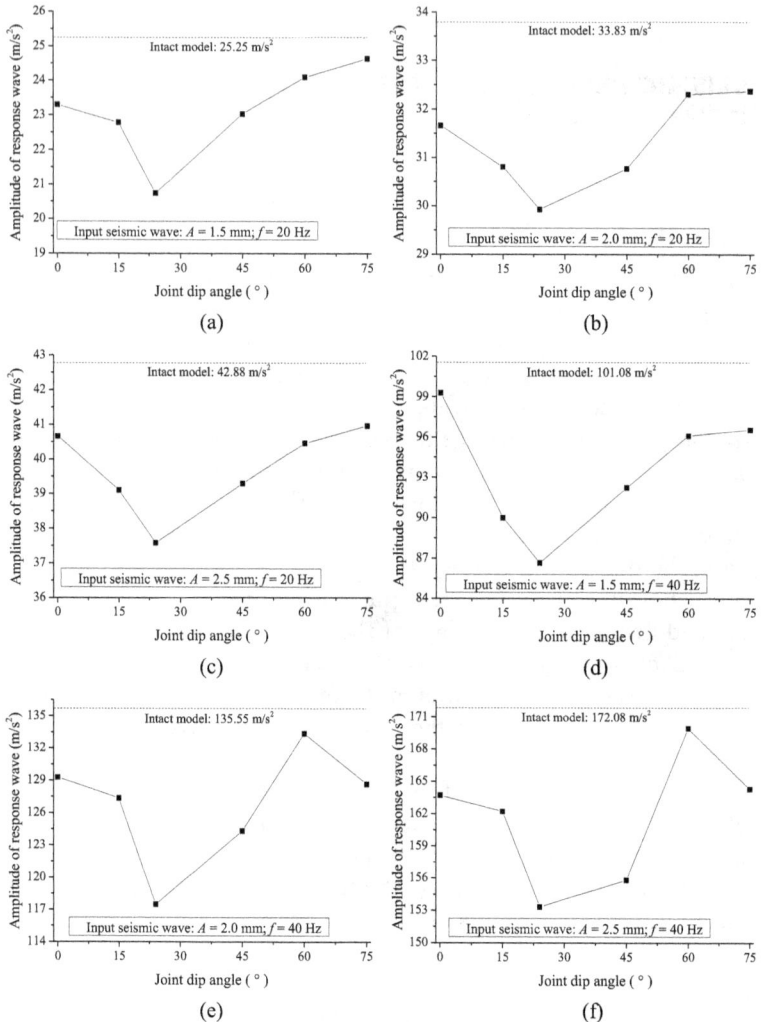

Figure 5.11 Relationship between the joint dip angles and the amplitudes of the response waves at the upper surfaces of Set-1 fractured models. (a) A = 1.5 mm and f = 20 Hz; (b) A = 2.0 mm and f = 20 Hz; (c) A = 2.5 mm and f = 20 Hz; (d) A = 1.5 mm and f = 40 Hz; (e) A = 2.0 mm and f = 40 Hz; (f) A = 2.5 mm and f = 40 Hz; (g) A = 1.5 mm and f = 50 Hz; (h) A = 2.0 mm and f = 50 Hz; (i) A = 2.5 mm and f = 50 Hz.

(g)

(h)

(i)

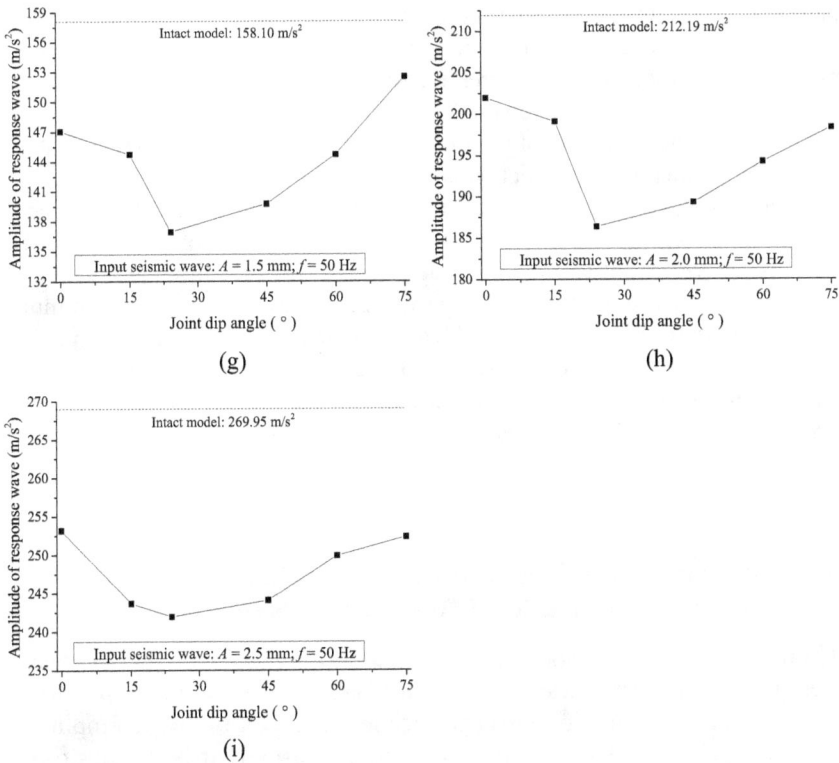

Figure 5.11 Continued

models. All the test results of the fractured models, as well as the intact model (for comparison) obtained under various dynamic loading conditions, are plotted in this figure (the amplitudes of input displacement waves are 1.5 mm, 2.0 mm, and 2.5 mm, respectively; and the frequencies of input waves are 20 Hz, 40 Hz, and 50 Hz, respectively). The test results indicate that although the amplitude and frequency of the input wave are different, the variation tendency of response wave amplitude with the change of joint dip angle exhibits good agreement. With the increase of joint dip angle (α) from 0° to 24°, the amplitude of the response wave at the model top decreases gradually and reaches its minimum value when the joint dip angle approaches 24°. After that, the amplitude of the response wave turns to increase with the continuous increase of the joint dip angle and reaches its maximum value when the joint dip angle is equal to 75° in most loading cases (see Figures 5.11a–d and 5.11g–i) or when the joint dip angle is equal to 60° in two loading cases (see Figures 5.11e and 5.11f).

The variation in wave amplitude may be due to the different deformation behavior of joints under dynamic loading. Generally, joint deformation is composed of two components: the tangential deformation along the joint plane that is caused by shear stresses, and the normal deformation induced by compressive or tensile stresses. When subjected to horizontal shear waves, the tangential deformation of the joint will decrease with the increase of joint dip angle, while the normal deformation of the joint will increase. If the joint dip angle is small (e.g., $\alpha = 0°$), a large slip deformation can be produced along the joint plane and consequently leads to a considerable deformation and wave amplitude at the model top. As the joint dip angle increases to around 24°, the amplitude of the response wave reaches its minimum, as a result that the tangential deformation of the joint decreases quickly. Afterward, the normal deformations of the joint (such as the joint closure and opening) make a major contribution to the total deformation of the fractured model. The increased normal deformation of joints in the cases of large dip angles (e.g., $\alpha = 60°$ and $\alpha = 75°$) enlarges the deformation and wave amplitude at the model top.

5.5.2 Attenuation Effect of Inclined Joints on the Amplitude of Response Wave

The test results of Set-1 models indicate that the amplitudes of response waves were reduced to some extent due to the presence of rock joints. To further investigate the attenuation effect of inclined joints on the wave amplitude, a comparison between the wave amplitudes measured at both sides of the inclined joint was carried out by taking two representative fractured models with the joint dip angles of 24° and 45° as examples, and the attenuation effect of the joint was estimated by using the attenuation ratio as defined by

$$p = \frac{a_{bottom} - a_{top}}{a_{bottom}} \times 100\% \qquad (5.2)$$

where p represents the attenuation ratio and a_{bottom} and a_{top} denote the amplitudes of acceleration waves measured at the bottom and the upper surface of a fractured model, respectively.

Figure 5.12 shows the amplitude attenuation ratios of fractured models as well as their variation tendency with the change of the amplitude of the input wave. It can be derived that with the increase of input wave amplitude, the attenuation ratio decreases continuously. This is because the increasing amplitude of the input wave raises the overall stress level in the model, and subsequently increases the stiffness of the joint to some extent, compared with the cases of low-amplitude input waves (the joint stiffness generally

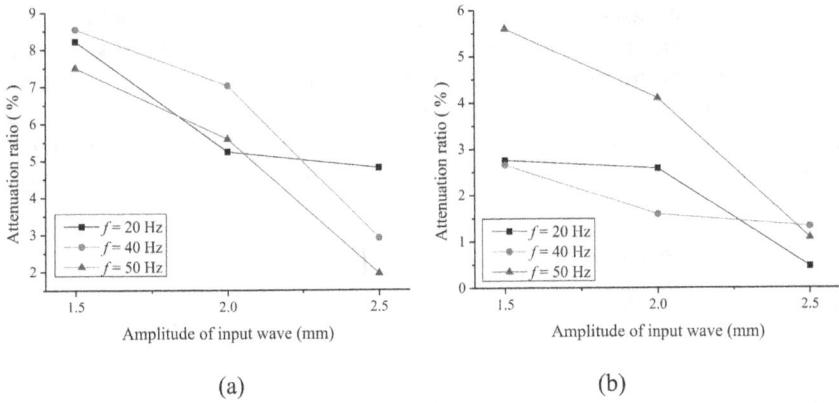

Figure 5.12 Variation of the amplitude attenuation ratio with the change of the amplitude of the input wave. (a) $\alpha = 24°$; (b) $\alpha = 45°$.

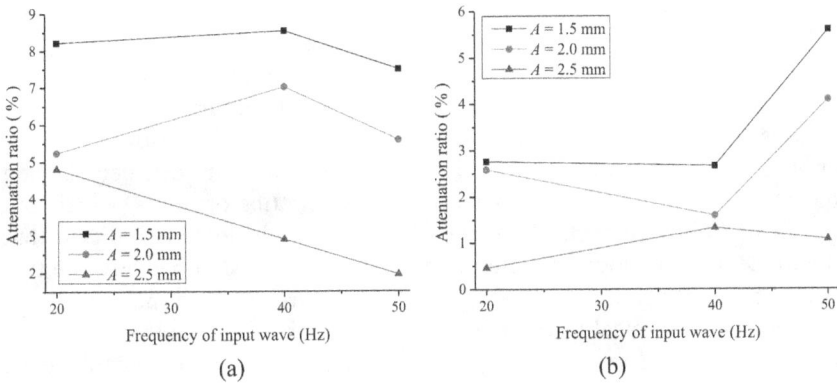

Figure 5.13 Variation of the amplitude attenuation ratio with the change of the frequency of the input wave. (a) $\alpha = 24°$; (b) $\alpha = 45°$.

increases with the increase of environmental stress, according to the nonlinear deformation theories of rock joints, such as the Barton-Bandis model). The joint with the increased stiffness exhibits a lower attenuation effect on the wave amplitude, and consequently, the attenuation ratio will decrease.

Similarly, the variation of the amplitude attenuation ratio with the change of input wave frequency was also analyzed, and the results are shown in Figure 5.13. The curves in this figure do not follow a certain law and exhibit some divergence characteristics. The divergence of variation tendency may be caused by two factors: (1) the increasing wave frequency leads to the increase of overall stress level in the model as well as the increase of joint stiffness, which subsequently causes the decrease of the attenuation ratio (similarly to the cases in Figure 5.12); and (2) the seismic wave with a high frequency will be attenuated severely in its amplitude due to the low-pass filter effect of the joint, which consequently leads to the increase of the attenuation ratio. It can be found that the two impact factors have opposite effects on the attenuation ratio, and make this problem complicated. Thereby, more future studies need to be carried out on this topic.

5.5.3 Failure Statues of Set-1 Models

In the dynamic centrifuge tests, the input seismic motions on the bottom of models generate large-amplitude acceleration waves (for the test models, the acceleration amplitude ranges from 23.69 m/s² to 246.74 m/s²; for the prototypes, the acceleration amplitude ranges from 0.47 m/s² to 4.93 m/s², according to the scaling laws mentioned before), which can subsequently induce the failure of models. Through analyzing the failure statuses of the fractured models, the failure behavior of prototype foundations under seismic loads can be inferred.

Figure 5.14 shows the failure statuses of the Set-1 models containing single joints after dynamic tests (the failure positions have been marked on the figures). It can be found that the failure of models occurs primarily in the regions near the pre-existing joints. When subjected to seismic loads, the pre-existing joints can induce localized stress concentration in their neighboring regions, especially in the regions adjacent to the tips of joints, which can cause the failure of rock-like materials. The test phenomena indicate that the model failure mostly resulted from the severe collision between the two parts of the model during the dynamic loading process, and the failure pattern belongs to typical brittle failure. It can be inferred from the test results that for the rock foundations, the regions where large joints or faults exist (especially the outcrop regions of joints and faults) need special attention, and some reinforcement measures should be performed to avoid the failure of rock mass induced by seismic loads affects the safety and stability of surface structures.

Figure 5.14 also indicates that the failure statuses of models are affected by the joint dip angle to some extent. For the cases of low dip angle joints (e.g., $\alpha = 0°$, $15°$, and $24°$, corresponding to Figure 5.14a, Figure 5.14b, and Figure 5.14c, respectively), the damage degree of models is relatively low, because the joint deformations caused by shear waves are primarily contributed by the slip deformation along the joint planes. With the increase of joint dip angle, the normal deformation of joints increases, and the collision between the two parts of the model can cause more severe damage to the models, as can be found in Figure 5.14e corresponding to the case of $\alpha = 60°$ and Figure 5.14f corresponding to the case of $\alpha = 75°$.

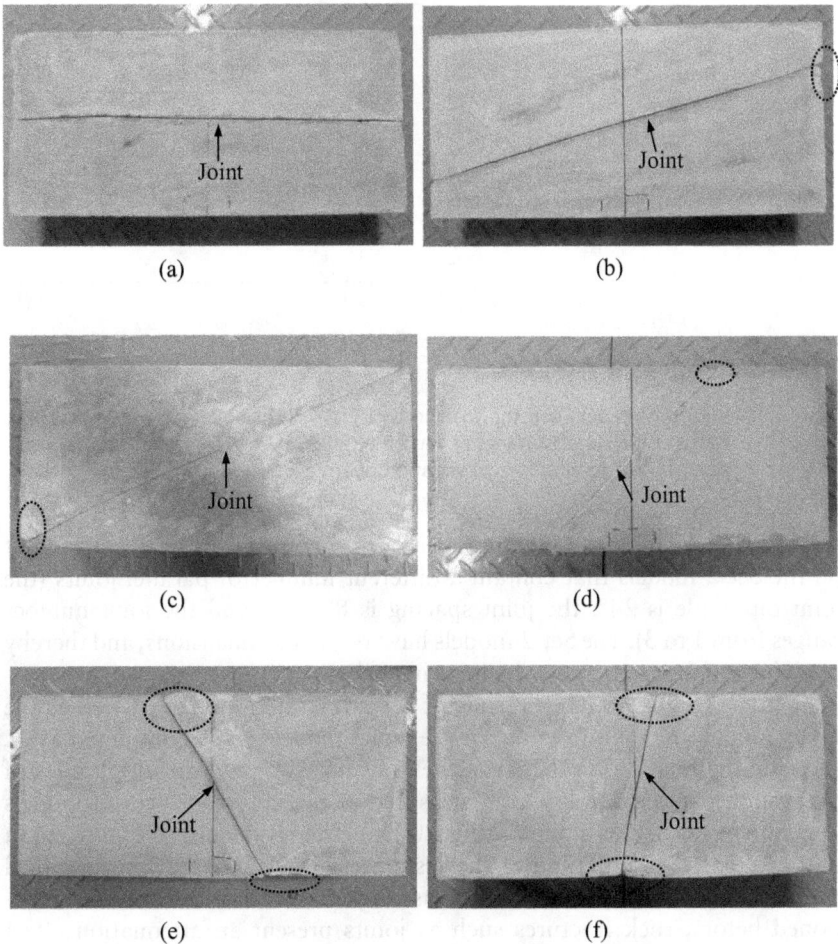

Figure 5.14 Failure statuses of Set-1 models after the dynamic centrifuge tests (the vertical lines in the middle of models were used to indicate the location of acceleration sensors). (a) $\alpha = 0°$; (b) $\alpha = 15°$; (c) $\alpha = 24°$; (d) $\alpha = 45°$; (e) $\alpha = 60°$; (f) $\alpha = 75°$.

5.6 SEISMIC RESPONSE OF TEST MODELS CONTAINING MULTIPLE JOINTS

The fracture (joint) intensity is an important geometrical parameter to describe the integrity of fractured rock masses, and its quantity concerns significantly the overall strength and deformation behavior of rock masses. To investigate the influence of joint intensity on the seismic behavior of

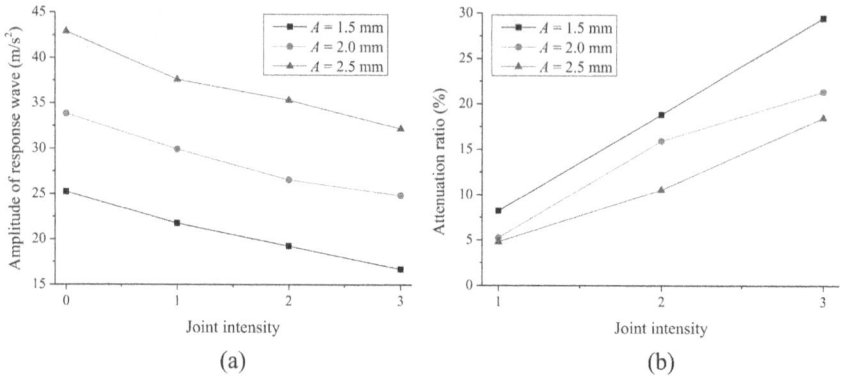

Figure 5.15 Relationship between the joint intensity and the amplitudes of the response waves at the upper surfaces of Set-2 fractured models (e.g., f = 20 Hz). (a) Variation of wave amplitudes; (b) variation of attenuation ratios.

fractured foundations, the related dynamic centrifuge tests were carried out on the Set-2 models that contain a different number of parallel joints (the joint dip angle is 24°; the joint spacing is 80 mm; and the joint number ranges from 1 to 3). The Set-2 models have the same dimensions, and thereby the joint number represents the joint intensity.

Figure 5.15 gives the test results of Set-2 fractured models as well as the intact model (the joint intensity of the intact model can be considered as 0), showing the relationship between the joint intensity and the amplitudes of the response acceleration waves measured at the upper surfaces of models (e.g., the frequency of input wave is 20 Hz). The test results in Figure 5.15a indicate that the amplitude of the response wave at the top of the model decreases significantly with the increase in joint intensity. As has been mentioned before, rock fractures such as joints present an attenuation effect on the wave amplitude, which is usually accompanied by the dissipation of dynamic energy. As the joint intensity increases, the attenuation effect becomes more obvious, consequently leading to a quick drop in the amplitude of the response wave. Figure 5.15b shows the corresponding amplitude attenuation ratios of Set-2 models, which were calculated by using Equation (5.2), indicating that the attenuation ratio increases continuously with the joint intensity. It can also be found that for the models containing multiple joints, the attenuation ratio decreases with the increase of input wave amplitude, due to the increase of joint stiffness under the higher stress level, as has been described before.

During the dynamic centrifuge tests, the high-amplitude input waves induced the failure of models. Figure 5.16 shows the failure statuses of Set-2 models (the failure positions have been marked on the figures), indicating

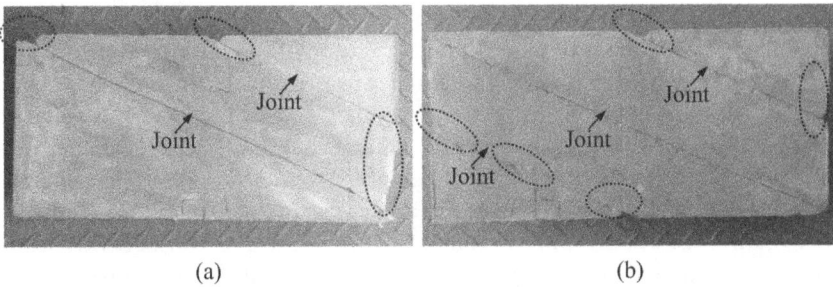

Figure 5.16 Failure statuses of Set-2 models after the dynamic centrifuge tests. (a) Model containing two joints; (b) model containing three joints.

that the model failure primarily occurs in the regions near the pre-existing joints, as a result of the localized stress concentration induced by joints. Compared with the cases of Set-1 models with single joints (see Figure 5.14), the damage degree of Set-2 models is more severe due to the combined effects of multiple joints. The test phenomena demonstrate that the rock foundation with a low integrity is more likely to be damaged under seismic loading, and thereby special attentions are required in engineering practices.

5.7 CONCLUSION

In this chapter, dynamic centrifuge model tests on rock-like models were conducted to investigate the seismic responses of fractured rock foundations to horizontal shear waves (sine waves). Before performing dynamic tests, a servo-controlled vibration apparatus composed of an electrohydraulic shaking table and a laminar model container was developed to apply seismic loads and to simulate the dynamic boundary conditions of models. Then dynamic centrifuge tests on two sets of fractured models were carried out, mainly focusing on the effects of joint dip angle and joint intensity on the overall seismic behavior of fractured foundations, respectively.

For the case of the intact model, the amplitude of the response wave measured at the model top is larger than that at the model bottom, indicating that the intact rock foundation has an amplification effect on the wave amplitude due to the deformation of the model under dynamic loads. The amplification effect becomes more obvious with the increase of frequency and amplitude of the input wave. By comparing the test results between the intact model and the fractured model, the attenuation effect of the joint on the wave amplitude and the time delay effect of the joint on the wave phase were confirmed.

The test results of the Set-1 models indicate that the joint dip angle has a significant effect on the overall seismic behavior of models. With the increase of joint dip angle from 0° to 24°, the amplitude of the response wave at the model top decreases gradually and reaches its minimum when $\alpha = 24°$. After that, the amplitude of the response wave turns to increase with the increase of joint dip angle and reaches its maximum when $\alpha = 75°$ in most loading cases. The variation in wave amplitude may be due to the different deformation behavior of joints. If the joint dip angle is small (e.g., $\alpha = 0°$), a large slip deformation can be produced along the joint plane and consequently leads to a considerable deformation and wave amplitude at the model top. When $\alpha = 24°$, the amplitude of the response wave reaches its minimum, as a result that the tangential deformation of the joint decreases quickly. For the cases of large dip angle joints ($\alpha = 60°$ or $75°$), the increased normal deformation of joints enlarges the deformation and wave amplitude at the model top. The attenuation effect of inclined joints is affected to some extent by the characteristics of the input wave. The attenuation ratio decreases with the increase of input wave amplitude, due to that the increased environmental stress caused by the high-amplitude wave can enhance the joint stiffness. The test results of the Set-2 models indicate that the amplitude of the response wave at the model top decreases significantly with the increase of joint intensity. As the joint intensity increases, the attenuation effect of joints becomes more obvious, consequently leading to a quick drop in the amplitude of the response wave.

The high-amplitude input wave induced the failure of fractured models. The failure of models primarily occurs in the regions near the pre-existing joints (especially in the regions adjacent to the joint tips), due to the localized stress concentration caused by joints. The failure statuses of models are affected by the joint dip angle to some extent. For the cases of low dip angle joints, the damage degree of models is relatively low, because the joint deformations caused by shear waves are primarily contributed by the slip deformation along joint planes. With the increase of joint dip angle, the normal deformation of joints increases, and the collision between the two parts of the model can cause more severe damage to the models. Additionally, with the increase of joint intensity, the damage to models becomes more severe.

Chapter 6

The Methods for Analyzing Seismic Response of Underground Structures

6.1 EXPERIMENTAL SEISMIC ANALYSIS

For an intuitive grasp of the seismic influence of earthquakes, porotype observation generally can provide the most authentic and effective post-earthquake results of tunnels. However, it cannot control the input mechanism of seismic waves and boundary conditions, nor can it actively change various factors to conduct a purposeful study on a certain phenomenon. If such information is required, the experimental seismic analysis must be undertaken to better investigate the seismic response. Two main types of experimental seismic analysis are available including field test with artificial seismic source and physical model test.

6.1.1 Field Test with the Artificial Seismic Source

Field test with artificial seismic source simulates the in-situ dynamic response of underground structures under seismic loading by artificial seismic source or blasting. Since there is no consideration for the effect of artificial truncation boundary and input mechanisms of seismic waves, it can provide much more accurate characteristics of the underground structure's dynamic response (Xue et al., 2013; Chen et al., 2017). Phillips and Luke (1991) conducted a tunnel dynamics experiment (TDE) to provide a dataset that included both measured ground motions and documented tunnel response. The experiment was designed and fielded 0.5 km from a recent underground nuclear explosion (UNE). The artificial TDE source could stimulate a tunnel response similar to what might be expected in the near-field region of a small to moderate earthquake. The experiment collected acceleration, permanent displacement, tunnel convergence, borehole observation, and photographs. Figure 6.1 illustrates the layout of experimental measurements in the TDE. The TDE produced a consistent picture of the dynamic response of the tunnel based on the various types of tunnel damage and measured ground motions. Moreover, the results of the TDE are consistent with case histories discussed in the literature (Spottiswoode & McGarr, 1975; McGarr et al., 1981; McGarr, 1981, 1982, 1983).

DOI: 10.1201/9781003401599-6

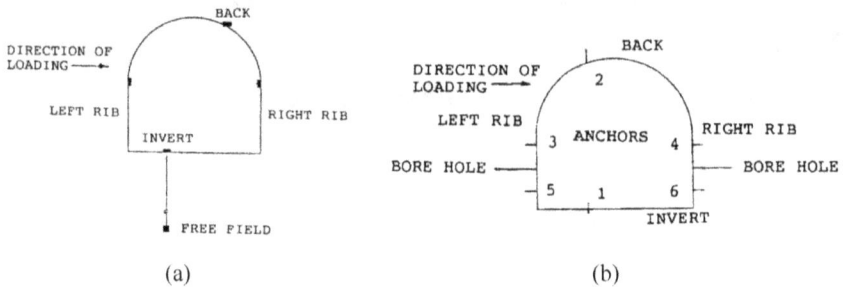

Figure 6.1 Layout of experimental measurements in the TDE. (a) Accelerometer location in the TDE; (b) layout of tunnel convergence measurements and exploratory boreholes.

The field test is seldom used since the effect of the nonlinear characteristic of structure and ground rupture on the seismic response cannot be well represented. Moreover, small artificial seismic force and high material and labor costs limit this method's use.

6.1.2 Physical Model Test

At present, the physical model test for the seismic response study of underground structures mainly includes the shaking table test and dynamic centrifuge test. The shaking table test is a method to study the seismic performance of an underground structure by placing a scaled-down structure model on the shaking table and applying sinusoidal or arbitrary excitation to the single direction or multiple directions of the shaking table. One representative shaking table equipment is illustrated in Figure 6.2. Many researchers have carried out relevant shaking table tests on seismic waves (e.g., magnitude, frequency), geographic conditions (e.g., fault, tunnel portal slope, tunnel depth), and structural characteristics of tunnel lining (e.g., lining form, dimension) on the seismic response of tunnels (e.g., Yakovlevich & Borisovna, 1978; Okamoto, 1984; Goto et al., 1988; Zhou, 1998; Iwatate et al., 2000; Che et al., 2002; Shen et al., 2008; Jiang et al., 2010a, 2011; Chen et al., 2010b, 2018; Xu et al., 2013; He et al., 2014; Gao et al., 2015; Xin et al., 2015, 2018; Moghadam et al., 2016; Wang et al., 2017; Bao et al., 2017). The dynamic centrifuge test increases the field acceleration to satisfy the physical similarity. One representative one is the medium-sized dynamic centrifuge apparatus at Nagasaki University, as illustrated in Figure 5.1. It has the advantage of simulating the dynamic response of the underground structure under real gravity conditions (Chen et al., 2017). It can provide a much more accurate description of the deformation and failure mechanism of the tunnel (Liu et al., 2008; Lanzano et al., 2012). The model test is always desirable to carry out to learn the actual seismic performance of the underground structures and to develop/check design theories (Ohtomo

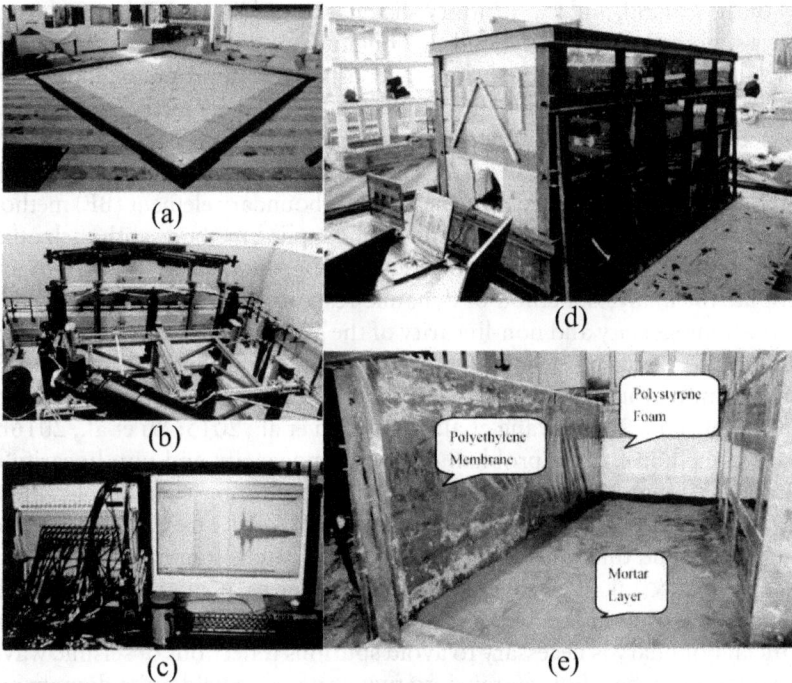

Figure 6.2 Shaking table and accessory facilities. (a) Shaking table; (b) loading controlling devices; (c) data logging apparatus; (d) model container with observation system; (e) boundary effect mitigation measures for the model container.

Source: Xin et al., 2018.

et al., 2001, 2003). In addition, it is helpful for the optimization and verification of the anti-seismic measures of the tunnel, thus providing the basis for the anti-seismic design. It has become an effective and intuitive method to study the seismic response of tunnels, yet there are some difficulties in the aspects of material selection for the lining structure, determining the similarity ratio, and dealing with the model boundary and scale effect. Moreover, for the dynamic centrifuge test, the wave passage effect cannot be well simulated, and the Coriolis effect existing in the experimental process cannot be avoided with the current technology, which restricts its further development (Chen et al., 2017).

6.2 DYNAMIC TIME-HISTORY ANALYSIS

Dynamic time-history analysis widely applied to analyze the seismic response of tunnels is by far the most acceptable method for the seismic analysis of underground structures. The method can efficiently describe the kinematic and inertial aspects of the ground-structure interaction, the complex geometry of the ground, and the arbitrary shape of tunnel lining. With time-history analysis, both the transversal and the longitudinal directions of the structure can be modeled and analyzed simultaneously using appropriate constitutive relationships and considering the complicated ground geometry, and the non-linear behavior of the ground and the structure (Pitilakis & Tsinidis, 2014). Several methods are available in the current literature: the finite element (FE) method, the finite difference (FD) method, the boundary element (BE) method, the discrete element (DE) method, and their coupling methods with each other.

The FE method is a way to discretize the ground-structure system and compute the analysis method of dynamic response. It can solve the problem of non-homogeneity and non-linearity of the ground and the lining structure. Research works using both the two-dimensional and three-dimensional FE models can be found in the current literature (e.g., Krauthammer & Chen, 1989; Ren et al., 2005; Wang et al., 2014; Wu et al., 2015; Yu et al., 2016b). As for a method for the problem of non-homogeneity and non-linearity of the ground-lining system, the FD method is more robust and applicable to the wider class of problems. Related analytical works using both the two-dimensional and three-dimensional FD models can be found in the literature (e.g., Hwang & Lu, 2007; Shen et al., 2014). Due to an artificially truncated boundary existing in the models of both the FE method and FD method, an artificial boundary is necessary to avoid spurious reflections of seismic waves by partitioning the whole system into two parts: near-field finite domain and far-field infinite domain. In current literature, several types of the artificial boundary could be found: Engquist-Majda boundary (Engquist & Majda, 1977, 1979), Bayliss-Turkel boundary (Bayliss & Turkel, 1980), Higdon boundary (Higdon, 1986, 1987), multi-transmitting boundary, viscous boundary, and spring-viscous boundary (White et al., 1977; Liu & Li, 2005).

The BE method can automatically satisfy the radiation condition of the far-field infinite domain without introducing an artificial boundary. It can transform a two-dimensional or three-dimensional integration problem into a one-dimensional or two-dimensional one. In this regard, a few works can be found in the literature (e.g., Besko, 1987; Zimmerman & Stern, 1993; Yu & Dravinski, 2010; Dravinski & Yu, 2011). The DE method proposed in the early 1970s by Cundall (1971) assumes that the rock mass is composed of rigid blocks. It can analyze the large deformation and instability process of the rock mass because of its unique advantages in dealing with fractures. Many works for the seismic response analysis of underground structures can be found in the literature (e.g., Zhang et al., 2006; Jia et al., 2013).

For a three-dimensional FE model, the computation work with local absorbing boundary conditions becomes prohibitively large for applications in the wide frequency range under consideration, as they require an extended finite element mesh together with a small element size. On the other hand, BE method partially solves such a problem as it reduces the discretization effort (Degrande et al., 2006). Moreover, the FE, FD, and DE methods are superior in the aspects of the complicated ground geometry, and the non-linear behavior of the ground and the structure, while the BE method is superior in solving homogeneous, linear infinite, and semi-infinite medium problems. Consequently, coupling methods of the FE, BE, and DE models are also performed in current literature (Jin et al., 2001; Degrande et al., 2006; Galvín et al., 2010; Costa et al., 2012).

6.3 SIMPLIFIED SEISMIC ANALYSIS

For practical design purposes, a simplified method considering the seismic effect is desirable. There are two common categories for seismic design and analysis methods. They are the force-based method and the deformation-based method (Wang, 1993; Pitilakis & Tsinidis, 2014).

The force-based method refers to the way that the seismic load is introduced in terms of equivalent or pseudo-static forces acting on the structure. It is much more available for aboveground structures since the structures are not only constrained by ground shaking but also experience an amplification of the shaking motions due to their vibratory characteristics (Wang, 1993). However, due to the easy application of the method, it is also quite common for underground structures in engineering practice (Mononobe & Matsuo, 1929; Schnabel et al., 1972; Greek Seismic Code, 2003). Contrary to the force-based method, the displacement-based method refers to the way that the seismic load is introduced in terms of seismic displacements. The seismic response of tunnel and underground structures is more sensitive to earthquake-induced deformation since they are constrained by the surrounding medium. The deformation-based method is much more available for tunnels and underground structures. The key point of the two methods is how to reasonably determine the quantitative relationship between seismic input, site conditions, structural form, and the magnitude and distribution of seismic force or seismic deformation.

Currently, several simple, rational, and practical methods are available in the literature for the seismic design of tunnels and large underground structures, such as the free-field deformation method, simplified equivalent static and dynamic method, simplified frame analysis method (Wang, 1993; Penzien, 2000), and mass-spring model method (e.g., Kawashima, 2000; Wang, 1993; Penzien, 2000; Hashash et al., 2001; ISO 23469, 2005; Hung et al., 2009; Pitilakis & Tsinidis, 2014).

6.3.1 Free-Field Deformation Method

The free-field deformation method can describe the seismic deformation of ground due to seismic waves without considering the tunnels or underground structures for practical purposes. Ground deformation is directly imposed on the structure to analyze the deformation of tunnels and underground structures. Complex characteristics and different transmission paths of seismic waves make the calculation difficult and rarely economical.

Therefore, the primary assumption on the seismic wavefield is necessary. The seismic wave is assumed to be plane waves with consistent amplitude along the entire tunnel, and wave scattering and complex three-dimensional wave propagation are not taken into consideration (Hashash et al., 2001). In the current literature, several simplified methods are available both in the longitudinal and transversal analysis corresponding to the three deformation modes illustrated in Section 1.4.1.

Newmark (1968) proposed the free-field deformation method to the ground strains by assuming a harmonic wave propagating with an incidence angle against the axis of the tunnel embedded in a homogeneous, isotropic, elastic medium based on the theory of wave propagation. Figure 6.3 shows the free-field ground deformation along the axis of the tunnel by a harmonic shear wave. Kuesel (1969) and Yeh (1974) extended the relations of Newmark to account for obliquely incident shear and Rayleigh wave. Besides, based on the method proposed by Newmark, St. John and Zahrah (1987) developed ground deformation solutions of ground axial and curvature deformation caused by compression, shear, and Rayleigh waves. Table 6.1 lists the solution of the free-field ground strains for the three types of seismic waves.

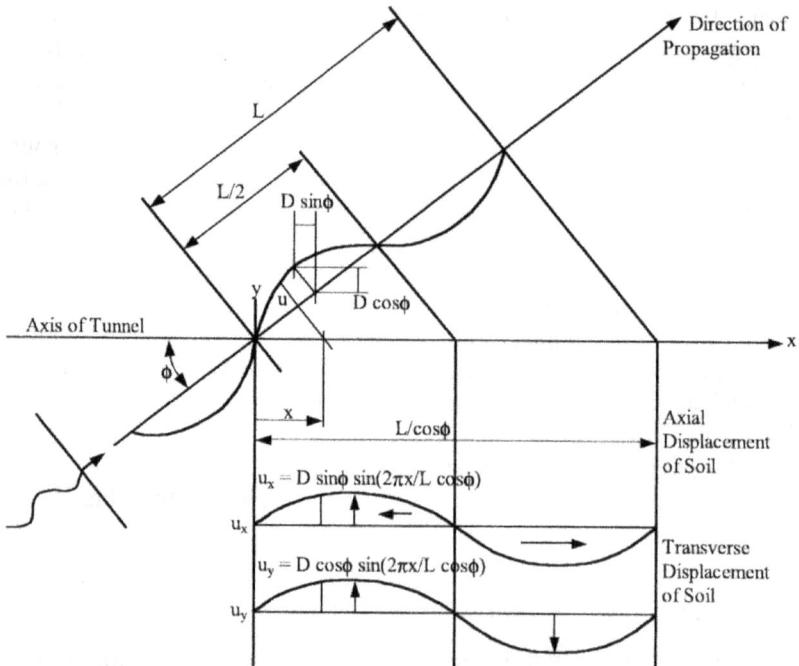

Figure 6.3 Free-field ground deformation along the axis of the tunnel by a harmonic shear wave.
Source: Hashash et al., 2001.

Table 6.1 Solution of the Free-Field Ground Strains for the Three Types of Seismic Wave

Wave Type		Longitudinal Strain	Normal Strain	Shear Strain	Curvature
Compression wave		$\varepsilon_l = \dfrac{V_P}{c_P}\cos^2\varphi$	$\varepsilon_n = \dfrac{V_P}{c_P}\sin^2\varphi$	$\gamma = \dfrac{V_P}{c_P}\sin\varphi\cos\varphi$	$\dfrac{1}{\rho} = \dfrac{a_P}{c_P^2}\sin\varphi\cos^2\varphi$
		$\varepsilon_{lmax} = \dfrac{V_P}{c_P}$ for $\varphi = 0°$	$\varepsilon_{nmax} = \dfrac{V_P}{c_P}$ for $\varphi = 90°$	$\gamma_{max} = \dfrac{V_P}{2c_P}$ for $\varphi = 45°$	$\dfrac{1}{\rho_{max}} = 0.385\dfrac{a_P}{c_P^2}$ for $\varphi = 35°16'$
Shear wave		$\varepsilon_l = \dfrac{V_S}{c_S}\sin\varphi\cos\varphi$	$\varepsilon_n = \dfrac{V_S}{c_S}\sin\varphi\cos\varphi$	$\gamma = \dfrac{V_S}{c_S}\cos^2\varphi$	$K = \dfrac{a_S}{c_S^2}\cos^3\varphi$
		$\varepsilon_{lmax} = \dfrac{V_S}{2c_S}$ for $\varphi = 45°$	$\varepsilon_{nmax} = \dfrac{V_S}{2c_S}$ for $\varphi = 45°$	$\gamma_{max} = \dfrac{V_S}{c_S}$ for $\varphi = 0°$	$K_{max} = \dfrac{a_S}{c_S^2}$ for $\varphi = 0°$
Rayleigh waves	Compressional component	$\varepsilon_l = \dfrac{V_{RP}}{c_R}\cos^2\varphi$	$\varepsilon_n = \dfrac{V_{RP}}{c_R}\sin^2\varphi$	$\gamma = \dfrac{V_{RP}}{c_R}\sin\varphi\cos\varphi$	$K = \dfrac{a_{RP}}{c_R^2}\sin\varphi\cos^2\varphi$
		$\varepsilon_{lmax} = \dfrac{V_{RP}}{C_R}$ for $\varphi = 0°$	$\varepsilon_{nmax} = \dfrac{V_{RP}}{C_R}$ for $\varphi = 90°$	$\gamma_{max} = \dfrac{V_{RP}}{2c_R}$ for $\varphi = 45°$	$K_{max} = 0.385\dfrac{a_{RP}}{c_R^2}$ for $\varphi = 35°16'$

(Continued)

Table 6.1 Continued

Wave Type	Longitudinal Strain	Normal Strain	Shear Strain	Curvature
Shear component	—	$\varepsilon_n = \dfrac{V_{RS}}{c_R}\sin\varphi$	$\gamma = \dfrac{V_{RS}}{c_R}\cos\varphi$	$K = \dfrac{\alpha_{RS}}{c_R^2}\cos^2\varphi$
	—	$\varepsilon_{nmax} = \dfrac{V_{RS}}{c_R}$ for $\varphi = 90°$	$\gamma_{max} = \dfrac{V_{RS}}{2c_R}$ for $\varphi = 0°$	$K_{max} = \dfrac{\alpha_{RS}}{c_R^2}$ for $\varphi = 0$

Notation:
V_P: peak particle velocity associated with P-wave
V_s: peak particle velocity associated with S-wave
α_P: peak particle acceleration associated with P-wave
α_S: peak particle acceleration associated with S-wave
c_P: apparent velocity of P-wave propagation
c_S: apparent velocity of S-wave propagation
V_{RP}: peak particle velocity associated with a compressional component of Rayleigh wave
V_{RS}: peak particle velocity associated with a shear component of Rayleigh wave
α_{RP}: peak particle acceleration associated with a compressional component of Rayleigh wave
α_{RS}: peak particle acceleration associated with a shear component of Rayleigh wave
c_R: apparent velocity of Rayleigh wave propagation
φ: angle of incidence of the wave concerning the tunnel axis

Source: After St. John & Zahrah, 1987.

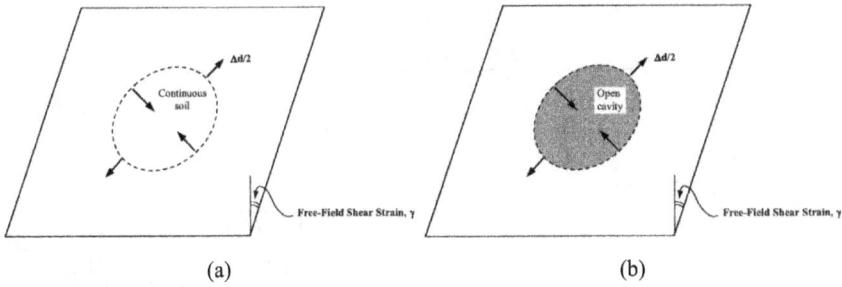

Figure 6.4 Free-field shear distortion of ground. (a) Non-perforated medium; (b) perforated ground.

Source: Wang, 1993.

Besides the axial and curvature deformation, the ovaling effect on tunnels and underground structures (see Figure 1.14) is also inevitable when the seismic wave propagates perpendicular to the axis of the tunnel for the seismic design and analysis for tunnels and underground structures. For the circular lining, there are two ways to describe the shear distortion of the surrounding medium (Wang, 1993), as illustrated in Figure 6.4. For the non-perforated surrounding medium (Figure 6.4a), the diametric strain of the tunnel lining is presented as:

$$\frac{\Delta d}{d} = \pm \frac{\gamma_{max}}{2} \qquad (6.1)$$

where d is the diameter of the tunnel and γ_{max} is the maximum free-field shear strain. For the perforated surrounding medium (Figure 6.4b), the diametric strain of the tunnel lining is expressed as:

$$\frac{\Delta d}{d} = \pm 2\gamma_{max}\left(1-v_m\right) \qquad (6.2)$$

where v_m is the Poisson's ratio of the surrounding medium.

On the other hand, the rectangular tunnels in rock and stiff soil also experience racking deformation due to the shear distortion of the surrounding medium. The racking deformation imposed on the structure is estimated based on the assumed medium deformations. One representative example of the medium deformation with depth is given in Figure 6.5a. The racking deformation could be taken as the difference between the medium deformation at the top (Point A in Figure 6.5b) and that at the bottom (Point B in Figure 6.5b) of the rectangular structures (St. John & Zahrah, 1987).

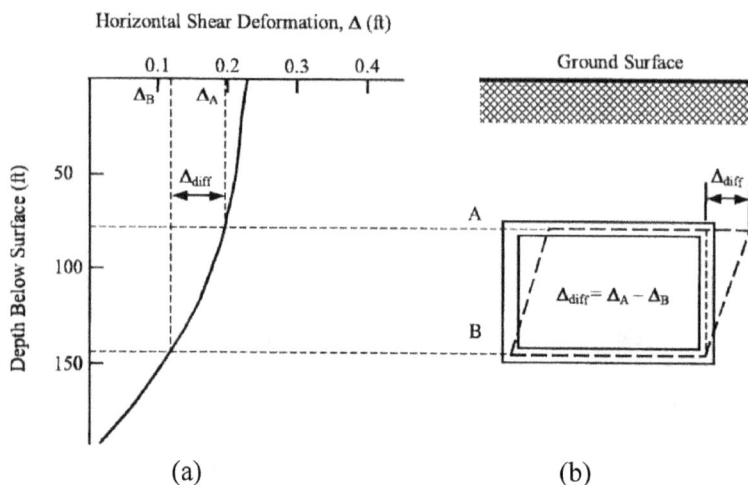

Horizontal Shear Deformation, Δ (ft)

Figure 6.5 Typical free-field racking deformation imposed on a buried rectangular frame. (a) Soil deformation profile; (b) racking deformation of a box structure.

Source: Wang, 1993.

The above-mentioned method is available for tunnels and underground structures which are relatively flexible compared to the surrounding medium since the tunnel structures in rock or stiff soils could deform according to the surrounding medium. It can provide a first-order estimate of the anticipated deformation of the structure (Hashash et al., 2001; Tao et al., 2015). The method has been applied successfully in several major transportation projects with the advantage of requiring the least number of inputs (Kuesel, 1969; Monsees & Merritt, 1988; Wang, 1993). However, such a method is not desirable for the situation where the stiff structures are buried in soft soil (Wang, 1993) since the tunnel resists, rather than conforms to the deformations imposed by the ground. Therefore, for such cases, the analysis of the tunnel-ground interaction effect was studied by employing the beam-on-elastic foundation approach (St. John & Zahrah, 1987) or modeling the underground structure as a cylindrical shell embedded in an elastic half-space (Datta et al., 1985; Wong et al., 1986; Liu et al., 1991; Luco & De Barros, 1994), and account for slippage at the soil-structure interface (O'Rourke & Hmadi, 1988; Ogawa & Koike, 2001).

6.3.2 Simplified Equivalent Static Analysis

Simplification of the seismic design and analysis of tunnels and underground structures subjected to earthquakes is a quite common and effective way for

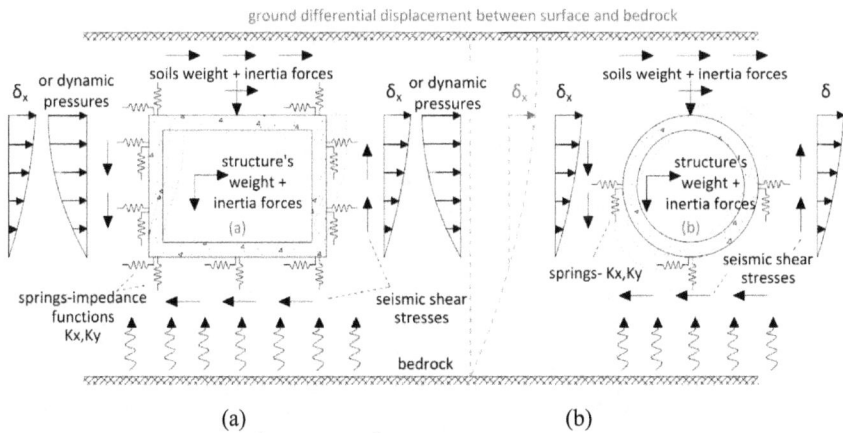

Figure 6.6 Simplified equivalent static method—equivalent forces. (a) Rectangular structure; (b) circular tunnel.

Source: Pitilakis & Tsinidis, 2014.

practical engineering purposes. One common model used for the simplified equivalent static analysis in the transversal direction is a frame-spring model (ISO 23469, 2005), as illustrated in Figure 6.6. Beam elements are introduced to model the structure and appropriate springs are applied to simulate the ground-structure interaction. The model is reduced to the simplest case without considering the ground-structure interaction when omitting the springs. Such a method categorizes the seismic loadings into three types: equivalent inertial static loadings, seismic shear stresses at the perimeter of the structure, and seismic earth pressures imposed on the sidewalls of the structure (Pitilakis & Tsinidis, 2014). Computation methods for each type of static loading are provided in current literature (Mononobe & Matsuo, 1929; Schnabel et al., 1972; Greek Seismic Code, 2003).

Even though this method can be used easily, there are still some significant shortcomings. For the fully embedded tunnels and underground structures, the exact magnitude and distribution of the seismic earth pressures remain unclear. Estimations based on guidelines for semi-embedded tunnels and underground structures may miscalculate the seismic earth pressures. Such misestimating may result in erroneous results. On the other hand, there is still a challenge for an accurate estimation of the seismic shear stresses at the perimeter of the tunnels and underground structures. Since the imperfect interaction between the structure and its surrounding medium may redistribute the stresses on the interface, such a simplified equivalent method cannot provide an excellent illustration of the actual situation (Pitilakis & Tsinidis, 2014).

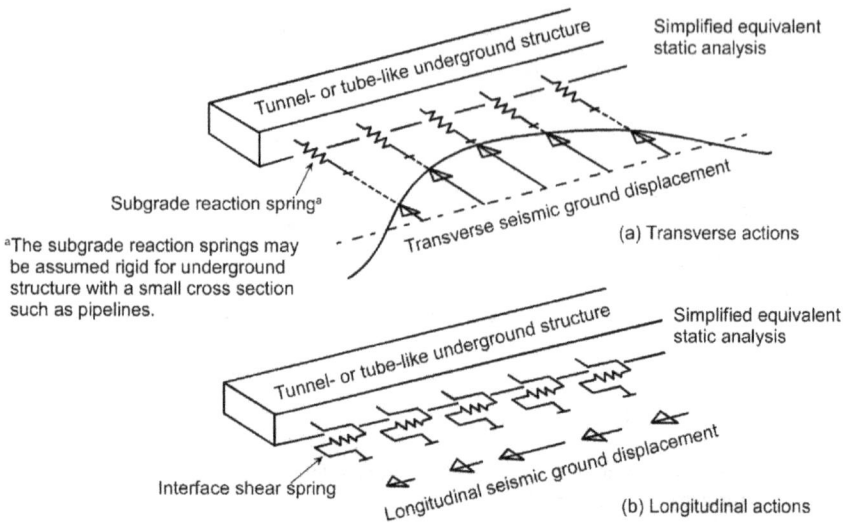

Figure 6.7 Simplified equivalent static analysis with spatial variation. (a) Transversal analysis; (b) longitudinal analysis.

Source: Iai, 2005.

The above-mentioned method is one way for seismic actions without spatial variation. Spatial variation for the ground motion is an inevitable aspect of tunnel seismic design and analysis due to their significant length. The seismic effect can be illustrated by simplifying the tunnel as a beam model on elastic foundations (i.e. springs), as illustrated in Figure 6.7. The equivalent static ground displacement is applied on the springs accounting for the spatial variation both in the transversal and longitudinal directions.

For both the seismic actions with or without the spatial variation, the accurate determination of impedance function (springs and dashpots) is of significant importance. Unfortunately, very few available formations are proposed in the current literature. By assuming a harmonic incidence wave, St. John and Zahrah (1987) provided an estimation for the transversal and longitudinal ground springs coefficients K_a and K_t, expressed as:

$$K_a = K_t = \frac{16 \times \pi \times G_m \times (1 - v_m)}{3 - 4v_m} \times \frac{D}{L} \tag{6.3}$$

where G_m is the ground shear modulus, v_m is the Poison's ratio of the ground, D is the diameter of the circular tunnel (or height of rectangular structure) and L is the harmonic wavelength. Besides, some other formulations are proposed by borrowing ideas from the elastic impedance functions of surface foundations or even retaining walls (e.g., Clough & Penzien, 1993; AFPS/AFTES, 2001; Vrettos, 2005). However, since underground structures have features that make their seismic behavior distinct from most surface foundations and retaining wall, most notably their complete enclosure in soil or rock, and their significant length, the application of these formulations may result in a large difference. Therefore, the accurate formulation of impedance functions for tunnels and underground structures is still an open issue for further work.

6.3.3 Simplified Frame Analysis Method (Wang, 1993)

Wang (1993) provided a step-by-step description of this method to calculate the actual racking deformation of the tunnel and underground structure, $\Delta_{structure}$, as follows:

$$\Delta_{structure} = R \times \Delta_{free\text{-}field} \tag{6.4}$$

where R is the racking coefficient and $\Delta_{free\text{-}field}$ is the free-field distortion. Using the estimation of the racking ratio, an equivalent force can be obtained and imposed on the underground structure in the forms of a concentrated force for the deeply buried tunnel (Figure 6.8a) and a pseudo-triangular pressure distribution for the shallow tunnel (Figure 6.8b). Each of the force forms has its advantages and disadvantages. The latter one could obtain a much more critical estimation of the moment capacity of the underground structure at its bottom joint than the former one, while the former one could provide a much more critical moment response at the roof-wall joint than the latter one. As a result, both models should be employed in the simplified frame analysis for design.

The simplified frame method can provide a good approximation of soil-structure interaction. On the other hand, reasonable accuracy in determining

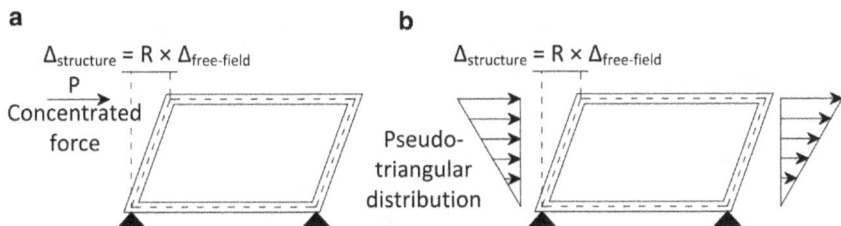

Figure 6.8 Simplified frame analysis models. (a) Concentrated force model; (b) triangular pressure distribution model.

Source: Pitilakis & Tsinidis, 2014.

the response of an underground structure can be obtained using this method. However, the precision is low for highly variable ground conditions and this method is not available for complicated subsurface ground profiles.

6.3.4 Mass-Spring Model Method (Kiyomiya, 1995)

Field observations have shown that the relative displacement along the axis of the tunnel during earthquakes is a primary factor in the design of the tunnel, especially for the immersed tunnel. One representative method to illustrate such a fact is the mass-spring model (Kiyomiya, 1995), as illustrated in Figure 6.9.

In the spring-mass model, the surface layer is divided into several slices perpendicular to the tunnel's axis. An equivalent mass-spring system is applied to simulate each slice. The ground slide mass is represented by the mass; a spring and a dashpot are applied to connect the mass to the base ground. The spring coefficient K_s is evaluated to guarantee that the natural period of the slide's first model of shear vibration is consistent with the natural period of the mass-spring system. Then along the axis of the tunnel, the adjacent masses are connected by using springs and dashpots in the longitudinal direction of the tunnel. The spring coefficient K_t has a relation with tension-compression resistance to the longitudinal relative deformation between the adjacent slice, or tangential resistance to the lateral relative deformation.

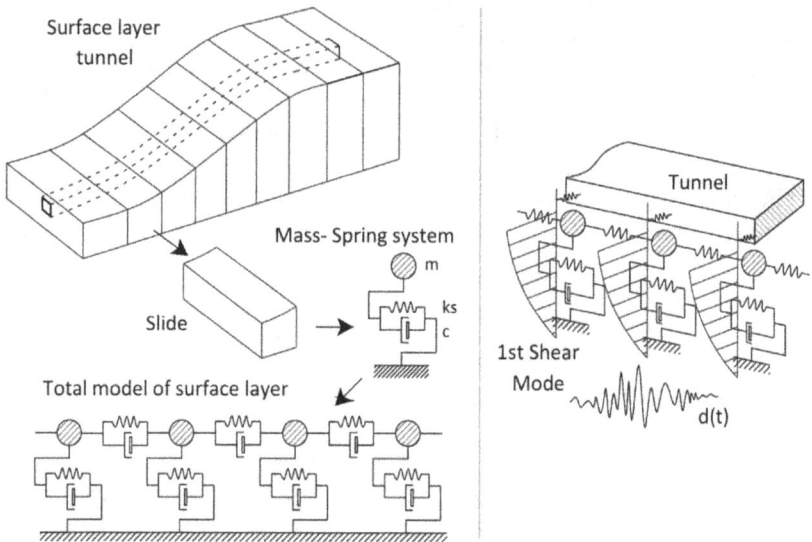

Figure 6.9 Mass-spring model method.

Source: Kiyomiya, 1995; Pitilakis & Tsinidis, 2014.

Through the general dynamic equilibrium equation expressed as Equation (6.5), the seismic response of the surrounding medium is obtained.

$$[M]\{\ddot{X}\}+[C]\{\dot{X}\}+[K]\{X\} = -[M]\{e\} \qquad (6.5)$$

where, $[M]$ is the matrix of the mass; $[C]$ is the matrix of damping; $[K]$ is the rigid matrix determined by spring coefficients K_t and K_s; $\{e\}$ is the input acceleration; $\{\ddot{X}\}$ is the acceleration vector of mass; $\{\dot{X}\}$ is the velocity of mass and $\{X\}$ is the displacement of the mass. The tunnel in this model is reduced to a beam supported by springs. After the determination of the ground deformation, the seismic response of the tunnels and underground structures could be obtained by imposing the ground deformation on one end of each spring.

6.4 ANALYTIC SEISMIC ANALYSIS

The analytical seismic analysis is one convenient and efficient way of providing direct quantitative insights into the physical mechanism. Wave theory is one popularly accepted method available for the analytic seismic analysis of underground structures. Several analytical methods are available in the current literature, such as the wave function expansion method (Pao & Mow, 1973; Liu et al., 2013), integral equation method, complex variable method (Liu et al., 1982), geometric ray theory (Pao et al., 1983; Li & Liu, 1987), and Fourier transform. Since the wave function expansion method is one of the most widely applied methods, a summary of the wave function expansion method is provided in this section.

Using the wave function expansion method, varieties of studies have taken into consideration plane or anti-plane stress and displacement of underground tunnels with or without lining embedded in a homogeneous, isotropic, and linear elastic full- or half-space (Pao & Mow, 1973; Smerzini et al., 2009; Davis et al., 2010; Lin et al., 2010; Liu & Wang, 2012; Liu et al., 2013). For a non-circular tunnel, such as a rectangular tunnel, semi-circular tunnel, or horseshoe-shaped tunnel which are common in engineering applications, the complex variable method and the weighted residual method are introduced to solve the problem caused by the scattering and diffraction of the seismic wave (Wang et al., 2005; Liu et al., 2016).

Tunnel excavation, especially using drilling and blasting, may result in a very irregular and rough excavation surface (Figure 6.10). Therefore, due to the existence of micro-cracks or interstitial media, the interface between the tunnel lining and its surrounding rock mass is not always perfect. The stress and displacement behaviors of tunnel lining are very dependent on the

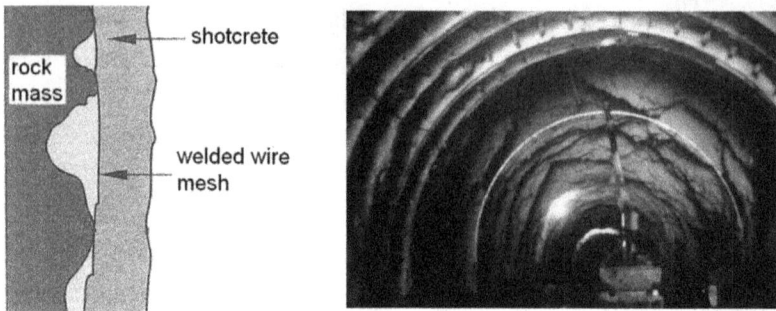

Figure 6.10 Illustration of the geometry of shotcrete lining with an irregular geometry of the tunnel boundary. (a) Sketch example of the irregular interface; (b) photograph of large over breaks induced by joint discontinuities around a road tunnel.
Source: Barpi & Peila, 2012.

surface status (Son & Cording, 2007). From the qualitative and quantitative points of view, the analytical results are different from the true ones if a correct interface is not placed around the tunnel lining. Recently, several imperfect interface models involving the elastic interface model (Yi et al., 2014; Fang et al., 2015, 2016) and time-dependent visco-elastic model (Fang & Jin, 2016, 2017) have been introduced.

For the convenience of derivation, the rock mass is usually assumed to be isotropic. In practical engineering, due to the mineral foliation in metamorphic rocks, stratification in sedimentary rocks, and discontinuities in the rock mass, many rocks have anisotropic characteristics (e.g., in their mechanical, thermal, seismic, and hydraulic properties). Generally speaking, rock anisotropy causes a non-uniform stress state and displacement around the boundary of an underground opening, creates a non-uniform depth of failure, and allows for stress-induced failure (near wall degradation) to possibly occur at relatively low-stress levels and under unexpected stress conditions (Bewick & Kaiser, 2009). Rock anisotropy is one of the most distinct features that must be considered in rock engineering disciplines. Moreover, a few studies have been devoted to investigating the anisotropy effect (e.g., Synge, 1956; Liu, 1988; Shi et al., 1996; Han & Liu, 1997; Červený & Pšenčík, 2005; Ting, 2010; Chen, 2012, 2015). Nevertheless, the aforementioned literature has failed to consider the influence of an imperfect interface on the dynamic response of an underground tunnel. In Chapters 8 and 9, two transversal analytic solutions are constructed to clarify the influences of rock anisotropy and imperfect interface on the seismic response of deep-buried and shallow tunnels, respectively.

The aforementioned literature focuses on transversal seismic performance. Longitudinal seismic analysis of long tunnels is equally important as transversal analysis. Several analytical methodologies based on structural dynamics are available for the longitudinal seismic analysis in the current literature. The ground-structure interaction effects are simulated by employing the beam-on-elastic foundation approach (St. John & Zahrah, 1987) or modeling the underground structure as a cylindrical shell embedded in an elastic half-space (Datta et al., 1985; Wong et al., 1986; Liu et al., 1991; Luco et al., 1994). On the other hand, slippage at the ground-structure interface is accounted for (O'Rourke & Hmadi, 1988; Ogawa & Koike, 2001). For more details on the method based on structural dynamics, the reader is referred to Section 6.3.1. In addition, the wave function expansion method using three-dimensional elastodynamics was employed to provide a full perspective of longitudinal seismic performance in long underground structures without considering the ground-structure interaction (Datta et al., 1985; Yasuda et al., 2014). Therefore, in Chapter 7, one longitudinal analytic solution is performed to clarify the effects of the imperfect interface on the seismic response of the deep-buried tunnel.

6.5 CONCLUSION

The chapter provides a short review of the available seismic research methods. Several issues that are still open and can significantly affect the performance and design of underground structures are also discussed.

Field test with artificial seismic source and physical model test are available for an intuitive grasp of the seismic influence of earthquakes, while they cannot control the input mechanism of seismic waves and boundary conditions, nor can they actively change various factors to conduct a purposeful study on a certain phenomenon. The dynamic time-history analysis is by far the most acceptable method for the seismic analysis of underground structures, which can efficiently describe the kinematic and inertial aspects of the ground-structure interaction, the complex geometry of the ground, and the arbitrary shape of tunnel lining. Currently, coupling methods of the FE, BE, and DE models with their own advantages are being performed for a three-dimensional seismic analysis, which is still an open issue. Currently, several simple, rational, and practical methods for simplified methods including the force-based method and the deformation-based method are available, such as the free-field deformation method, simplified equivalent static and dynamic method, simplified frame analysis method, and mass-spring model method. A convenient and efficient way of providing direct quantitative insights into the physical mechanism is the analytical seismic analysis. Wave theory is one popularly accepted method available. The effects of complex ground-structure interaction, ground anisotropic characteristics, and arbitrary shape of tunnel lining are being discussed in a simplified way using several mathematical methodologies, which is still an open issue.

Chapter 7

Analytic Analysis for Longitudinal Seismic Response

7.1 GOVERNING EQUATIONS

A deep underground tunnel subjected to a seismic wave is considered as illustrated in Figure 7.1. For the generalization of the results, a typical circular cross-section of the tunnel was adopted. The tunnel is embedded in a homogeneous, isotropic, and linearly elastic rock mass. The tunnel lining is also assumed to be homogeneous, isotropic, and linearly elastic. The Lamé constants, density, and Poisson's ratio are λ_1, μ_1, ρ_1, and υ_1 for the rock mass and λ_2, μ_2, ρ_2, and υ_2 for the tunnel lining. It is assumed that the tunnel lining is infinitely long and continuous. The inner and outer radii of the lining are denoted by R_1 and R_2, respectively. A harmonic compression P wave with circular frequency ω propagates in an incidence angle φ_{inc} along the tunnel axis. The tunnel axis passes through the vertical propagation plane x-z.

The displacement U satisfies the motion equation of the elastic medium when ignoring the body force effect as Equation (7.1) (Achenbach, 1973; Graff, 1973).

$$(\lambda + \mu)\nabla\nabla \cdot U + \mu\nabla^2 U = \rho\frac{\partial^2 U}{\partial^2 t} \tag{7.1}$$

where ∇ is the gradient and ∇^2 is the Laplacian, λ and μ are Lamé constants, and ρ is density. To simulate the traveling wave along the tunnel, the x-dependence of U will be taken as $U(r,\theta,z)\,e^{i(kz-\omega t)}$ in the cylindrical polar coordinate system r-θ-z, where k is the wave number of the incident wave along the tunnel, ω is the circular frequency, and t is time. Since the present model is in a steady state, the time t can be separated as an independent variable by the exponential factor $e^{-i\omega t}$ (Kubenko et al., 1978). The solution of Equation (7.1) can be written with one scalar (longitudinal wave potential φ) potential and two vector potentials (shear potentials ψ and χ) as Equation (7.2).

DOI: 10.1201/9781003401599-7

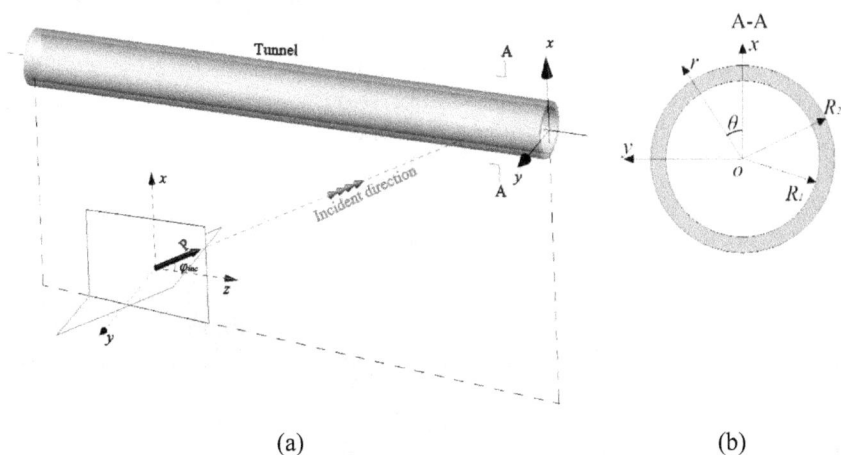

(a)   (b)

Figure 7.1 Three-dimensional geometry model (a) and cross-section (b) of a circular tunnel under a seismic wave.

$$U = \nabla\varphi + \nabla \times (\psi e_z) + \frac{1}{k_s}\nabla \times \nabla \times (\chi e_z) \qquad (7.2)$$

where k_s is the shear wave number determined by $k_s = \omega / c_s$ with $c_s = \sqrt{\mu/\rho}$.

For the harmonic incident wave, the potentials φ, ψ, and χ satisfy the following Helmholtz equation in the form of Equation (7.3).

$$\nabla^2\varphi + k_p^2\nabla^2\varphi = 0$$
$$\nabla^2\psi + k_s^2\nabla^2\psi = 0 \qquad (7.3)$$
$$\nabla^2\chi + k_p^2\nabla^2\chi = 0$$

where k_p is the P wave number determined by $k_p = \omega / c_p$ with $c_p = \sqrt{(\lambda+2\mu)/\rho}$.

7.2 WAVE FIELDS IN THE ISOTROPIC ROCK MEDIUM

7.2.1 Seismic Wave Potentials

The incident wave potential φ satisfying the conditions at infinity is expressed in the form of Equation (7.4).

$$\varphi_1^{inc} = \varphi_0 \sum_{n=0}^{\infty} \varepsilon_n i^n J_n(\alpha_1 r)\cos n\theta\, e^{i(\gamma_p z - \omega t)} \qquad (7.4)$$

where φ_0 is the amplitude of the incident compression wave, $\varepsilon_0 = 1$ and $\varepsilon_n = 2$ for $n \geq 1$, $i^2 = -1$, $J_n(\cdot)$ is the nth Bessel function of the first kind, α_1 is the apparent wave number of the compression wave in the x-direction, and γ_p is the apparent wave number of the compression wave in the z-direction. The apparent wave numbers α_1 and γ_p are defined as

$$\left. \begin{array}{l} \alpha_1 = k_{p(1)} \sin \varphi_{inc}^P \\ \gamma_p = k_{p(1)} \cos \varphi_{inc}^P \end{array} \right\} \tag{7.5}$$

where, $k_{p(1)}$ and $k_{s(1)}$ are the wave number for the compression and shear waves in the surrounding rock medium, and φ_{inc}^P is the compression wave incidence angle.

When the incident waves run into the tunnel, the waves are scattered multiple times at the interface of the tunnel and its surrounding rock medium. The scattered wave fields in the surrounding rock medium are given by

$$\begin{array}{l} \varphi_1^{sc} = \sum_{n=0}^{\infty} A_n H_n^{(1)}(\alpha_1 r) \cos n\theta \, e^{i(\gamma z - \omega t)} \\[2mm] \psi_1^{sc} = \sum_{n=0}^{\infty} B_n H_n^{(1)}(\beta_1 r) \sin n\theta \, e^{i(\gamma z - \omega t)} \\[2mm] \chi_1^{sc} = \sum_{n=0}^{\infty} C_n H_n^{(1)}(\beta_1 r) \cos n\theta \, e^{i(\gamma z - \omega t)} \end{array} \tag{7.6}$$

where A_n, B_n, and C_n are the mode coefficients of scattered waves, $H_n^{(1)}$ is the nth Hankel function of the first kind, which denotes the outgoing wave, β_1 is the apparent wave number of the shear wave in the x-direction, and α_1, β_1, and γ depend on the type of incident wave as defined in Equations (7.5) and (7.7).

$$\left. \begin{array}{l} \gamma = \gamma_P = k_{P(1)} \cos \varphi_{inc}^P \\ \beta_1 = \sqrt{k_{s(1)}^2 - \gamma^2} \end{array} \right\} \tag{7.7}$$

where $k_{s(1)}$ is the shear wave number in the surrounding rock medium.

7.2.2 Seismic Stresses and Displacements

The displacements and stresses in the isotropic rock medium can be solved using three-dimensional elastodynamic analysis (see Appendix A). Therefore, in the present model as illustrated in Figure 7.1, the stresses in the isotropic medium resulting from the incident and scattered waves are reached by Equations (7.8) to (7.13).

$$\sigma_{rr(1)} = \frac{2\mu_1}{r^2}\sum_{n=0}^{\infty}\left(\varepsilon_n i^n \varphi_0 X_{11}^{(1)} + A_n X_{11}^{(3)} + B_n X_{12}^{(3)} + C_n X_{13}^{(3)}\right)\cos n\theta\, e^{i(\gamma z - \omega t)} \qquad (7.8)$$

$$\sigma_{\theta\theta(1)} = \frac{2\mu_1}{r^2}\sum_{n=0}^{\infty}\left(\varepsilon_n i^n \varphi_0 X_{21}^{(1)} + A_n X_{21}^{(3)} + B_n X_{22}^{(3)} + C_n X_{23}^{(3)}\right)\cos n\theta\, e^{i(\gamma z - \omega t)} \qquad (7.9)$$

$$\sigma_{zz(1)} = \frac{2\mu_1}{r^2}\sum_{n=0}^{\infty}\left(\varepsilon_n i^n \varphi_0 X_{31}^{(1)} + A_n X_{31}^{(3)} + B_n X_{32}^{(3)} + C_n X_{33}^{(3)}\right)\cos n\theta\, e^{i(\gamma z - \omega t)} \qquad (7.10)$$

$$\sigma_{r\theta(1)} = \frac{2\mu_1}{r^2}\sum_{n=0}^{\infty}\left(\varepsilon_n i^n \varphi_0 X_{41}^{(1)} + A_n X_{41}^{(3)} + B_n X_{42}^{(3)} + C_n X_{43}^{(3)}\right)\sin n\theta\, e^{i(\gamma z - \omega t)} \qquad (7.11)$$

$$\sigma_{rz(1)} = \frac{2\mu_1}{r^2}\sum_{n=0}^{\infty}\left(\varepsilon_n i^n \varphi_0 X_{51}^{(1)} + A_n X_{51}^{(3)} + B_n X_{52}^{(3)} + C_n X_{53}^{(3)}\right)\cos n\theta\, e^{i(\gamma z - \omega t)} \qquad (7.12)$$

$$\sigma_{\theta z(1)} = \frac{2\mu_1}{r^2}\sum_{n=0}^{\infty}\left(\varepsilon_n i^n \varphi_0 X_{61}^{(1)} + A_n X_{61}^{(3)} + B_n X_{62}^{(3)} + C_n X_{63}^{(3)}\right)\sin n\theta\, e^{i(\gamma z - \omega t)} \qquad (7.13)$$

The terms $X_{sq}^{(b)}$ (s = 1, 2, 3, 4, 5, 6; q = 1, 2, 3 and b = 1, 3) are presented in Appendix B.

The displacements in the isotropic medium resulting from the incident and scattered waves are given by Equations (7.14) to (7.16).

$$u_{r(1)} = \frac{1}{r}\sum_{n=0}^{\infty}\left(\varepsilon_n i^n \varphi_0 X_{71}^{(1)} + A_n X_{71}^{(3)} + B_n X_{72}^{(3)} + C_n X_{73}^{(3)}\right)\cos n\theta\, e^{i(\gamma z - \omega t)} \qquad (7.14)$$

$$u_{\theta(1)} = \frac{1}{r}\sum_{n=0}^{\infty}\left(\varepsilon_n i^n \varphi_0 X_{81}^{(1)} + A_n X_{81}^{(3)} + B_n X_{z2}^{(3)} + C_n X_{z3}^{(3)}\right)\sin n\theta\, e^{i(\gamma z - \omega t)} \qquad (7.15)$$

$$u_{z(1)} = \frac{1}{r}\sum_{n=0}^{\infty}\left(\varepsilon_n i^n \varphi_0 X_{91}^{(1)} + A_n X_{91}^{(3)} + B_n X_{92}^{(3)} + C_n X_{93}^{(3)}\right)\cos n\theta\, e^{i(\gamma z - \omega t)} \qquad (7.16)$$

The terms $X_{sq}^{(b)}$ (s = 7, 8, 9; q = 1, 2, 3; and b = 1, 3) are presented in Appendix B.

7.3 WAVE FIELDS IN THE ISOTROPIC TUNNEL LINING

7.3.1 Seismic Wave Potentials

The refracted wave potentials being confined inside the concrete lining are expressed as Equation (7.17).

$$\varphi_2^{rr} = \sum_{n=0}^{\infty} D_n H_n^{(2)}(\alpha_2 r)\cos n\theta e^{i(\gamma z-\omega t)}$$

$$\psi_2^{rr} = \sum_{n=0}^{\infty} E_n H_n^{(2)}(\beta_2 r)\sin n\theta e^{i(\gamma z-\omega t)} \qquad (7.17)$$

$$\chi_2^{rr} = \sum_{n=0}^{\infty} F_n H_n^{(2)}(\beta_2 r)\cos n\theta e^{i(\gamma z-\omega t)}$$

where D_n, E_n, and F_n are the mode coefficients of refracted waves, $H_n^{(2)}$ is the nth Hankel function of the second kind which denotes the ingoing wave, and α_2 and β_2 are the apparent wave number in the concrete lining defined as Equation (7.18).

$$\alpha_2 = \sqrt{k_{p(2)}^2 - \gamma^2}$$

$$\beta_2 = \sqrt{k_{s(2)}^2 - \gamma^2} \qquad (7.18)$$

where $k_{p(2)}$ and $k_{s(2)}$ are the wave number for the compression and shear waves in the concrete lining.

Meanwhile, reflected waves propagate outwards from the lining inner boundary. The displacement fields of reflected waves are given as Equation (7.19).

$$\varphi_2^{rf} = \sum_{n=0}^{\infty} G_n H_n^{(1)}(\alpha_2 r)\cos n\theta e^{i(\gamma z-\omega t)}$$

$$\psi_2^{rf} = \sum_{n=0}^{\infty} M_n H_n^{(1)}(\beta_2 r)\sin n\theta e^{i(\gamma z-\omega t)} \qquad (7.19)$$

$$\chi_2^{rf} = \sum_{n=0}^{\infty} N_n H_n^{(1)}(\beta_2 r)\cos n\theta e^{i(\gamma z-\omega t)}$$

where G_n, M_n, and N_n are the mode coefficients of the reflected waves.

The total wave fields in the isotropic rock medium are produced by the superposition of the incident waves and the scattered waves as Equation (7.20).

$$\varphi_1^{total} = \varphi_1^{inc} + \varphi_1^{sc}$$

$$\psi_1^{total} = \psi_1^{sc} \qquad (7.20)$$

$$\chi_1^{total} = \chi_1^{sc}$$

The total wave fields in the concrete lining are produced by the superposition of the refracted waves and reflected waves as Equation (7.21).

$$\varphi_2^{\text{total}} = \varphi_2^{rr} + \varphi_2^{rf}$$
$$\psi_2^{\text{total}} = \psi_2^{rr} + \psi_2^{rf} \qquad (7.21)$$
$$\chi_2^{\text{total}} = \chi_2^{rr} + \chi_2^{rf}$$

7.3.2 Seismic Stresses and Displacements

The displacements and stresses in the isotropic concrete lining can be also obtained based on the three-dimensional elastodynamic analysis (see Appendix A). Therefore, in the present model as illustrated in Figure 7.1, the lining stresses resulting from the reflected and refracted waves are given as Equations (7.22) to (7.27).

$$\sigma_{rr(2)} =$$
$$\frac{2\mu_2}{r^2} \sum_{n=0}^{\infty} \left(D_n X_{11}^{(4)} + F_n X_{11}^{(3)} + M_n X_{12}^{(4)} + E_n X_{12}^{(3)} + G_n X_{13}^{(4)} + G_n X_{13}^{(3)} \right) \cos n\theta\, e^{i(\gamma z - \omega t)} \qquad (7.22)$$

$$\sigma_{\theta\theta(2)} =$$
$$\frac{2\mu_2}{r^2} \sum_{n=0}^{\infty} \left(D_n X_{21}^{(4)} + F_n X_{21}^{(3)} + M_n X_{22}^{(4)} + E_n X_{22}^{(3)} + G_n X_{23}^{(4)} + G_n X_{23}^{(3)} \right) \cos n\theta\, e^{i(\gamma z - \omega t)} \qquad (7.23)$$

$$\sigma_{zz(2)} =$$
$$\frac{2\mu_2}{r^2} \sum_{n=0}^{\infty} \left(D_n X_{31}^{(4)} + F_n X_{31}^{(3)} + M_n X_{32}^{(4)} + E_n X_{32}^{(3)} + G_n X_{33}^{(4)} + G_n X_{33}^{(3)} \right) \cos n\theta\, e^{i(\gamma z - \omega t)} \qquad (7.24)$$

$$\sigma_{r\theta(2)} =$$
$$\frac{2\mu_2}{r^2} \sum_{n=0}^{\infty} \left(D_n X_{41}^{(4)} + F_n X_{41}^{(3)} + M_n X_{42}^{(4)} + E_n X_{42}^{(3)} + G_n X_{43}^{(4)} + G_n X_{43}^{(3)} \right) \sin n\theta\, e^{i(\gamma z - \omega t)} \qquad (7.25)$$

$$\sigma_{rZ(2)} =$$
$$\frac{2\mu_2}{r^2} \sum_{n=0}^{\infty} \left(D_n X_{51}^{(4)} + F_n X_{51}^{(3)} + M_n X_{52}^{(4)} + E_n X_{52}^{(3)} + G_n X_{53}^{(4)} + G_n X_{53}^{(3)} \right) \cos n\theta\, e^{i(\gamma z - \omega t)} \qquad (7.26)$$

$$\sigma_{\theta z(2)} =$$
$$\frac{2\mu_2}{r^2} \sum_{n=0}^{\infty} \left(D_n X_{61}^{(4)} + F_n X_{61}^{(3)} + M_n X_{62}^{(4)} + E_n X_{62}^{(3)} + G_n X_{63}^{(4)} + G_n X_{63}^{(3)} \right) \sin n\theta\, e^{i(\gamma z - \omega t)} \qquad (7.27)$$

The terms $X_{sq}^{(b)}$ ($s = 1, 2, 3, 4, 5, 6$; $q = 1, 2, 3$; and $b = 1, 3$) are presented in Appendix B.

The lining displacements resulting from the reflected and refracted waves are given by Equations (7.28) to (7.30).

$$u_{r(2)} =$$
$$\frac{1}{r} \sum_{n=0}^{\infty} \left(D_n X_{71}^{(4)} + F_n X_{71}^{(3)} + M_n X_{72}^{(4)} + E_n X_{72}^{(3)} + G_n X_{73}^{(4)} + G_n X_{73}^{(3)} \right) \cos n\theta\, e^{i(\gamma z - \omega t)} \qquad (7.28)$$

$$u_{\theta(2)} =$$

$$\frac{1}{r}\sum_{n=0}^{\infty}\left(D_n X_{81}^{(4)} + F_n X_{81}^{(3)} + M_n X_{81}^{(4)} + E_n X_{82}^{(3)} + G_n X_{83}^{(4)} + G_n X_{83}^{(3)}\right)\sin n\theta\, e^{i(\gamma z - \omega t)} \qquad (7.29)$$

$$u_{z(2)} =$$

$$\frac{1}{r}\sum_{n=0}^{\infty}\left(D_n X_{91}^{(4)} + F_n X_{91}^{(3)} + M_n X_{92}^{(4)} + E_n X_{92}^{(3)} + G_n X_{93}^{(4)} + G_n X_{93}^{(3)}\right)\cos n\theta\, e^{i(\gamma z - \omega t)} \qquad (7.30)$$

The terms $X_{sq}^{(b)}(s = 7, 8, 9;\ q = 1, 2, 3;$ and $b = 1, 3)$ are presented in Appendix B.

7.4 BOUNDARY CONDITION

As aforementioned, the ground-lining interaction due to the imperfect contact between the tunnel lining and its surrounding medium is inevitable and would result in different seismic behaviors (Son & Cording, 2007). Analytic results would be different from the true ones if a correct interface is not placed around the tunnel lining. The present study introduced an elastic model to analyze the interface influence. Its validity has been verified through theoretical (Yi et al., 2014; Fang et al., 2015, 2016) and experimental results (Honarvar & Sinclair, 1998), as well as numerical simulation (Lombard & Piraux, 2006). With the elastic model, tractions at the inner boundary $(r = R_1)$ of the tunnel lining are free described as expressed in Equation (7.31).

$$\sigma_{rr(2)}\big|_{r=R_1} = 0$$

$$\sigma_{r\theta(2)}\big|_{r=R_1} = 0 \qquad (7.31)$$

$$\sigma_{rz(2)}\big|_{r=R_1} = 0$$

At the lining's outer boundary $(r = R_2)$, tractions are continuous, but displacements are discontinuous across the interface expressed as in Equation (7.32).

$$\sigma_{rr(2)}\big|_{r=R_2} = \sigma_{rr(1)}\big|_{r=R_2}, \quad \left(u_{r(2)} - u_{r(1)}\right)\big|_{r=R_2} = \frac{\sigma_{rr(1)}\big|_{r=R_2}}{k_r}$$

$$\sigma_{r\theta(2)}\big|_{r=R_2} = \sigma_{r\theta(1)}\big|_{r=R_2}, \quad \left(u_{\theta(2)} - u_{\theta(1)}\right)\big|_{r=R_2} = \frac{\sigma_{r\theta(1)}\big|_{r=R_2}}{k_\theta} \qquad (7.32)$$

$$\sigma_{rz(2)}\big|_{r=R_2} = \sigma_{rz(2)}\big|_{r=R_2}, \quad \left(u_{z(2)} - u_{z(1)}\right)\big|_{r=R_2} = \frac{\sigma_{rz(1)}\big|_{r=R_2}}{k_z}$$

where k_r, k_θ, and k_z are the imperfect interface's radial, circumferential, and axial stiffness. If $k_r \to \infty$, $k_\theta \to \infty$, and $k_z \to \infty$, the imperfect interface will be reduced to the perfect one. The interface stiffness is often determined from the rock mass data using a simple relationship that is derived from the Winkler theory (Barpi & Peila, 2012). Oreste (2007) also proposed a more refined formulation with a hyperbolic constitutive relationship. However, some empirical parameters for the interface stiffness used by Fang et al. (2016) were adopted due to the difficulty of determining these parameters from the in-situ investigation. With such boundary conditions, a set of algebra equations can be obtained. Then the unknown coefficients can be obtained by solving the algebra equations.

Numerical results have shown that the mode with $n = 0$ is axial deformation with tension and compression, the mode with $n = 1$ is curvature deformation, and the mode with $n = 2$ is an ovaling deformation of the cross-section (Yasuda et al., 2014). The three modes could represent the three primary deformations of the tunnel due to seismic shaking: axial compression and extension, longitudinal bending, and ovaling/racking (Pitilakis & Tsinidis, 2014), as illustrated in Figure 1.15. Therefore, the three conditions ($n = 0, 1, 2$) are taken into consideration in the following analysis.

7.5 FUNDAMENTAL SOLUTION

To analyze the dynamic response around a deep underground tunnel subjected to seismic shaking, three dimensionless dynamic stress concentration factors (DSCF) are introduced as Equations (7.33) to (7.35).

$$\sigma_{\theta\theta}^* = \left| \frac{\sigma_{\theta\theta}}{\sigma_0} \right| \tag{7.33}$$

$$\sigma_{zz}^* = \left| \frac{\sigma_{zz}}{\sigma_0} \right| \tag{7.34}$$

$$\sigma_{\theta z}^* = \left| \frac{\sigma_{\theta z}}{\sigma_0} \right| \tag{7.35}$$

where σ_0 is the maximum magnitude of the incident stress defined as $\sigma_0 = (\lambda_1 + 2\mu_1)k_{p1}^2 \varphi_0$.

To illustrate the three-dimensional dynamic behavior of a mountain tunnel, case studies on parametric influence are presented. We focused on the influences of the incidence angle and the imperfect interface stiffness on the stress distribution at the lining inner side ($r = R_1$). Three dimensionless stiffness parameters of the interface are defined as $k_r^* = k_r R_2/\mu_1$, $k_\theta^* = k_\theta R_2/\mu_1$, and $k_z^* = k_z R_2/\mu_1$. For convenience, it is assumed that $k^* = k_r^* = k_\theta^* = k_z^*$.

7.5.1 Case I: Incidence Angle Influence

For illustration of the incidence angle influence, three incidence angles of 15°, 45°, and 75° are employed. Detailed properties of the rock mass and tunnel lining are listed in Table 7.1, based on the conditions encountered in the Tawarayama tunnel in Kumamoto Prefecture, Japan. Figure 7.2 shows the

Table 7.1 Input Data for the Material Properties of the Model

Item		Value
Properties of Rock Mass		
Young's modulus	E_1 (Gpa)	13.08
Poisson's ratio	υ_1	0.3
Density	ρ_1 (kg/m³)	2200
Properties of Tunnel Lining		
Young's modulus	E_2 (Gpa)	22.00
Poisson's ratio	υ_2	0.2
Density	ρ_2 (kg/m³)	2450
Thickness	t (m)	0.50

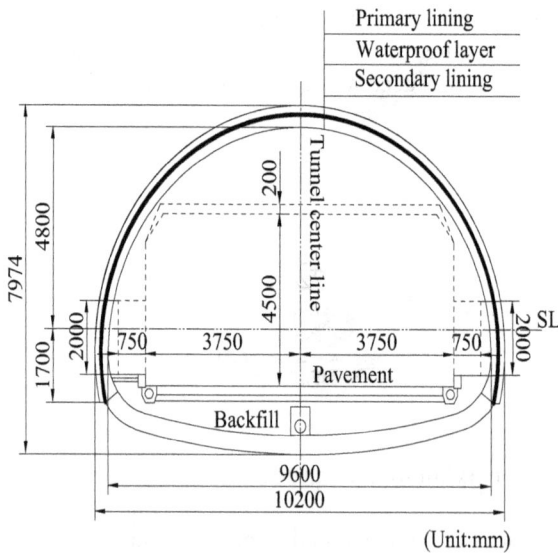

Figure 7.2 The horseshoe-shaped cross-section in the Tawarayama tunnel.

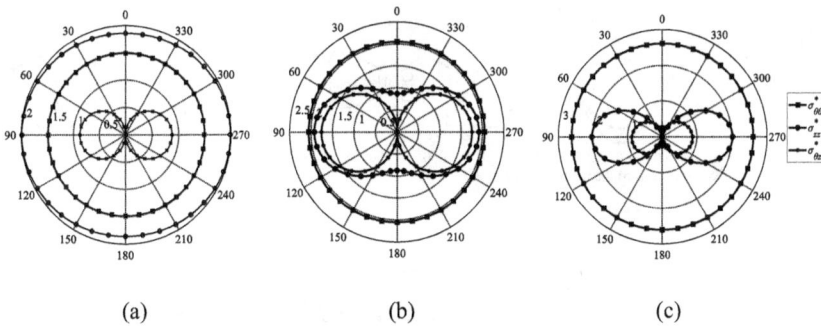

(a)   (b)   (c)

Figure 7.3 Distribution of stress amplitude at the lining inner side considering incidence angle. (a) Incidence angle of 15°; (b) incidence angle of 45°; (b) incidence angle of 75°.

horseshoe cross-section shape of the tunnel. The inner radius of the crown is 4.8 m and the thickness of the tunnel lining is 0.5 m. Here, the inner radius of the crown is adopted to represent the inner radius of the circular cross-section. The frequency of the incident wave is 3 Hz. Here, the interface was assumed to be perfect.

Figure 7.3 illustrates the DSCF distribution at the lining inner side considering the incidence angle. The incidence angle has a great influence on the scattering of seismic waves. In the case with a shallow incidence angle of 15° (nearly parallel to the axis of the tunnel lining), the axial stress σ_{zz} is dominant. The axial stress σ_{zz} and the shear stress $\sigma_{\theta z}$ show their maximums at $\theta = 90°$ and 270°, while the circumferential stress $\sigma_{\theta\theta}$ is substantially uniform along the entire lining wall surface. In the case with a slightly larger incidence angle of 45°, the circumferential stress $\sigma_{\theta\theta}$ becomes dominant. The maximums of the axial stress σ_{zz} and the shear stress $\sigma_{\theta z}$ also occur at $\theta = 90°$ and 270°. The circumferential stress $\sigma_{\theta\theta}$ also distributes uniformly. In the case with the incidence angle of 75°, the stress distribution is the same as that in the case with the incidence angle of 45°. In addition, the axial stress σ_{zz} at $\theta = 0°$ changes dramatically as the incidence angle increases.

The DSCF distribution of the axial stress, circumferential stress, and shear stress along the tunnel lining at $\theta = 90°$ and 270° are illustrated in Figures 7.4 and 7.5, respectively. Due to the harmonic incident wave, the axial stress σ_{zz}, circumferential stress $\sigma_{\theta\theta}$, and shear stress $\sigma_{\theta z}$ distribute along the tunnel axis periodically. On the other hand, the incidence angle influences the wavelength of the stresses. In the case with a shallow incidence angle (e.g., 15°), the wavelength of the stresses is about 1100 m. With the increase of incidence angle, the wavelength of the stress turns out to be 1500 m for the incidence angle of 45°, and 4000 m for the incidence angle of 75°.

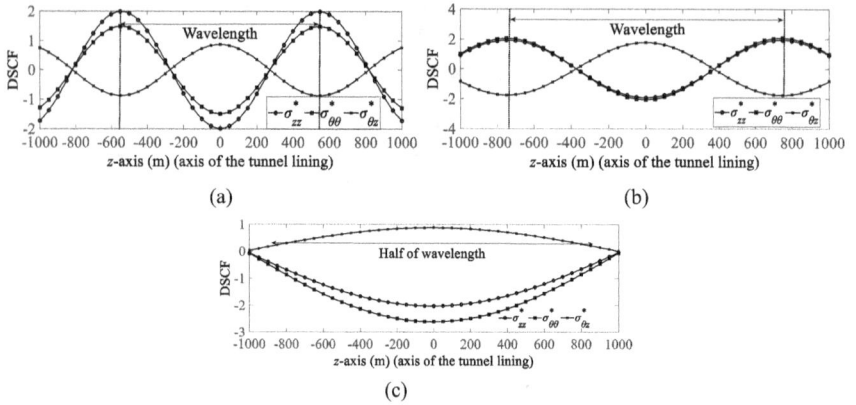

Figure 7.4 DSCF distribution along the lining axis ($\theta = 90°$). (a) Incidence angle of 15°; (b) incidence angle of 45°; (c) incidence angle of 75°.

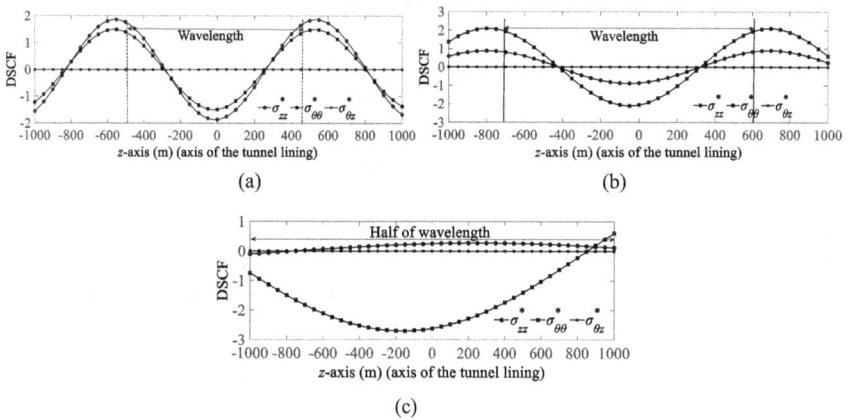

Figure 7.5 DSCF distribution along the lining axis ($\theta = 0°$). (a) Incidence angle of 15°; (b) incidence angle of 45°; (c) incidence angle of 75°.

To illustrate the influence of incidence angle on the seismic response of tunnel lining subjected to compression wave, the axial stress, circumferential stress, and shear stress at $\theta = 0°$, 45°, and 90° are investigated. Figure 7.6 shows the stress variation with incidence angle. The axial stress dominates with a compression wave propagating in a shallow incidence angle along the lining axis. It indicates that axial deformation with tension and compression is predominant. With the increase of incidence angle, the axial stress σ_{zz}

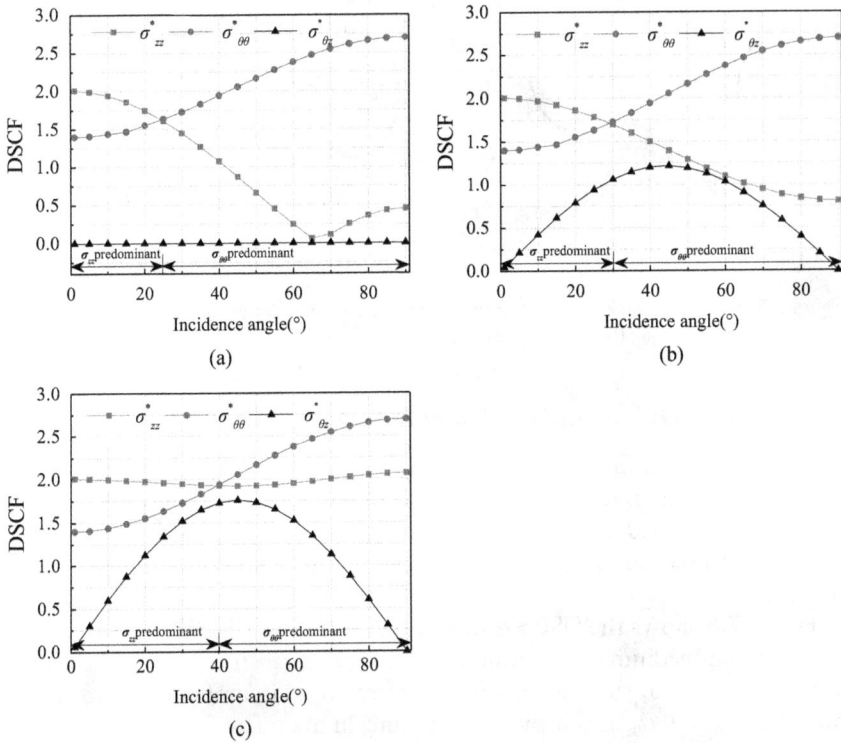

Figure 7.6 Stress variation with incident angle. (a) $\theta = 0°$; (b) $\theta = 45°$; (c) $\theta = 90°$.

decreases especially at $\theta = 0°$ and $45°$, while the circumferential stress $\sigma_{\theta\theta}$ increases. After a critical incidence angle, the circumferential stress becomes dominant, which indicates that curvature deformation becomes predomi-nant. The critical incidence angles are different at different locations of the cross-section. They are $25°$ at $\theta = 0°$, $30°$ at $\theta = 45°$, and $40°$ at $\theta = 90°$, respectively.

As different deformation modes occur with different incidence angles, the potential damage will be different. Figure 7.7 presents the typical potential damage to the tunnel lining subjected to compression waves. In the case of a shallow incidence angle (Figure 7.7a), the circumferential crack and another failure (e.g., construction joint/pavement failure) could occur due to the dominant axial deformation with tension and compression. In the case of a deep incidence angle (Figure 7.7b), longitudinal cracks along the tunnel axis are likely to occur at $\theta = 90°$ and $270°$ due to the superposition of axial deformation and curvature deformation.

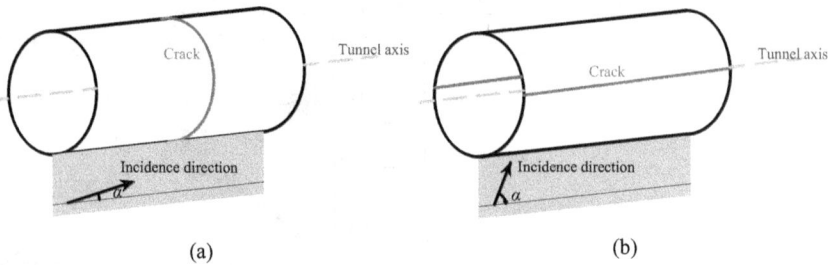

Figure 7.7 Potential crack patterns concerning different incidence angles. (a) Shallow incidence angle; (b) steep incidence angle.

7.5.2 Case II: Imperfect Interface Influence

To illustrate the influence of the interaction between the surrounding rock medium and lining, four sets of stiffness parameters are adopted which are $k^* = 50.0, 10.0, 1.0,$ and 0.5, respectively. A shallow incidence angle of $15°$ is taken into consideration. Other geometric and material parameters refer to Case I.

Figure 7.8 shows the DSCF distribution at the lining inner side with a soft surrounding medium considering different interface stiffness k^*. In the case with k^* being 0.5, the DSCFs of axial stress σ_{zz}, circumferential stress $\sigma_{\theta\theta}$, and shear stress $\sigma_{\theta z}$ reach their maximum. In a similar fashion to the case with the perfect interface (Figure 7.3), the axial stress σ_{zz} and the shear stress $\sigma_{\theta z}$ become maximum at $\theta = 90°$ and $270°$, while the circumferential stress $\sigma_{\theta\theta}$ along the entire lining wall surface is substantially uniform. In the case with k^* being 1.0, the DSCFs of axial stress σ_{zz}, circumferential stress $\sigma_{\theta\theta}$, and shear stress $\sigma_{\theta z}$ decrease compared to the case with k^* being 0.5. But the maximums of the axial stress σ_{zz} and the shear stress $\sigma_{\theta z}$ also occur at $\theta = 90°$ and $270°$. The circumferential stress $\sigma_{\theta\theta}$ also distributes along the entire lining wall surface uniformly. In the cases with k^* being 10 and 50, the DSCFs of axial stress σ_{zz}, circumferential stress $\sigma_{\theta\theta}$, and shear stress $\sigma_{\theta z}$ continue to decrease and the distribution of stresses keep consistent with those in the cases with k^* being 0.5 and 1.0. The seismic stress concentration becomes larger with a worse interface between the tunnel lining and its surrounding rock mass. Large stress concentration is more likely to induce lining concrete spalling, even lining structure collapse along with the fault influence.

As the maximum values of circumferential stress $\sigma_{\theta\theta}$ and shear stress $\sigma_{\theta z}$ occur at the position of $\theta = 90°$ and $270°$, the stress variation with the interface stiffness at the position of $\theta = 90°$ are considered. Without loss of generality, stresses at the positions of $\theta = 0°$ and $45°$ are also investigated. Figure 7.9 illustrates the DSCF variation of stresses at $\theta = 0°, 45°$ and $90°$

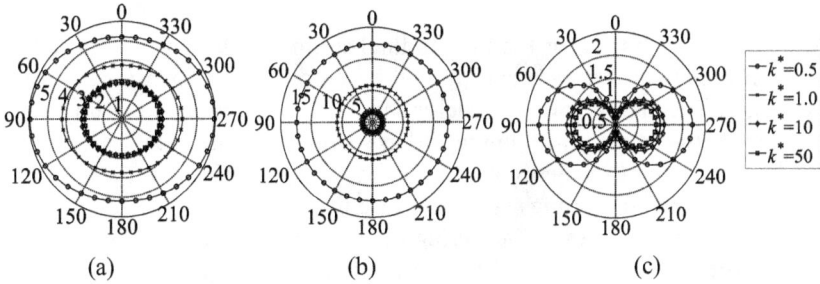

Figure 7.8 Distribution of stress amplitude at the lining inner side considering different interface stiffness k^*. (a) Axial stress; (b) circumferential stress; (c) shear stress.

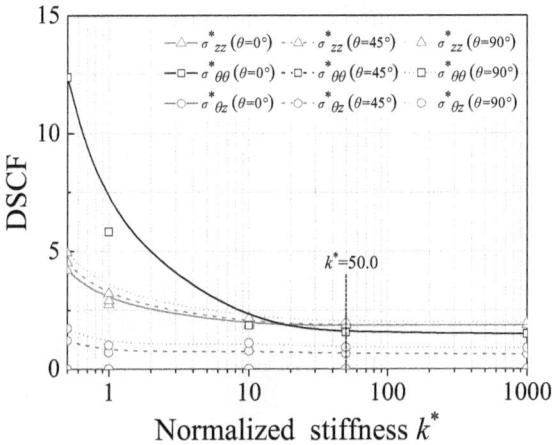

Figure 7.9 Influence of the interaction between the surrounding rock medium and tunnel lining.

with different interface stiffness k^*. As discussed before, increasing interface stiffness contributes to the decrease of the DSCFs. However, they tend to become stable as the interface stiffness increases. Taking the axial stress as one example, the decrease percentages at $\theta = 0°$, 45° and 90° are 34.7%, 34.8%, and 35.0% with the interface stiffness increasing from 0.5 to 1.0, while the decrease percentages are 0.09%, 3.6%, and 6.7% with interface stiffness increasing from 10 to 50. As a limit analysis, the case of k^* being 10^3 is compared to the case of k^* being 50. It can be seen that DSCFs are much the same, regardless of axial stress, circumferential stress, and shear stress. Thus, with a much larger interface stiffness, the interface influence can be neglected to consider the interface perfect in the numerical analysis.

7.6 CONCLUSION

An analytical solution for the longitudinal seismic performance of a tunnel with an imperfect interface is proposed based on three-dimensional elasto-dynamics. The elastic spring model is introduced to simulate the imperfect interface between the tunnel lining and its surrounding medium. Case studies on the influence of incidence angle and imperfect interface on the dynamic stress of the medium-lining system are discussed.

The numerical analysis considering the wave propagation obtained some fundamental deformation and failure modes. Seismic waves with different incidence angles induce varying patterns of lining damage. A seismic wave with a shallow incidence angle concerning the axis of the tunnel lining causes a circumferential crack, and a seismic wave with a steep incidence angle causes a longitudinal crack along the tunnel axis. The imperfect interface between the tunnel lining and its surrounding rock medium contributes to the large dynamic stress concentration of the tunnel lining, which influences the stability of the lining-ground system.

Chapter 8

Analytic Analysis for Cross-Sectional Seismic Response

Part I—Deep-Buried Tunnel

8.1 GOVERNING EQUATIONS

A deep underground tunnel with a horseshoe-shaped lining is embedded in a homogeneous, anisotropic, and linearly elastic rock medium. An anti-plane shear wave (SH wave) propagates into the tunnel lining with an incidence angle. Such a problem can be simplified as a plane strain one considering the model's geometry and boundary conditions. Figure 8.1 illustrates the vertical cross-section of the essentially two-dimensional problem. Two coordinate systems are adopted: one Cartesian coordinate system (x, y) in the Z plane and one polar coordinate system (r, θ) in the ζ plane having a common origin at the tunnel center with the Cartesian coordinate system. The z-axis denotes the tunnel lining axis direction, not being plotted in Figure 8.1.

For anisotropic elastic materials, 36 elastic constants are required to describe their properties according to the generalized Hooke's law. Since a plane strain problem with an anti-plane SH wave incidence is considered, elastic constants C_{44}, C_{45}, and C_{55} are enough to describe such a problem. The constraints $C_{44} > 0$, and $C_{44}C_{55} - C_{45}^2 > 0$ should be incorporated to satisfy the positive strain energy density condition. Besides, the density of the surrounding rock medium is characterized by ρ_1. The tunnel lining with outer radius R_1 and inner radius R_2 is assumed to be homogeneous, isotropic, linearly elastic material with properties characterized by Lamé constants λ_2, μ_2, and density ρ_2. The imperfect interface between the tunnel lining and its surrounding rock medium is considered.

In a fixed rectangular coordinate system (x, y, z), let u_x, u_y, and u_z be the displacements in directions x, y, and z. For the anti-plane deformation, only the out-of-plane displacement u_z is not zero, that is,

$$u_x = u_y = 0; u_z = W(x, y, t) \tag{8.1}$$

where W is the out-of-plane displacement and t is the time variable. The governing equation in the anisotropic mass in the Z plane is Equation (8.2) (Liu, 1988).

DOI: 10.1201/9781003401599-8

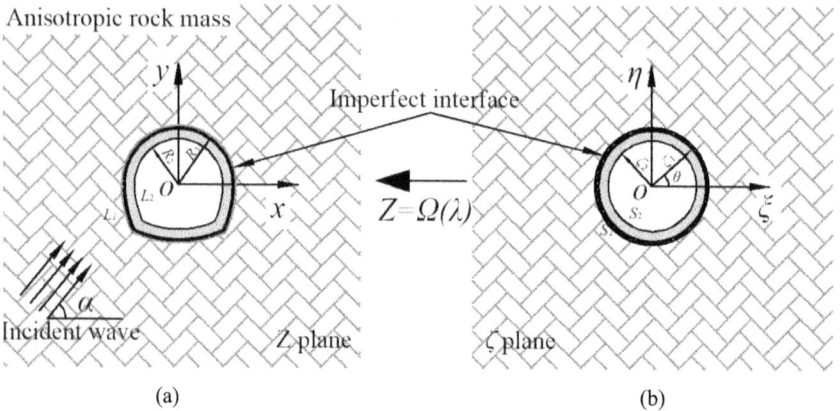

Figure 8.1 A horseshoe-shaped tunnel with an imperfect interface under an anti-plane shear wave in the z-plane (a) and its mapping ring region in the ζ plane (b).

$$\left(C_{55}-C_{44}+2iC_{45}\right)\frac{\partial^2 W}{\partial z^2}+2\left(C_{44}+C_{55}\right)\frac{\partial^2 W}{\partial z\partial\bar{z}}+\left(C_{55}-C_{44}-2iC_{45}\right)\frac{\partial^2 W}{\partial\bar{z}^2}=\rho_1\frac{\partial^2 W}{\partial t^2} \quad (8.2)$$

where $z = x+iy$ and $\bar{z} = x-iy$ are a complex variable and its complex conjugate.

To express the analytical solution of wave fields around the tunnel with different cross-section shapes, the conforming mapping method of a complex function is employed as Equation (8.3).

$$Z=\Omega(\zeta)=c\left(\zeta+\sum_{m=0}^{N}c_m\zeta^{-m}\right),\zeta=\xi+i\eta=\rho e^{i\theta} \quad (8.3)$$

where c and c_m are coefficients of the complex function, ξ and η are the rectangular coordinate axes in the ζ plane, ρ and θ denote the polar coordinates in the ζ plane and $i^2 = -1$. The mapping function transforms the boundaries L_1 and L_2 (Figure 8.1a) in the Z plane into two concentric circles S_1 and S_2 (Figure 8.1b) with ζ_2 (<1) and unit radius (ζ_1) in the ζ plane.

Submitting Equation (8.3) into Equation (8.2), the governing equation becomes Equation (8.4).

$$\left(C_{55}-C_{44}+2iC_{45}\right)\frac{1}{\Omega'(\zeta)}\frac{\partial}{\partial\zeta}\left(\frac{1}{\Omega'(\zeta)}\frac{\partial W}{\partial\zeta}\right)+2\left(C_{44}+C_{55}\right)\frac{1}{\Omega'(\zeta)}\frac{\partial}{\partial\zeta}\left(\frac{1}{\overline{\Omega'(\zeta)}}\frac{\partial W}{\partial\bar\zeta}\right)+$$

$$\left(C_{55}-C_{44}-2iC_{45}\right)\frac{1}{\overline{\Omega'(\zeta)}}\frac{\partial}{\partial\bar\zeta}\left(\frac{1}{\overline{\Omega'(\zeta)}}\frac{\partial W}{\partial\bar\zeta}\right)=\rho_1\frac{\partial^2 W}{\partial t^2} \quad (8.4)$$

Additional variables on the complex function are introduced as Equation (8.5).

$$\left. \begin{aligned} \chi &= \frac{1}{2}[(1-i\gamma)z + (1+i\gamma)\bar{z}] \\ \bar{\chi} &= \frac{1}{2}[(1-i\bar{\gamma})z + (1+i\bar{\gamma})\bar{z}] \end{aligned} \right\} \tag{8.5}$$

where γ is a complex constant determined by $\gamma = -\dfrac{C_{45}}{C_{44}} + \dfrac{i\sqrt{C_{55}C_{44} - C_{45}^2}}{C_{44}}$ and $\bar{\gamma}$ is the complex conjugate of γ.

Then, considering the steady solution, that is, $W = we^{-i\omega t}$ (ω is the circular frequency of the SH wave, and w is the anti-plane deformation ignoring the time variable t), and substituting Equation (8.5) into Equation (8.4), the governing equation in the ζ plane is rewritten as Equation (8.6).

$$\frac{\partial^2 w}{\partial \chi \partial \bar{\chi}} = \left(\frac{ik_1}{2}\right)^2 w \tag{8.6}$$

where $k_1 = \dfrac{\omega}{c_1}$ with $c_1 = \dfrac{\mu_1}{\rho_1}$ and $\mu_1 = \dfrac{C_{44}C_{55} - C_{45}^2}{C_{44}}$. It is noted that the time-dependent term $e^{-i\omega t}$ of the incident wave and the following wave fields is omitted.

8.2 WAVE FIELDS IN THE ANISOTROPIC ROCK MASS

8.2.1 Seismic Wave Potentials

The anti-plane SH wave with incidence angle α is expanded in a complex form in the ζ plane as Equation (8.7).

$$w^{(in)} = w_0 \sum_{n=-\infty}^{\infty} i^n J_n\left(k_{in}|\Omega(\zeta)|\right) \left[\frac{\Omega(\zeta)}{|\Omega(\zeta)|}\right]^n e^{-in\alpha} \tag{8.7}$$

where w_0 denotes an amplitude of the incident SH wave, k_{in} is the wave number in the rock mass determined by $\omega = k_{in}c_{in}$ with

$$c_{in} = \left[\frac{\left(C_{55}\cos^2\alpha + 2C_{45}\sin\alpha\cos\alpha + C_{44}\sin^2\alpha\right)}{\rho_1}\right]^{\frac{1}{2}}, \text{ and } J_n(\cdot) \text{ is the } n\text{th Bessel}$$

function of the first kind.

Following Equation (8.6), the scattered wave due to an SH wave incidence around the tunnel lining is written as Equation (8.8).

$$w_1^{(sc)} = \sum_{n=-\infty}^{\infty} A_n H_n^{(1)}\left(k_1 \mid \chi \mid\right)\left[\frac{\chi}{\mid \chi \mid}\right]^n \tag{8.8}$$

where A_n is an uncertain coefficient of the scattered wave around tunnel lining, $H_n^{(1)}(\cdot)$ is the Hankel function of the first kind and the nth order denoting the outgoing wave.

8.2.2 Seismic Stresses

Using Hooke's law, the relationships between stress and strain in the rectangular coordinate are

$$\tau_{xz} = C_{55}\frac{\partial w}{\partial x} + C_{45}\frac{\partial w}{\partial y}$$
$$\tau_{yz} = C_{45}\frac{\partial w}{\partial x} + C_{44}\frac{\partial w}{\partial y} \tag{8.9}$$

By using the relations $z = x + iy$ and $\bar{z} = x - iy$, stresses in complex plane can be written as Equation (8.10).

$$\tau_{xz} = \left(C_{55} + iC_{45}\right)\frac{\partial w}{\partial z} + \left(C_{55} - iC_{45}\right)\frac{\partial w}{\partial \bar{z}}$$
$$\tau_{yz} = \left(C_{45} + iC_{44}\right)\frac{\partial w}{\partial z} + \left(C_{45} - iC_{44}\right)\frac{\partial w}{\partial \bar{z}} \tag{8.10}$$

In polar coordinates, the stresses in Equation (8.10) are written as Equation (8.11).

$$\tau_{rz} = \frac{1}{2}\left[\left(C_{55} + C_{44}\right)\frac{\partial w}{\partial z} + \left(C_{55} - C_{44} - 2iC_{45}\right)\frac{\partial w}{\partial \bar{z}}\right]e^{i\theta}$$
$$+ \frac{1}{2}\left[\left(C_{55} - C_{44} + 2iC_{45}\right)\frac{\partial w}{\partial z} + \left(C_{55} + C_{44}\right)\frac{\partial w}{\partial \bar{z}}\right]e^{-i\theta}$$
$$\tau_{\theta z} = \frac{1}{2}\left[i\left(C_{55} + C_{44}\right)\frac{\partial w}{\partial z} + \left(2C_{45} - i(C_{55} - C_{44})\right)\frac{\partial w}{\partial \bar{z}}\right]e^{i\theta}$$
$$+ \frac{1}{2}\left[\left(2C_{45} + i(C_{44} - C_{55})\right)\frac{\partial w}{\partial z} - i\left(C_{55} + C_{44}\right)\frac{\partial w}{\partial \bar{z}}\right]e^{-i\theta} \tag{8.11}$$

The stresses in ζ plane are expressed as Equations (8.12) and (8.13).

$$\tau_{rz} = \frac{1}{4}\left\{\begin{array}{l} \left[(C_{55}+C_{44})(1-i\gamma)+(C_{55}-C_{44}-2iC_{45})(1+i\gamma)\right]\dfrac{\partial w}{\partial\zeta} \\[2mm] +\left[(C_{55}+C_{44})(1-i\bar{\gamma})+(C_{55}-C_{44}-2iC_{45})(1+i\bar{\gamma})\right]\dfrac{\partial w}{\partial\bar{\zeta}} \end{array}\right\}e^{i\theta}$$

$$+\frac{1}{4}\left\{\begin{array}{l} \left[(C_{55}-C_{44}+2iC_{45})(1-i\gamma)+(C_{55}+C_{44})(1+i\gamma)\right]\dfrac{\partial w}{\partial\zeta} \\[2mm] +\left[(C_{55}-C_{44}+2iC_{45})(1-i\bar{\gamma})+(C_{55}+C_{44})(1+i\bar{\gamma})\right]\dfrac{\partial w}{\partial\bar{\zeta}} \end{array}\right\}e^{-i\theta} \tag{8.12}$$

$$\tau_{\theta z} = \frac{1}{4}\left\{\begin{array}{l} \left[i(C_{55}+C_{44})(1-i\gamma)+(2C_{45}+i(C_{55}-C_{44}))(1+i\gamma)\right]\dfrac{\partial w}{\partial\zeta} \\[2mm] +\left[i(C_{55}+C_{44})(1-i\bar{\gamma})+(2C_{45}+i(C_{55}-C_{44}))(1+i\bar{\gamma})\right]\dfrac{\partial w}{\partial\bar{\zeta}} \end{array}\right\}e^{i\theta}$$

$$+\frac{1}{4}\left\{\begin{array}{l} \left[(2C_{45}+i(C_{44}-C_{55}))(1-i\gamma)-i(C_{55}+C_{44})(1+i\gamma)\right]\dfrac{\partial w}{\partial\zeta} \\[2mm] +\left[(2C_{45}+i(C_{44}-C_{55}))(1-i\bar{\gamma})-i(C_{55}+C_{44})(1+i\bar{\gamma})\right]\dfrac{\partial w}{\partial\bar{\zeta}} \end{array}\right\}e^{-i\theta} \tag{8.13}$$

By using Equations (8.12) and (8.13), the stresses in ζ plane resulting from the incident wave in Equation (8.7) are expressed as Equations (8.14) and (8.15).

$$\tau_{rz,1}^{(in)} = \frac{C_{55}k_{in}w_0}{4}\left\{\begin{array}{l} \left[\begin{array}{l}(1+\beta)\displaystyle\sum_{n=-\infty}^{\infty} i^n J_{n-1}(k_{in}|\Omega(\zeta)|)\left[\dfrac{\Omega(\zeta)}{|\Omega(\zeta)|}\right]^{n-1}e^{-in\alpha} - \\[4mm] (1-\beta-2i\kappa)\displaystyle\sum_{n=-\infty}^{\infty} i^n J_{n+1}(k_{in}|\Omega(\zeta)|)\left[\dfrac{\Omega(\zeta)}{|\Omega(\zeta)|}\right]^{n+1}e^{-in\alpha}\end{array}\right]\dfrac{\zeta}{R}\dfrac{\Omega'(\zeta)}{|\Omega'(\zeta)|} \\[12mm] +\left[\begin{array}{l}(1-\beta+2i\kappa)\displaystyle\sum_{n=-\infty}^{\infty} i^n J_{n-1}(k_{in}|\Omega(\zeta)|)\left[\dfrac{\Omega(\zeta)}{|\Omega(\zeta)|}\right]^{n-1}e^{-in\alpha} - \\[4mm] (1+\beta)\displaystyle\sum_{n=-\infty}^{\infty} i^n J_{n+1}(k_{in}|\Omega(\zeta)|)\left[\dfrac{\Omega(\zeta)}{|\Omega(\zeta)|}\right]^{n+1}e^{-in\alpha}\end{array}\right]\dfrac{\bar{\zeta}}{R}\dfrac{\overline{\Omega'(\zeta)}}{|\Omega'(\zeta)|} \end{array}\right\} \tag{8.14}$$

$$\tau_{\theta z,1}^{(in)} = \frac{C_{55}k_{in}w_0}{4}\left\{\begin{array}{l} \left[\begin{array}{l}i(1+\beta)\displaystyle\sum_{n=-\infty}^{\infty} (i)^n J_{n-1}(k_{in}|\Omega(\zeta)|)\left[\dfrac{\Omega(\zeta)}{|\Omega(\zeta)|}\right]^{n-1}e^{-in\alpha} - \\[4mm] (2\kappa+i(1-\beta))\displaystyle\sum_{n=-\infty}^{\infty} (i)^n J_{n+1}(k_{in}|\Omega(\zeta)|)\left[\dfrac{\Omega(\zeta)}{|\Omega(\zeta)|}\right]^{n+1}e^{-in\alpha}\end{array}\right]\dfrac{\zeta}{R}\dfrac{\Omega'(\zeta)}{|\Omega'(\zeta)|} \\[12mm] +\left[\begin{array}{l}(2\kappa-i(1-\beta))\displaystyle\sum_{n=-\infty}^{\infty} (i)^n J_{n-1}(k_{in}|\Omega(\zeta)|)\left[\dfrac{\Omega(\zeta)}{|\Omega(\zeta)|}\right]^{n-1}e^{-in\alpha} - \\[4mm] +i(1+\beta)\displaystyle\sum_{n=-\infty}^{\infty} (i)^n J_{n+1}(k_{in}|\Omega(\zeta)|)\left[\dfrac{\Omega(\zeta)}{|\Omega(\zeta)|}\right]^{n+1}e^{-in\alpha}\end{array}\right]\dfrac{\bar{\zeta}}{R}\dfrac{\overline{\Omega'(\zeta)}}{|\Omega'(\zeta)|} \end{array}\right\} \tag{8.15}$$

where $\beta = C_{44}/C_{55}$ and $\kappa = C_{45}/C_{55}$ represent the rock anisotropy degree. It is assumed that C_{44} is less than C_{55}, that is $\beta < 1$. Then, with a higher κ and smaller β, the medium has a higher anisotropy degree. Since the elastic parameters C_{55}, C_{44}, and C_{45} satisfy the positive definite condition of the elastic matric (that is the positive strain energy density condition), β and κ should meet the condition $\beta - \kappa^2 > 0$. The rock medium is isotropic in case of $\beta = 1.0$ and $\kappa = 0.0$, orthotropic in the case of $\beta \neq 1.0$ and $\kappa = 0$ and anisotropic in the case of $\beta \neq 1.0$ and $\kappa \neq 0$.

By using Equations (8.12) and (8.13), the stresses resulting from the scattered wave in Equation (8.8) are expressed as Equations (8.16) and (8.17).

$$
\tau_{rz,1}^{(sc)} = \frac{C_{55}k_1}{4} \left\{
\begin{bmatrix}
(a+ic)\sum_{n=-\infty}^{\infty}A_n H_{n-1}^{(1)}(k_1|\chi|)\left[\frac{\chi}{|\chi|}\right]^{n-1} - \\
(b-ic)\sum_{n=-\infty}^{\infty}A_n H_{n+1}^{(1)}(k_1|\chi|)\left[\frac{\chi}{|\chi|}\right]^{n+1}
\end{bmatrix}\frac{\zeta}{R}\frac{\Omega'(\zeta)}{|\Omega'(\zeta)|}
+ \begin{bmatrix}
(b+ic)\sum_{n=-\infty}^{\infty}A_n H_{n-1}^{(1)}(k_1|\chi|)\left[\frac{\chi}{|\chi|}\right]^{n-1} - \\
(a-ic)\sum_{n=-\infty}^{\infty}A_n H_{n+1}^{(1)}(k_1|\chi|)\left[\frac{\chi}{|\chi|}\right]^{n+1}
\end{bmatrix}\frac{\bar{\zeta}}{R}\frac{\overline{\Omega'(\zeta)}}{|\Omega'(\zeta)|}
\right\}
\tag{8.16}
$$

$$
\tau_{\theta z,1}^{(sc)} = \frac{C_{55}k_1}{4} \left\{
\begin{bmatrix}
(-c+ia)\sum_{n=-\infty}^{\infty}A_n H_{n-1}^{(1)}(k_1|\chi|)\left[\frac{\chi}{|\chi|}\right]^{n-1} - \\
(c+ib)\sum_{n=-\infty}^{\infty}A_n H_{n+1}^{(1)}(k_1|\chi|)\left[\frac{\chi}{|\chi|}\right]^{n+1}
\end{bmatrix}\frac{\zeta}{R}\frac{\Omega'(\zeta)}{|\Omega'(\zeta)|}
+ \begin{bmatrix}
(c-ib)\sum_{n=-\infty}^{\infty}A_n H_{n-1}^{(1)}(k_1|\chi|)\left[\frac{\chi}{|\chi|}\right]^{n-1} + \\
(c+ia)\sum_{n=-\infty}^{\infty}A_n H_{n+1}^{(1)}(k_1|\chi|)\left[\frac{\chi}{|\chi|}\right]^{n+1}
\end{bmatrix}\frac{\bar{\zeta}}{R}\frac{\overline{\Omega'(\zeta)}}{|\Omega'(\zeta)|}
\right\}
\tag{8.17}
$$

where $a = \sqrt{C_{55}C_{44} - C_{45}^2}\left(1 + \dfrac{\sqrt{C_{55}C_{44} - C_{45}^2}}{C_{44}}\right)$, $b = -\sqrt{C_{55}C_{44} - C_{45}^2}\left(1 - \dfrac{\sqrt{C_{55}C_{44} - C_{45}^2}}{C_{44}}\right)$ and

$c = \dfrac{C_{45}\sqrt{C_{55}C_{44} - C_{45}^2}}{C_{44}}$.

8.3 WAVE FIELDS IN THE ISOTROPIC TUNNEL LINING

8.3.1 Seismic Wave Potentials

The reflected wave that propagates outwards from the lining inner boundary is expressed in Equation (8.18).

$$w_2^{(rf)} = \sum_{n=-\infty}^{\infty} B_n H_n^{(1)}\left(k_2\,|\Omega(\zeta)|\right)\left[\frac{\Omega(\zeta)}{|\Omega(\zeta)|}\right]^n \tag{8.18}$$

where B_n is an uncertain coefficient of the reflected wave in the tunnel lining and k_2 is the wave number in the tunnel lining determined by $\omega = k_2 c_2$ with $c_2 = \sqrt{\dfrac{u_2}{\rho_2}}$.

The refracted wave being confined into the tunnel lining is expressed in Equation (8.19).

$$w_2^{(rr)} = \sum_{n=-\infty}^{\infty} C_n H_n^{(2)}\left(k_2\,|\Omega(\zeta)|\right)\left[\frac{\Omega(\zeta)}{|\Omega(\zeta)|}\right]^n \tag{8.19}$$

where C_n is an uncertain coefficient of the refracted wave in the tunnel lining, $H_n^{(2)}(\cdot)$ is the Hankel function of the second kind and the nth order denoting the ingoing wave.

8.3.2 Seismic Stresses

By using Equations (8.12) and (8.13), the stresses resulting from the reflected wave in Equation (8.18) are expressed as Equations (8.20) and (8.21).

$$\tau_{rz,2}^{(rf)} = \frac{\mu_2 k_2}{2}\left\{
\begin{aligned}
&\sum_{n=-\infty}^{\infty} B_n H_{n-1}^{(1)}\left(k_2|\Omega(\zeta)|\right)\left[\frac{\Omega(\zeta)}{|\Omega(\zeta)|}\right]^{n-1}\frac{\zeta}{R}\frac{\Omega'(\zeta)}{|\Omega'(\zeta)|} \\
&-\sum_{n=-\infty}^{\infty} B_n H_{n+1}^{(1)}\left(k_2|\Omega(\zeta)|\right)\left[\frac{\Omega(\zeta)}{|\Omega(\zeta)|}\right]^{n+1}\frac{\overline{\zeta}}{R}\frac{\overline{\Omega'(\zeta)}}{|\Omega'(\zeta)|}
\end{aligned}
\right\} \tag{8.20}$$

$$\tau_{\theta z,2}^{(rf)} = \frac{i\mu_2 k_2}{2}\left\{
\begin{aligned}
&\sum_{n=-\infty}^{\infty} B_n H_{n-1}^{(1)}\left(k_2|\Omega(\zeta)|\right)\left[\frac{\Omega(\zeta)}{|\Omega(\zeta)|}\right]^{n-1}\frac{\zeta}{R}\frac{\Omega'(\zeta)}{|\Omega(\zeta)|} \\
&+\sum_{n=-\infty}^{\infty} B_n H_{n+1}^{(1)}\left(k_2|\Omega(\zeta)|\right)\left[\frac{\Omega(\zeta)}{|\Omega(\zeta)|}\right]^{n+1}\frac{\overline{\zeta}}{R}\frac{\overline{\Omega'(\zeta)}}{|\Omega'(\zeta)|}
\end{aligned}
\right\} \tag{8.21}$$

By using Equations (8.12) and (8.13), the stresses resulting from the refracted wave in Equation (8.19) are written as Equations (8.22) and (8.23).

$$
\tau_{rz,2}^{(rr)} = \frac{\mu_2 k_2}{2} \left\{ \begin{array}{l} \displaystyle\sum_{n=-\infty}^{\infty} C_n H_{n-1}^{(2)}\left(k_2 |\Omega(\varsigma)|\right) \left[\frac{\Omega(\varsigma)}{|\Omega(\varsigma)|}\right]^{n-1} \frac{\varsigma}{R} \frac{\Omega'(\varsigma)}{|\Omega'(\varsigma)|} - \\[4mm] \displaystyle\sum_{n=-\infty}^{\infty} C_n H_{n+1}^{(2)}\left(k_2 |\Omega(\varsigma)|\right) \left[\frac{\Omega(\varsigma)}{|\Omega(\varsigma)|}\right]^{n+1} \frac{\overline{\varsigma}}{R} \frac{\overline{\Omega'(\varsigma)}}{|\Omega'(\varsigma)|} \end{array} \right\} \tag{8.22}
$$

$$
\tau_{\theta z,2}^{(rr)} = \frac{i\mu_2 k_2}{2} \left\{ \begin{array}{l} \displaystyle\sum_{n=-\infty}^{\infty} C_n H_{n-1}^{(2)}\left(k_2 |\Omega(\varsigma)|\right) \left[\frac{\Omega(\varsigma)}{|\Omega(\varsigma)|}\right]^{n-1} \frac{\varsigma}{R} \frac{\Omega'(\varsigma)}{|\Omega'(\varsigma)|} + \\[4mm] \displaystyle\sum_{n=-\infty}^{\infty} C_n H_{n+1}^{(2)}\left(k_2 |\Omega(\varsigma)|\right) \left[\frac{\Omega(\varsigma)}{|\Omega(\varsigma)|}\right]^{n+1} \frac{\overline{\varsigma}}{R} \frac{\overline{\Omega'(\varsigma)}}{|\Omega'(\varsigma)|} \end{array} \right\} \tag{8.23}
$$

8.4 BOUNDARY CONDITION

The total wave fields in the anisotropic rock mass are produced by the superposition of the incident wave and the scattered wave expressed as Equation (8.24).

$$
w_1^{(t)} = w^{(in)} + w_1^{(sc)} \tag{8.24}
$$

The total wave fields in the isotropic tunnel lining are produced by the superposition of the refracted wave and reflected wave expressed as Equation (8.25).

$$
w_2^{(t)} = w_2^{(rr)} + w_2^{(rf)} \tag{8.25}
$$

The same elastic model as discussed in Chapter 3 is introduced to analyze the imperfect interface influence. With the elastic model, tractions at the lining outer boundary ($R = R_1$, L_1 in Figure 8.1a) are continuous, but displacements are discontinuous across the interface, described as Equation (8.26).

$$
\left. \begin{array}{l} w_1^{(t)}\Big|_{R=R1} - w_2^{(t)}\Big|_{R=R_1} = \dfrac{\tau_{rz,1}^{(t)}}{k}\bigg|_{R=R_1} \\[4mm] \tau_{rz,1}^{(t)}\Big|_{R=R_1} = \tau_{rz,2}^{(t)}\Big|_{R=R_1} \end{array} \right\} \tag{8.26}
$$

where k is the imperfect interface stiffness, $\tau_{rz,1}^{(t)} = \tau_{rz,1}^{(in)} + \tau_{rz,1}^{(sc)}$, and $\tau_{rz,2}^{(t)} = \tau_{rz,2}^{(rr)} + \tau_{rz,2}^{(rf)}$.

At the lining inner boundary $(R = R_2, L_2$ in Figure 8.1a), tractions are zero expressed as Equation (8.27).

$$\tau_{rz,2}^{(t)}\Big|_{R=R_2} = 0 \qquad (8.27)$$

where $\tau_{rz,2}^{(t)} = \tau_{rz,2}^{(rr)} + \tau_{rz,2}^{(rf)}$.

Substituting Equations (8.24) and (4.25) into Equations (8.26) and (8.27), it can be obtained that

$$\sum_{m=1}^{3} \sum_{n=-\infty}^{\infty} K_n^{mj} X_n^m = Q^j \ (j = 1, 2, 3) \qquad (8.28)$$

where $X_n^1 = A_n$, $X_n^2 = B_n$, $X_n^3 = C_n$, K_n^{mj} and Q^j are presented in Appendix C.

By multiplying both sides of Equation (8.28) with the orthogonality of $e^{-is\theta}$ $(s = 0, \pm1, \pm2, \pm3 \ldots)$ and integrating over the interval $[-\pi, \pi]$, the unknown coefficients A_n, B_n, and C_n can be determined straightforwardly by solving a set of the infinite linear algebraic system with the expression as Equation (8.29).

$$\sum_{m=1}^{3} \sum_{n=-\infty}^{\infty} K_{ns}^{mj} X_n^m = Q_s^j \ (j = 1, 2, 3) \qquad (8.29)$$

where $K_{ns}^{mj} = \dfrac{1}{2\pi} \int_{-\pi}^{\pi} K_n^{mj} e^{-is\theta} d\theta$ and $Q_s^j = \dfrac{1}{2\pi} \int_{-\pi}^{\pi} Q^j e^{-is\theta} d\theta$.

8.5 FUNDAMENTAL SOLUTION

To analyze the influences of rock anisotropy and imperfect interface on the dynamic response around a deep underground tunnel subjected to an anti-plane SH wave, the dimensionless dynamic stress concentration factor (DSCF) is introduced. According to the definition of DSCF by Pao and Mow (1973), DSCF of the circumferential stress around the tunnel is written as Equation (8.30).

$$DSCF = \tau_{\theta z,p}^* = \left| \frac{\tau_{\theta z,p}^{(t)}}{\tau_0} \right| \quad \begin{matrix} (p = 1 \text{ for the rock mass and} \\ p = 2 \text{ for the tunnel lining}) \end{matrix} \qquad (8.30)$$

where, τ_0 is the maximum magnitude of the incident stress defined as $\tau_0 = C_{55} k_{in} w_0$, and $\tau_{\theta z,1}^{(t)} = \tau_{\theta z,1}^{(in)} + \tau_{\theta z,1}^{(sc)}$, $\tau_{\theta z,2}^{(t)} = \tau_{\theta z,2}^{(rr)} + \tau_{\theta z,2}^{(rf)}$.

The circumferential stress along the interface of the anisotropic rock mass and the tunnel lining is as Equation (8.31).

$$
\begin{aligned}
\tau_{\theta z,1}^{(t)} = \frac{C_{55}k_{in}w_0}{4} &\left\{
\begin{bmatrix}
\left[i(1+\beta) \displaystyle\sum_{n=-\infty}^{\infty} i^n J_{n-1}\left(k_{in}\left|\Omega(\varsigma_1)\right|\right)\left[\dfrac{\Omega(\varsigma_1)}{\left|\Omega(\varsigma_1)\right|}\right]^{n-1} e^{-in\alpha} \right. \\[2ex]
\left. -i(1-\beta-2i\kappa) \displaystyle\sum_{n=-\infty}^{\infty} i^n J_{n+1}\left(k_{in}\left|\Omega(\varsigma_1)\right|\right)\left[\dfrac{\Omega(\varsigma_1)}{\left|\Omega(\varsigma_1)\right|}\right]^{n+1} e^{-in\alpha} \right] \dfrac{\varsigma_1}{R_1}\dfrac{\Omega'(\varsigma_1)}{\left|\Omega'(\varsigma_1)\right|} \\[3ex]
+\left[-i(1-\beta+2i\kappa) \displaystyle\sum_{n=-\infty}^{\infty} i^n J_{n-1}\left(k_{in}\left|\Omega(\varsigma_1)\right|\right)\left[\dfrac{\Omega(\varsigma_1)}{\left|\Omega(\varsigma_1)\right|}\right]^{n-1} e^{-in\alpha} \right. \\[2ex]
\left. +i(1+\beta) \displaystyle\sum_{n=-\infty}^{\infty} i^n J_{n+1}\left(k_{in}\left|\Omega(\varsigma_1)\right|\right)\left[\dfrac{\Omega(\varsigma_1)}{\left|\Omega(\varsigma_1)\right|}\right]^{n+1} e^{-in\alpha} \right] \dfrac{\overline{\varsigma_1}}{R_1}\dfrac{\overline{\Omega'(\varsigma_1)}}{\left|\Omega'(\varsigma_1)\right|}
\end{bmatrix}
\right\} \\[4ex]
+\frac{C_{55}k_1}{4} &\left\{
\begin{bmatrix}
\left[(-c+ia) \displaystyle\sum_{n=-\infty}^{\infty} A_n H_{n-1}^{(1)}\left(k_1\left|\chi(\varsigma_1)\right|\right)\left[\dfrac{\chi(\varsigma_1)}{\left|\chi(\varsigma_1)\right|}\right]^{n-1} \right. \\[2ex]
\left. -(c+ib) \displaystyle\sum_{n=-\infty}^{\infty} A_n H_{n+1}^{(1)}\left(k_1\left|\chi(\varsigma_1)\right|\right)\left[\dfrac{\chi(\varsigma_1)}{\left|\chi(\varsigma_1)\right|}\right]^{n+1} \right] \dfrac{\varsigma_1}{R_1}\dfrac{\Omega'(\varsigma_1)}{\left|\Omega'(\varsigma_1)\right|} \\[3ex]
+\left[(c+ib) \displaystyle\sum_{n=-\infty}^{\infty} A_n H_{n-1}^{(1)}\left(k_1\left|\chi(\varsigma_1)\right|\right)\left[\dfrac{\chi(\varsigma_1)}{\left|\chi(\varsigma_1)\right|}\right]^{n-1} \right. \\[2ex]
\left. +(c+ia) \displaystyle\sum_{n=-\infty}^{\infty} A_n H_{n+1}^{(1)}\left(k_1\left|\chi(\varsigma_1)\right|\right)\left[\dfrac{\chi(\varsigma_1)}{\left|\chi(\varsigma_1)\right|}\right]^{n+1} \right] \dfrac{\overline{\varsigma_1}}{R_1}\dfrac{\overline{\Omega'(\varsigma_1)}}{\left|\Omega'(\varsigma_1)\right|}
\end{bmatrix}
\right\}
\end{aligned}
\tag{8.31}
$$

The circumferential stresses of the lining's outer ($g = 1$) and inner ($g = 2$) sides are written as Equation (8.32).

$$
\begin{aligned}
\tau_{\theta z,2}^{(t),g} = \frac{i\mu_2 k_2}{2} &\left\{
\begin{bmatrix}
\displaystyle\sum_{n=-\infty}^{\infty} B_n H_{n-1}^{(1)}\left(k_2\left|\Omega(\varsigma_g)\right|\right)\left[\dfrac{\Omega(\varsigma_g)}{\left|\Omega(\varsigma_g)\right|}\right]^{n-1} \\[2ex]
+\displaystyle\sum_{n=-\infty}^{\infty} C_n H_{n-1}^{(2)}\left(k_2\left|\Omega(\varsigma_g)\right|\right)\left[\dfrac{\Omega(\varsigma_g)}{\left|\Omega(\varsigma_g)\right|}\right]^{n-1}
\end{bmatrix} \dfrac{\overline{\varsigma_g}}{R_g}\dfrac{\overline{\Omega'(\varsigma_g)}}{\left|\Omega'(\varsigma_g)\right|} \right. + \frac{i\mu_2 k_2}{2} \\[4ex]
&\left. \begin{bmatrix}
\displaystyle\sum_{n=-\infty}^{\infty} B_n H_{n+1}^{(1)}\left(k_2\left|\Omega(\varsigma_g)\right|\right)\left[\dfrac{\Omega(\varsigma_g)}{\left|\Omega(\varsigma_g)\right|}\right]^{n+1} \\[2ex]
+\displaystyle\sum_{n=-\infty}^{\infty} C_n H_{n-1}^{(2)}\left(k_2\left|\Omega(\varsigma_g)\right|\right)\left[\dfrac{\Omega(\varsigma_g)}{\left|\Omega(\varsigma_g)\right|}\right]^{n+1}
\end{bmatrix} \dfrac{\overline{\varsigma_g}}{R_g}\dfrac{\Omega'(\varsigma_g)}{\left|\Omega'(\varsigma_g)\right|} \right\}
\end{aligned}
\tag{8.32}
$$

Two dimensionless variables are used for simplification: the incident wave number $k_a^* = k_{in}R$, and the interface stiffness parameter $k^* = kR_2/C_{55}$. In the following analysis of the chapter, the truncation number is tested to be 5 for the low frequency and 7 for the high frequency. Liu also pointed out that a smaller truncated number could be adopted when the wave frequency is smaller (Liu & Wang, 2012; Liu et al., 2013). To validate this dynamic model, a comparison with the existing approaches in the work of Fang et al. (2016) is discussed. Here, the rock mass is isotropic, the interface is perfect, and the tunnel is circular. The same parameters for tunnel lining and surrounding rock in Table 8.1 are used. The rock mass is reduced to isotropic by defining $\beta = 1.0$, $\kappa = 0.0$. A lower dimensionless wave number $k_a^* = 0.1$ and a higher dimensionless wave number $k_a^* = 1.0$ are taken into consideration, which are the same as those in the work of Fang and Jin (2016). After trial and error, the dimensionless stiffness $k^* = 10^4$ is enough to guarantee the interface is approximately perfect with the present parameters. Figure 8.2 shows the comparisons of the present solutions with those by Fang et al. (2016) for circular tunnels under horizontal SH wave in the isotropic rock mass. The reduced model is consistent with that of Fang et al. (2016). Since distributions of the dynamic stresses around the circular tunnel lining are perfectly symmetrical about the incident direction, the maximum dynamic stresses $DSCF_{max\_upper}$ and $DSCF_{max\_lower}$ near both ends of the vertical line of the incident direction are taken into consideration in the following analysis, as illustrated in Figure 8.3.

Table 8. 1 Coefficients in Conformal Mapping Function

R_1 (m)	R_2 (m)	$\bar{\rho}$	c_1	c_2	c_3	c_4	c_5	c_6
5.3	4.8	0.93	0.00240	−0.04721i	−0.04766	0.02038i	−0.00333	0.00977i

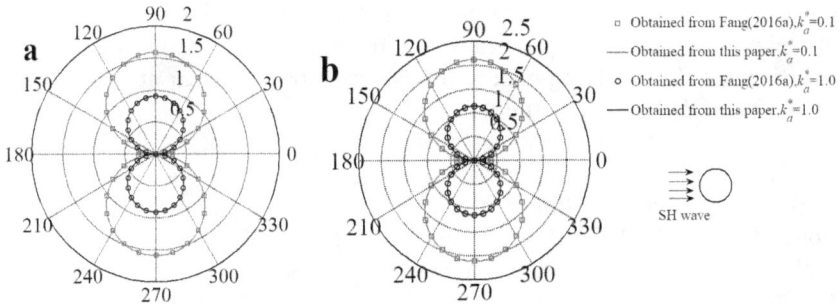

Figure 8.2 Comparisons of the present solutions with those by Fang (2016a) for a circular tunnel under horizontal SH-wave in the isotropic rock mass. (a) DSCF of the inner side of the tunnel lining; (b) DSCF of the surrounding rock mass at the interface.

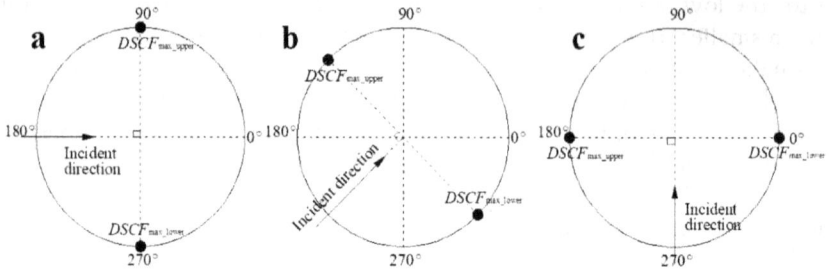

Figure 8.3 Schematic of the monitoring area of the tunnel lining subjected to horizontal incident wave (a), inclined incident wave (b), and vertical incident wave (c).

8.5.1 Case I: Rock Anisotropy Influence

For illustration of the rock anisotropy influence, two types of anisotropy degree, that is, lower degree of anisotropy (LDA) with $\beta = 0.8$, $\kappa = 0.2$ and higher degree of anisotropy (HDA) with $\beta = 0.5$, $\kappa = 0.5$, are employed. Figure 8.4 shows the mapping result and the actual dimension of the horseshoe cross-section in the Tawarayama tunnel. The inner radius of the crown is 8.8 m, and the thickness of the tunnel lining is 0.50 m. Coefficients in the conformal mapping function (Equation (8.3)) for the horseshoe-shaped tunnel lining are depicted in Table 8.1. The same parameters for tunnel lining and surrounding rock in Table 8.1 are used. Here, the elastic constant C_{55} is assumed to be equal to the elastic modulus of the Andesite lava (Shi et al., 1996). The interface between the tunnel lining and its surrounding rock/soil medium is assumed to be perfect ($k^* = 10^4$).

Figures 8.5–8.7 present the influence of the anisotropy parameters on the circumferential stresses in the lining and medium with different incidence angles. When rock mass is isotropic ($\beta = 1.0$, $\kappa = 0.0$), the DSCF distributions are bilaterally symmetric (the black line in Figure 8.5). Like the circular tunnel (Figure 8.2), the lining maximum dynamic stresses occur at the position of $\theta = 90°$ ($DSCF_{max\_upper}$) and $\theta = 270°$ ($DSCF_{max\_lower}$).

When the rock mass becomes anisotropic ($\beta \neq 1.0$, $\kappa \neq 0.0$), the DSCF distributions become asymmetric (the blue and red lines in Figure 8.5). In the LDA rock mass, the lining maximum dynamic stresses occur at the positions of $\theta = 94°$ ($DSCF_{max\_upper}$) and $\theta = 278°$ ($DSCF_{max\_lower}$). In the HDA rock mass, the lining maximum dynamic stresses occur at the positions of $\theta = 98°$ ($DSCF_{max\_upper}$) and $\theta = 286°$ ($DSCF_{max\_lower}$). In the cases of inclined and vertical incident SH wave (Figures 8.6 and 8.7), the DSCF distributions in the anisotropic rock mass also become asymmetric by comparison to those in the isotropic rock mass. Corresponding maximum dynamic stress

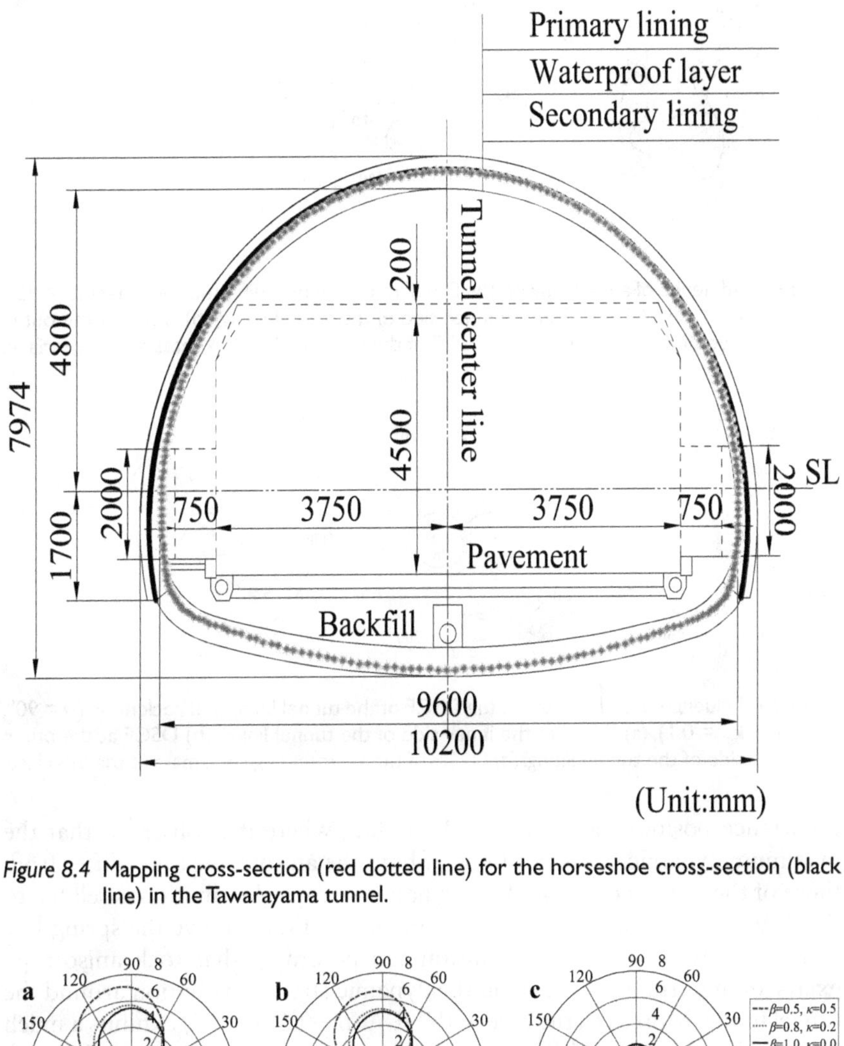

(Unit:mm)

Figure 8.4 Mapping cross-section (red dotted line) for the horseshoe cross-section (black line) in the Tawarayama tunnel.

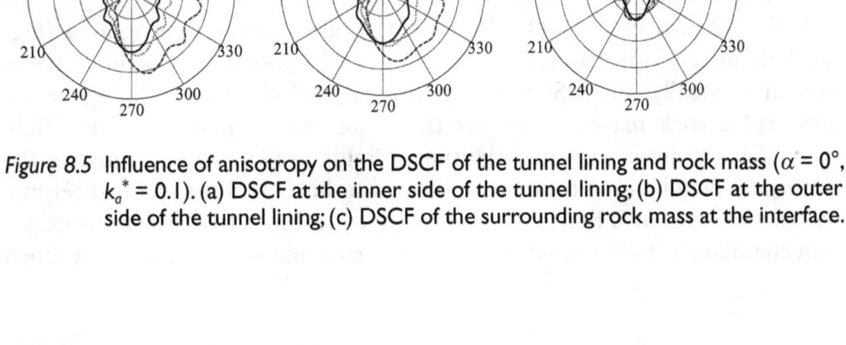

Figure 8.5 Influence of anisotropy on the DSCF of the tunnel lining and rock mass (α = 0°, k_a^* = 0.1). (a) DSCF at the inner side of the tunnel lining; (b) DSCF at the outer side of the tunnel lining; (c) DSCF of the surrounding rock mass at the interface.

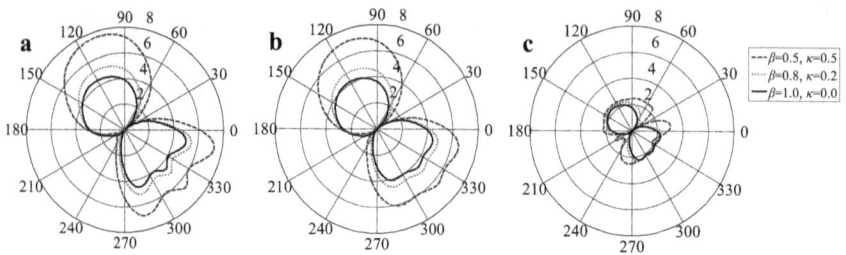

Figure 8.6 Influence of anisotropy on the DSCF of the tunnel lining and rock mass ($\alpha = 45°$, $k_a^* = 0.1$). (a) DSCF at the inner side of the tunnel lining; (b) DSCF at the outer side of the tunnel lining; (c) DSCF of the surrounding rock mass at the interface.

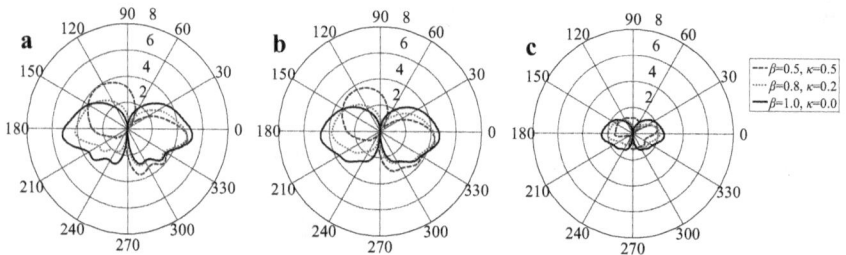

Figure 8.7 Influence of anisotropy on the DSCF of the tunnel lining and rock mass ($\alpha = 90°$, $k_a^* = 0.1$). (a) DSCF at the inner side of the tunnel lining; (b) DSCF at the outer side of the tunnel lining; (c) DSCF of the surrounding rock mass at the interface.

occurrence positions are listed in Table 8.2, where it is observed that the maximum dynamic stress positions of both the anisotropic cases differ from those of the isotropic one, and the dynamic stress redistribution is related to the degree of rock anisotropy, especially in the lining above the spring line (θ from 0° to 180°). The conclusion can be drawn that rock anisotropy exerts an important influence on the dynamic stress distribution around the tunnel lining. Moreover, the case with HDA ($\beta = 0.5$, $\kappa = 0.5$) exhibits a much larger value due to the larger difference of rock characteristics in different directions compared to the LDA case ($\beta = 0.5$, $\kappa = 0.2$).

Table 8.2 also lists the DSCF maximum near the monitoring area (Figure 8.4) in the anisotropic and isotropic rock masses. When the wave is excited vertically, the DSCF maximums around the tunnel lining in the anisotropic rock mass are smaller than those in the isotropic one. While the DSCF maximums around the tunnel lining are larger than those in the isotropic one when the tunnel is subjected to horizontal or inclined seismic waves. One reason is that rock anisotropy contributes to different propagation characteristics in each direction of the rock mass. Furthermore, a much

Table 8.2 Maximum of DSCF in the Anisotropic and Isotropic Rock Mass

Incident angle	Rock Anisotropy	$DSCF_{max\_upper}$ (L_1 side)	$DSCF_{max\_lower}$ (L_1 side)	$DSCF_{max\_upper}$ (L_2 side)	$DSCF_{max\_lower}$ (L_2 side)	θ_{max_upper} /°	θ_{max_lower} /°
$\alpha = 0°$	$\beta = 0.5$, $\kappa = 0.5$	6.63	5.57	7.12	6.20	98	286
	$\beta = 0.8$, $\kappa = 0.2$	5.29	4.30	5.70	4.76	94	278
	$\beta = 1.0$, $\kappa = 0.0$	5.07	4.06	5.46	4.48	90	270
$\alpha = 45°$	$\beta = 0.5$, $\kappa = 0.5$	7.10	6.40	7.57	6.64	107	320
	$\beta = 0.8$, $\kappa = 0.2$	4.86	4.82	5.17	5.00	116	319
	$\beta = 1.0$, $\kappa = 0.0$	4.17	4.36	4.44	4.53	123	320
$\alpha = 90°$	$\beta = 0.5$, $\kappa = 0.5$	3.68	4.35	3.93	4.86	129	352
	$\beta = 0.8$, $\kappa = 0.2$	3.80	4.32	4.18	4.81	183	354
	$\beta = 1.0$, $\kappa = 0.0$	4.70	4.70	5.21	5.21	187	355

larger maximum value can be obtained by choosing a much higher degree of anisotropy in the cases of the horizontal and inclined wave.

8.5.2 Case II: Imperfect Interface Influence

Three sets of stiffness parameters are selected and they are $k^* = 50.0, 5.0$, and 0.5, respectively, for the discussion of imperfect interface influence. Other parameters are the same as those in Case I.

Figures 8.8 and 8.9 illustrate the influence of the imperfect interface on the dynamic response of the horseshoe-shaped tunnel under the vertical incident wave with low frequency ($k_a^* = 0.1$). In a similar fashion to the circular tunnel, the maximum dynamic stresses around the tunnel occur at the positions near the top of the vault and the bottom of the invert. The interface influence at these positions is the greatest and increasing interface stiffness contributes to the decrease of the dynamic stress. Near the positions $\theta = 0°$ and $\theta = 180°$, the interface influence is not obvious on the dynamic stress. By comparing Figure 8.8 with Figure 8.9, it can be observed that a much higher degree of rock anisotropy brings an increase in the interface influence, especially at the invert.

Figures 8.10 and 8.11 illustrate the influence of the imperfect interface on the dynamic response of the horseshoe tunnel under the vertical incident wave with high frequency ($k_a^* = 1.0$). Dynamic stresses become slightly more

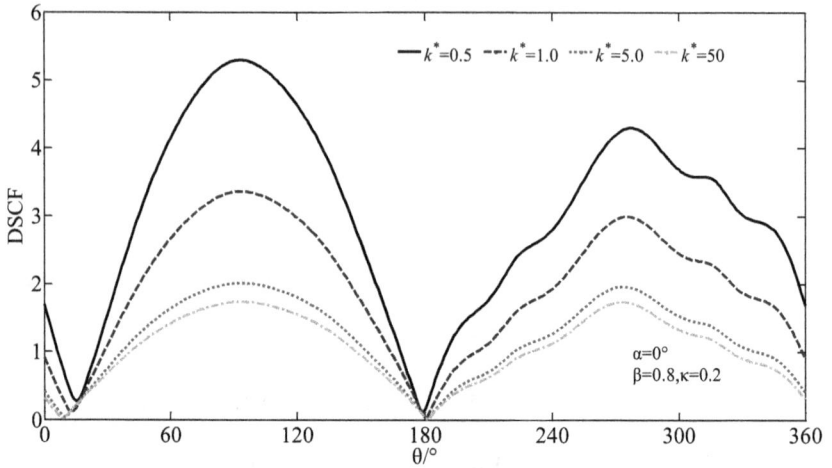

(a)

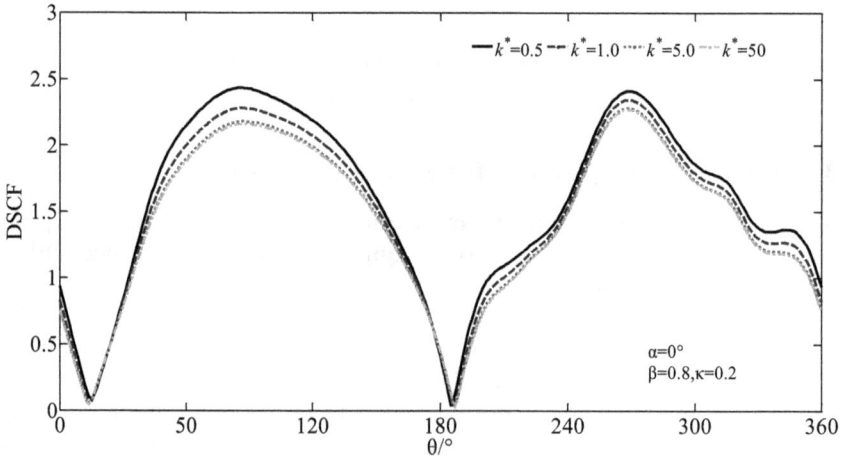

(b)

Figure 8.8 Influence of interface stiffness on DSCF at the outer side of tunnel lining (a) and rock mass at the interface (b) ($k_a^* = 0.1$, $\beta = 0.8$, $\kappa = 0.2$).

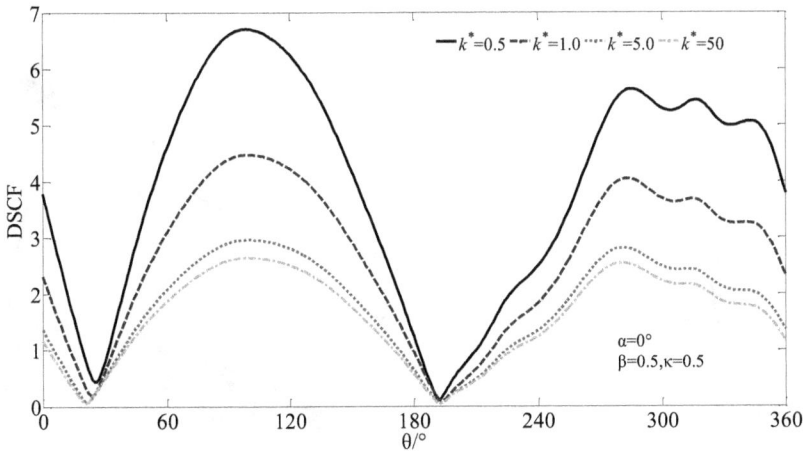

(a)

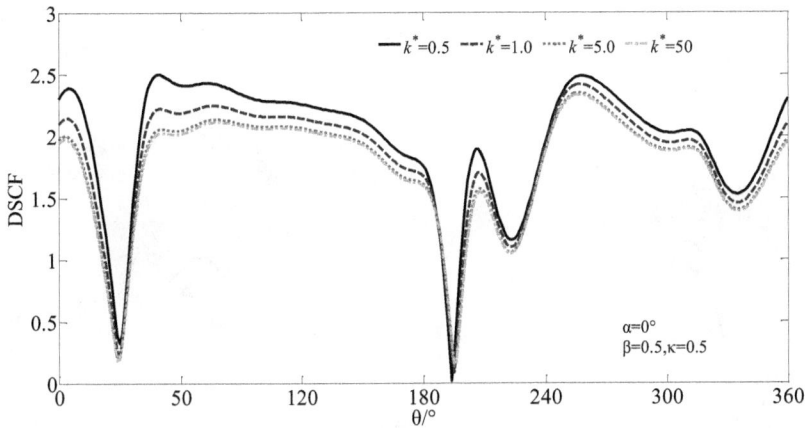

(b)

Figure 8.9 Influence of interface stiffness on DSCF at the outer side of tunnel lining (a) and rock mass at the interface (b) ($k_a^* = 0.1, \beta = 0.5, \kappa = 0.5$).

complicated with several peaks due to the high-frequency loading and the rock anisotropy. The interface influence in the region of high frequency on the dynamic stress increases significantly compared to that in the region of low frequency. From Figures 8.8–8.11, it can be observed that the interface influence is much more significant if the k^* is less than 1.0, as observed by Fang et al. (2016). Furthermore, the interface influence increases with the anisotropic characteristic of the rock mass.

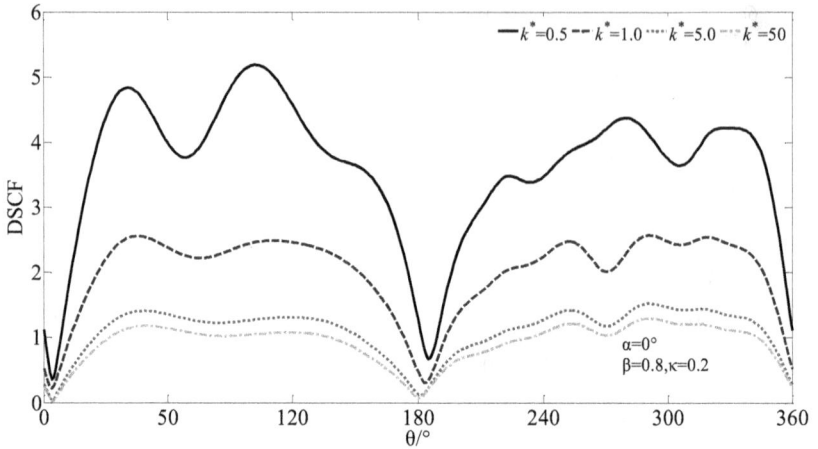

(a)

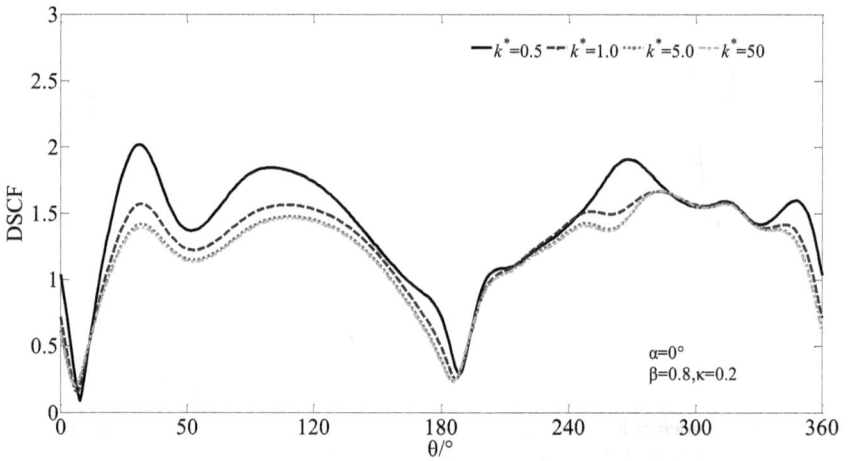

(b)

Figure 8.10 Influence of interface stiffness on DSCF of the outer side of tunnel lining (a) and rock mass at the interface (b) ($k_a^* = 1.0$, $\beta = 0.8$, $\kappa = 0.2$).

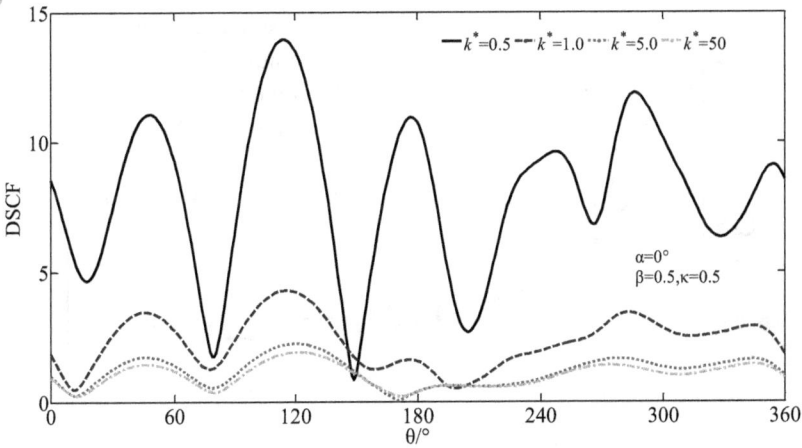

(a)

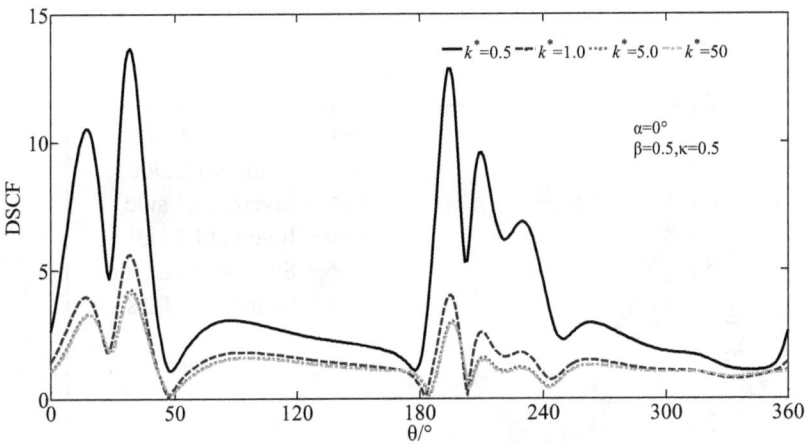

(b)

Figure 8.11 Influence of interface stiffness on DSCF of the outer side of tunnel lining (a) and rock mass at the interface (b) ($k_a^* = 1.0, \beta = 0.5, \kappa = 0.5$).

The lining DSCF variations at $\theta = 45°$ (spandrel), $\theta = 90°$ (vault) and $\theta = 270°$ (invert) with the interface stiffness k^* are presented in Figure 8.12. The DSCFs on the vault, spandrel, and invert are high when the stiffness is low. Increasing interface stiffness contributes to decreasing DSCF. As the stiffness increases, the dynamic stress concentration decreases greatly to a

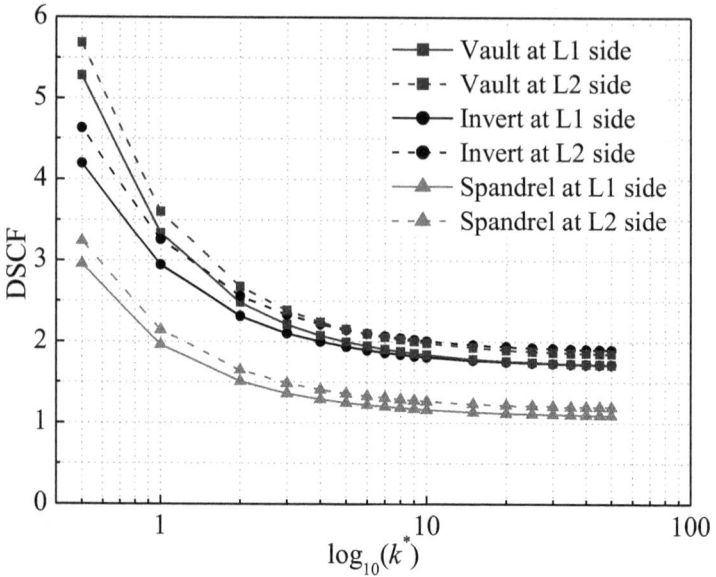

(a)

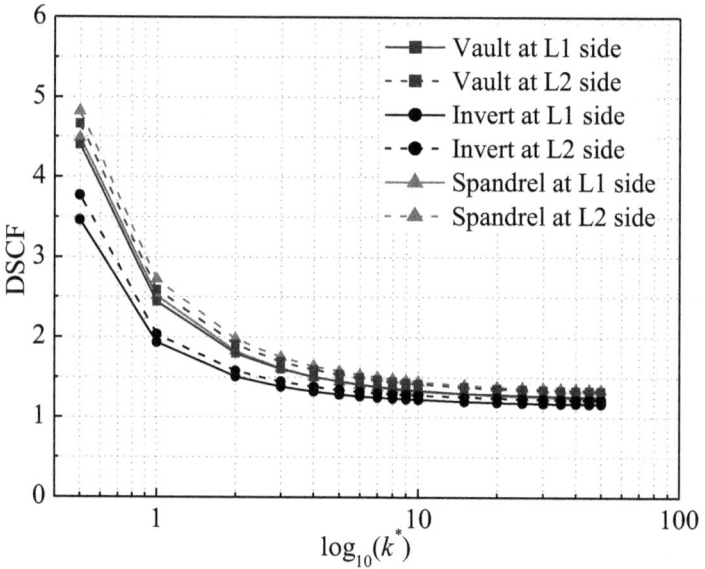

(b)

Figure 8.12 Variation of DSCF along with the interface stiffness constants k^* (β = 0.8, κ = 0.2). (a) k_a^* = 0.1; (b) k_a^* = 1.0.

steady value. Taking Figure 8.12a as an illustration, the decreasing percentages at the vault and invert of the lining inner side are 65.1% and 56.6% with the interface stiffness increasing from 0.5 to 10, while the decreasing percentages are 6.8% and 5.4% with interface stiffness increasing from 10 to 50. It can be concluded that the influence of interface stiffness decreases gradually along with the interface stiffness increasing. Thus, with a much larger interface stiffness, the interface influence can be neglected to make the interface perfect. This observation also corroborates the selection of k^* in the validation of the present analytical method.

8.6 CONCLUSION

An analytical solution for the cross-sectional seismic performance of a deep-buried tunnel in an anisotropic rock mass is proposed based on the wave function expansion method. Case studies on the influences of rock anisotropy and imperfect interface as well as the incidence angle and wave frequency are discussed. Some important conclusions are as follows.

Rock anisotropy is important in the scattering of a seismic wave. The rock anisotropy changes the shape of the DSCF distribution, unlike the symmetric distribution in the isotropic rock mass. The case with HDA exhibits much more serious asymmetry than that with LDA. Besides, different propagation characteristics in each direction due to the rock anisotropy contribute to a different DSCF maximum variation with the rock anisotropy.

Increasing interface stiffness contributes to the decrease of DSCF. When the high-frequency wave is excited, the shapes of DSCF become slightly more complicated with several peaks, compared with the smooth shape under a low-frequency incident wave. The influence of the interface between the tunnel lining and its surrounding rock mass weakens gradually along with the increase of the interface stiffness. Nevertheless, a much higher degree of rock anisotropy brings an increase in the interface influence, especially at the invert.

Chapter 9

Analytic Analysis for Cross-Sectional Seismic Response

Part II—Shallow-Buried Tunnel

9.1 GOVERNING EQUATIONS

A cross-section computation model of a shallow-buried lined circular tunnel and the adopted coordinated systems are shown in Figure 9.1. Herein, an infinitely long shallow-buried tunnel is embedded in a homogeneous, anisotropic, and linearly elastic medium. An anti-plane shear wave (SH wave) propagates into the lining's cross-section with an incidence angle α. As discussed in Chapter 8, such a shallow-buried tunnel problem can also be simplified as a plane strain one considering the model's geometry and boundary conditions. Two coordinate systems are adopted: one Cartesian coordinate system (x, y) in the Z plane and one polar coordinate system (r, θ) in the ζ plane having a common origin at the center with the Cartesian coordinate system. The z-axis denotes the tunnel's axis not being plotted in Figure 9.1.

As discussed in Chapter 8, three elastic constants C_{44}, C_{45}, and C_{55} are enough to describe the medium deformation. The density of the surrounding rock/soil medium is characterized by ρ_1. The tunnel lining with outer radius R_1 and inner radius R_2 is assumed to be homogeneous, isotropic, and linearly elastic material with properties characterized by Lamé constants λ_2, μ_2, and density ρ_2. The axis of the tunnel is at a depth h below the ground surface. The imperfect interface between the tunnel lining and its surrounding soil/rock medium is taken into consideration.

The governing equation in the anisotropic medium in the Z plane of complex space is the same as Equation (8.1) in Chapter 8. To meet both the zero-stress boundary condition on the ground surface and the Sommerfeld radiation condition at infinity, one complex variable χ and its conjugate complex variable $\bar{\chi}$ in Equation (8.5) are also introduced. The steady solution of the anti-plane deformation is considered, and time t could be separated as an independent variable by the exponential factor $e^{-i\omega t}$ which is also omitted in the following analysis of the current chapter. The governing equation for an anisotropic half-space in case of an incident SH wave is the same as Equation (8.6).

DOI: 10.1201/9781003401599-9

Ground surface

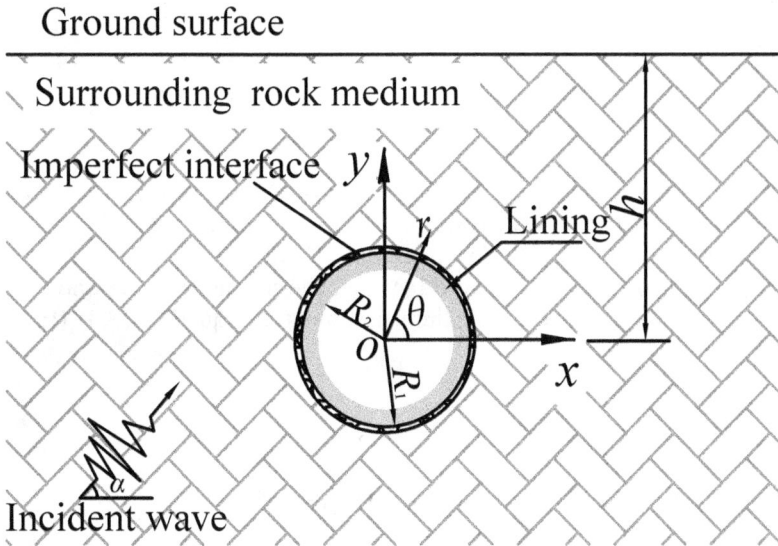

Figure 9.1 A shallow-lined circular tunnel with an imperfect interface under an anti-plane shear wave.

9.2 WAVE FIELDS IN THE ANISOTROPIC ROCK MASS

9.2.1 Seismic Wave Potentials

The anti-plane SH wave $w^{(in)}$ with incidence angle α is expressed as Equation (9.1).

$$w^{(in)} = w_0 e^{\frac{ik_{in}}{2}\left[(z+ih)\zeta_{in}+(\bar{z}-ih)\bar{\zeta}_{in}\right]} \tag{9.1}$$

where w_0 is the amplitude of the incident SH wave, k_{in} is the wave number of the incident wave in the anisotropic medium, and $\zeta_{in} = e^{i\alpha}$ and $\bar{\zeta}_{in} = e^{-i\alpha}$. The wave number can be determined as Equation (9.2) (Liu, 1988).

$$\omega = k_{in} c_{in}$$
$$c_{in} = \left[\left(C_{55}\cos^2\alpha + 2C_{45}\sin\alpha\cos\alpha + C_{44}\sin^2\alpha\right)/\rho_1\right]^{\frac{1}{2}}\Big\} \tag{9.2}$$

The reflected wave $w_1^{(r)}$ resulting from the horizontal surface is written as Equation (9.3).

$$w_1^{(r)} = w_0 e^{\frac{ik_r}{2}\left[(z+ih)\zeta_r + (\bar{z}-ih)\bar{\zeta}_r\right]}$$

(9.3)

where k_r is the wave number of the reflected wave in the medium determined by $\omega_r = k_r c_r$ (c_r is the propagation velocity of the reflected wave). The complex variables ζ_r, $\bar{\zeta}_r$, and the propagation velocity of the reflected wave c_r are dependent on the relationship between the incidence angle and elastic constants of the anisotropic medium described as Equation (9.4) (Han & Liu, 1997).

$$\left.\begin{array}{l} \zeta_r = e^{-i\alpha_r} \\ \bar{\zeta}_r = e^{i\alpha_r} \\ \tan\alpha_r = \tan\alpha - 2C_{45}/C_{44} \\ c_r = \left[\dfrac{\left(C_{55}\cos^2\alpha_r + 2C_{45}\sin\alpha_r\cos\alpha_r + C_{44}\sin^2\alpha_r\right)}{\rho_1}\right]^{\frac{1}{2}} \end{array}\right\}(\text{when } \tan\alpha > 2C_{45}/C_{44})$$

(9.4a)

$$\left.\begin{array}{l} \zeta_r = e^{i\alpha_r} \\ \bar{\zeta}_r = e^{-i\alpha_r} \\ \tan\alpha_r = -\left(\tan\alpha - \dfrac{2C_{45}}{C_{44}}\right) \\ c_r = \left[\dfrac{\left(C_{55}\cos^2\alpha_r - 2C_{45}\sin\alpha_r\cos\alpha_r + C_{44}\sin^2\alpha_r\right)}{\rho_1}\right]^{\frac{1}{2}} \end{array}\right\}(\text{when } \tan\alpha < 2C_{45}/C_{44})$$

(9.4b)

Following Equation (8.6), the scattered wave $w_1^{(sc)}$ around the tunnel lining is written as Equation (9.5).

$$w_1^{(sc)} = \sum_{n=-\infty}^{\infty} A_n \left\{ H_n^{(1)}(k_1|\chi|)\left[\frac{\chi}{|\chi|}\right]^n + H_n^{(1)}(K_1|\chi'|)\left[\frac{\chi'}{|\chi'|}\right]^n \right\}$$

(9.5)

where A_n is an uncertain coefficient of the scattered wave around tunnel lining, $H_n^{(1)}(\cdot)$ is the Hankel function of the first kind and the nth order denoting the outgoing wave and $\chi' = \frac{1}{2}[(1-i\gamma)(z+i2h)+(1+i\gamma)(z-i2h)]$.

9.2.2 Seismic Stresses

Using Equations (8.12) and (8.13), the stresses arising from the incident wave (Equation (9.1)) are presented as Equations (9.6) and (9.7).

$$\tau_{rz,1}^{(in)} = \frac{ik_{in}w^{(in)}}{4} \left\{ \begin{array}{l} \left[\left(C_{55}+C_{44}\right)\zeta_{in} + \left(C_{55}-C_{44}-2iC_{45}\right)\bar{\zeta}_{in} \right]e^{i\theta} + \\ \left[\left(C_{55}-C_{44}+2iC_{45}\right)\zeta_{in} + \left(C_{55}+C_{44}\right)\bar{\zeta}_{in} \right]e^{-i\theta} \end{array} \right\} \tag{9.6}$$

$$\tau_{\theta z,1}^{(in)} = \frac{ik_{in}w^{(in)}}{4} \left\{ \begin{array}{l} \left[i\left(C_{55}+C_{44}\right)\zeta_{in} + \left(2C_{45}+i\left(C_{55}-C_{44}\right)\right)\bar{\zeta}_{in} \right]e^{i\theta} + \\ \left[\left(2C_{45}-i\left(C_{55}-C_{44}\right)\right)\zeta_{in} - i\left(C_{55}+C_{44}\right)\bar{\zeta}_{in} \right]e^{-i\theta} \end{array} \right\} \tag{9.7}$$

Using Equations (8.12) and (8.13), the stresses due to the reflected wave (Equation (9.3)) could be obtained as Equations (9.8) and (9.9).

$$\tau_{rz,1}^{(r)} = \frac{ik_{r}w_{1}^{(r)}}{4} \left\{ \begin{array}{l} \left[\left(C_{55}+C_{44}\right)\zeta_{r} + \left(C_{55}-C_{44}-2iC_{45}\right)\bar{\zeta}_{r} \right]e^{i\theta} + \\ \left[\left(C_{55}-C_{44}+2iC_{45}\right)\zeta_{r} + \left(C_{55}+C_{44}\right)\bar{\zeta}_{r} \right]e^{-i\theta} \end{array} \right\} \tag{9.8}$$

$$\tau_{\theta z,1}^{(r)} = \frac{ik_{r}w_{1}^{(r)}}{4} \left\{ \begin{array}{l} \left[i\left(C_{55}+C_{44}\right)\zeta_{r} + \left(2C_{45}+i\left(C_{55}-C_{44}\right)\right)\bar{\zeta}_{r} \right]e^{i\theta} + \\ \left[\left(2C_{45}-i\left(C_{55}-C_{44}\right)\right)\zeta_{r} - i\left(C_{55}+C_{44}\right)\bar{\zeta}_{r} \right]e^{-i\theta} \end{array} \right\} \tag{9.9}$$

Using Equations (8.12) and (8.13), the stresses arising from the scattered wave (Equation (9.5)) could be obtained as Equations (9.10) and (9.11).

$$\tau_{rz,1}^{(sc)} = \frac{c_{55}k_{1}}{4} \sum_{n=-\infty}^{\infty} A_{n} \left\{ \begin{array}{l} \left[\left(a+ic\right)\left(F_{n-1}-F_{n+1}'\right) - \left(b-ic\right)\left(F_{n+1}-F_{n-1}'\right) \right]e^{i\theta} + \\ \left[\left(b+ic\right)\left(F_{n-1}-F_{n+1}'\right) - \left(a-ic\right)\left(F_{n+1}-F_{n-1}'\right) \right]e^{-i\theta} \end{array} \right\} \tag{9.10}$$

$$\tau_{\theta z,1}^{(sc)} = \frac{c_{55}k_{1}}{4} \sum_{n=-\infty}^{\infty} A_{n} \left\{ \begin{array}{l} \left[\left(-c+ia\right)\left(F_{n-1}-F_{n+1}'\right) - \left(c+ib\right)\left(F_{n+1}-F_{n-1}'\right) \right]e^{i\theta} + \\ \left[\left(c-ib\right)\left(F_{n-1}-F_{n+1}'\right) + \left(c+ia\right)\left(F_{n+1}-F_{n-1}'\right) \right]e^{-i\theta} \end{array} \right\} \tag{9.11}$$

where $F_n = H_n^{(1)}\left(k_1 |\chi|\right)\left[\dfrac{\chi}{|\chi|}\right]^{n}$, $F_n' = H_n^{(1)}\left(k_1 |\chi'|\right)\left[\dfrac{\chi'}{|\chi'|}\right]^{-n}$, $a = \sqrt{\beta-\kappa^2}\left(1+\sqrt{\dfrac{1}{\beta}-\dfrac{\kappa^2}{\beta^2}}\right)$,

$b = -\sqrt{\beta-\kappa^2}\left(1-\sqrt{\dfrac{1}{\beta}-\dfrac{\kappa^2}{\beta^2}}\right)$, $c = \kappa\sqrt{\beta-\kappa^2}/\beta$, $\beta = C_{44}/C_{55}$, and $\kappa = C_{45}/C_{55}$.

9.3 WAVE FIELDS IN THE ISOTROPIC TUNNEL LINING

9.3.1 Seismic Wave Potentials

The reflected wave $w_2^{(rf)}$ that propagates outwards from the lining's inner boundary is expressed as Equation (9.12).

$$w_2^{(rf)} = \sum_{n=-\infty}^{\infty} B_n H_n^{(1)}\left(k_2 \mid z \mid\right)\left[\frac{z}{\mid z \mid}\right]^n \tag{9.12}$$

where B_n is an uncertain coefficient of the reflected wave in the tunnel lining and k_2 is the wave number in the isotropic lining determined by $k_2 = \omega / c_2$ with $c_2 = \sqrt{u_2 / \rho_2}$.

The refracted wave $w_2^{(rr)}$ being confined into the tunnel lining is expressed as Equation (9.13).

$$w_2^{(rr)} = \sum_{n=-\infty}^{\infty} C_n H_n^{(2)}\left(k_2 \mid z \mid\right)\left[\frac{z}{\mid z \mid}\right]^n \tag{9.13}$$

where C_n is an uncertain coefficient of the refracted wave in the tunnel lining and $H_n^{(2)}(\cdot)$ is the Hankel function of the second kind and the nth order denoting the ingoing wave.

9.3.2 Seismic Stresses

Using Equations (8.12) and (8.13), the stresses arising from the reflected wave (Equation (9.12)) could be obtained as Equations (9.14) and (9.15).

$$\tau_{rz,2}^{(rf)} = \frac{\mu_2 k_2 B_n}{2}\left\{\sum_{n=-\infty}^{\infty} H_{n-1}^{(1)}\left(K_2 \mid z \mid\right)\left[\frac{z}{\mid z \mid}\right]^{n-1} e^{i\theta} - \sum_{n=-\infty}^{\infty} H_{n+1}^{(1)}\left(k_2 \mid z \mid\right)\left[\frac{z}{\mid z \mid}\right]^{n+1} e^{-i\theta}\right\} \tag{9.14}$$

$$\tau_{\theta z,2}^{(rf)} = \frac{i\mu_2 K_2 B_n}{2}\left\{\sum_{n=-\infty}^{\infty} H_{n-1}^{(1)}\left(k_2 \mid z \mid\right)\left[\frac{z}{\mid z \mid}\right]^{n-1} e^{i\theta} + \sum_{n=-\infty}^{\infty} H_{n+1}^{(1)}\left(k_2 \mid z \mid\right)\left[\frac{z}{\mid z \mid}\right]^{n+1} e^{-i\theta}\right\} \tag{9.15}$$

Using Equations (8.12) and (8.13), the stresses arising from the refracted wave (Equation (9.13)) could be obtained as Equations (9.16) and (9.17).

$$\tau_{rz,2}^{(rr)} = \frac{\mu_2 k_2 C_n}{2}\left\{\sum_{n=-\infty}^{\infty} H_{n-1}^{(2)}\left(k_2 \mid z \mid\right)\left[\frac{z}{\mid z \mid}\right]^{n-1} e^{i\theta} - \sum_{n=-\infty}^{\infty} H_{n+1}^{(2)}\left(k_2 \mid z \mid\right)\left[\frac{z}{\mid z \mid}\right]^{n+1} e^{-i\theta}\right\} \tag{9.16}$$

$$\tau_{\theta z,2}^{(rr)} = \frac{i\mu_2 k_2 C_n}{2} \left\{ \sum_{n=-\infty}^{\infty} H_{n-1}^{(2)} (k_2 |z|) \left[\frac{z}{|z|} \right]^{n-1} e^{i\theta} + \sum_{n=-\infty}^{\infty} H_{n+1}^{(2)} (k_2 |z|) \left[\frac{z}{|z|} \right]^{n+1} e^{-i\theta} \right\} \quad (9.17)$$

9.4 BOUNDARY CONDITION

The total wavefield $w_1^{(t)}$ confined in the anisotropic medium is the superposition of the incident wavefield, reflected wave field, and scattered wave field as Equation (9.18).

$$w_1^{(t)} = w^{(in)} + w_1^{(r)} + w_1^{(sc)} \quad (9.18)$$

The total wavefield $w_2^{(t)}$ confined in the isotropic tunnel lining is the superposition of the refracted wave field and the reflected wavefield as Equation (9.19).

$$w_2^{(t)} = w_2^{(rr)} + w_2^{(rf)} \quad (9.19)$$

The same elastic model as discussed in Chapters 7 and 8 is introduced to analyze the imperfect interface influence. Tractions at the lining's outer boundary ($R = R_1$, L_1 in Figure 9.1) are continuous, and displacements are discontinuous across the interface (Equation (9.20)).

$$\left. \begin{array}{l} w_1^{(t)} \Big|_{R=R_1} - w_2^{(t)} \Big|_{R=R_1} = \frac{\tau_{rz,1}^{(t)}}{k} \Big|_{R=R_1} \\ \\ \tau_{rz,1}^{(t)} \Big|_{R=R_1} = \tau_{rz,2}^{(t)} \Big|_{R=R_1} \end{array} \right\} \quad (9.20)$$

where k is the stiffness of the imperfect interface, $\tau_{rz,1}^{(t)} = \tau_{rz,1}^{(in)} + \tau_{rz,1}^{(r)} + \tau_{rz,1}^{(sc)}$ and $\tau_{rz,2}^{(t)} = \tau_{rz,2}^{(rr)} + \tau_{rz,2}^{(rf)}$. Details on the stiffness parameter selection refer to Chapter 7.

At the lining's inner boundary ($R = R_2$, L_2 in Figure 9.1), tractions are zero (Equation (9.21)).

$$\tau_{rz,2}^{(t)} \Big|_{R=R_2} = 0 \quad (9.21)$$

where $\tau_{rz,2}^{(t)} = \tau_{rz,2}^{(rr)} + \tau_{rz,2}^{(rf)}$.

Substituting the stresses into Equations (9.20) and (9.21), Equation (9.22) can be derived.

$$\sum_{m=1}^{3} \sum_{n=-\infty}^{\infty} K_n^{mj} X_n^m = Q^j \quad (j=1, 2, 3) \tag{9.22}$$

where $X_n^1 = A_n$, $X_n^2 = B_n$, $X_n^3 = C_n$, K_n^{mj} and Q^j are presented in Appendix D.

Multiplying both sides of Equation (9.22) with the orthogonality of $e^{-is\theta}$ ($s = 0, \pm1, \pm2, \pm3 \ldots$) and integrating over the interval $[-\pi, \pi]$, the unknown constants A_n, B_n, and C_n can be obtained through solving a set of the infinite linear algebraic systems as Equation (9.23).

$$\sum_{m=1}^{3} \sum_{n=-\infty}^{\infty} K_{ns}^{mj} X_n^m = Q_s^j \quad (j = 1, 2, 3) \tag{9.23}$$

where $K_{ns}^{mj} = \dfrac{1}{2\pi} \int_{-\pi}^{\pi} K_n^{mj} e^{-is\theta} d\theta$ and $Q_s^j = \dfrac{1}{2\pi} \int_{-\pi}^{\pi} Q^j e^{-is\theta} d\theta$.

9.5 FUNDAMENTAL SOLUTION

To present the dynamic response of the half-space under excitation of the anti-plane SH wave, normalized surface displacement $|w^*|$ and dimensionless frequency η of the incident wave are adopted as Equations (9.24) and (9.25).

$$w^* = \left| \frac{w_1^{(t)}}{w_0} \right| \tag{9.24}$$

$$\eta = 2R_2 / \lambda = \omega R_2 / \pi c_1 \tag{9.25}$$

As described in Chapters 7 and 8, to analyze the dynamic response around a shallow-buried tunnel subjected to the anti-plane SH wave, a dimensionless dynamic stress concentration factor (DSCF) is also introduced. DSCF of the tangential stress $\tau_{\theta z,p}^*$ around the tunnel is expressed as Equation (9.26).

$$DSCF = \tau_{\theta z,p}^* = \left| \frac{\tau_{\theta z,p}^{(t)}}{\tau_0} \right| \quad (p = 1 \text{ for the medium and } p = 2 \tag{9-26}$$

for the tunnel lining)

where $\tau_{\theta z,1}^{(t)} = \tau_{\theta z,1}^{(in)} + \tau_{\theta z,1}^{(r)} + \tau_{\theta z,1}^{(sc)}$ and $\tau_{\theta z,2}^{(t)} = \tau_{\theta z,2}^{(rr)} + \tau_{\theta z,2}^{(rf)}$.

To obtain a general solution, some dimensionless parameters are defined. The influencing factors are normalized as the medium anisotropy degree

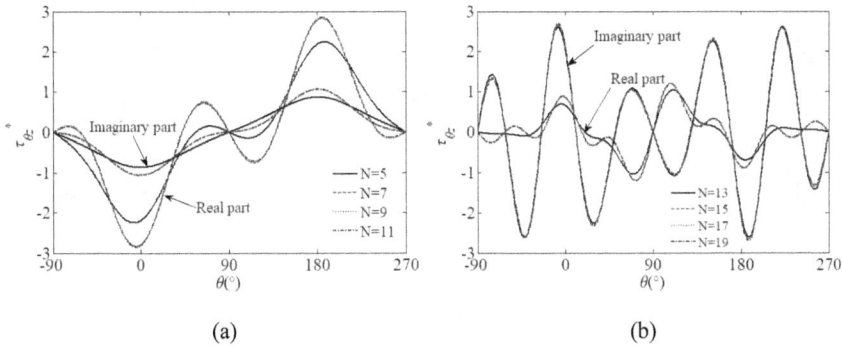

(a)   (b)

Figure 9.2 Convergence test. (a) $\eta = 0.5$ (lower frequency); (b) $\eta = 1.5$ (higher frequency).

$\beta = C_{44}/C_{55}$ and $\kappa = C_{45}/C_{55}$ (same as the definition in Chapter 8), and the interface stiffness parameter $k^* = kR_1/C_{55}$.

Figure 9.2 illustrates the tangential stress along the lining's inner side with different wave frequencies by different truncation numbers N for testing the convergence of the current method. The geometric and material parameters of the system are chosen as $h = 1.5R_1$, $R_1 = 5.0$ m, $\mu_1 = 3.26 \times 10^7$ Pa, $\rho_1 = 1932$ kg/m^3, $\mu_2 = 0.35\mu_1$, and $\rho_2 = 2200$ kg/m^3. Usually, earthquake waves have a frequency of less than 20 Hz (Allred et al., 2008). A lower dimensionless frequency $\eta = 0.5$ (seismic wave frequency of 6.46 Hz) and a higher dimensionless frequency $\eta = 1.5$ (seismic wave frequency of 19.48 Hz) are taken into consideration for the convergence test. The incident SH wave was deemed to excite vertically at an incidence angle of $\alpha = \pi/2$. Figure 9.2 indicates that both real and imaginary parts of $\tau_{\theta z,p}^*$ have good convergence as the truncation number N increases. It should be noted that the truncation number N is related to the frequency of the incident wave, and a smaller truncated number can be adopted when the wave frequency is smaller. To guarantee the efficiency of computation time and accuracy, the truncation can provide practically adequate results after $N = 11$ in the case of dimensionless frequency 0.5 and after $N = 19$ in the case of dimensionless frequency 1.5.

Besides, the current solution is simplified, and compared to the available complete results for a vertically incident SH wave in the research work of Lee and Trifunac (1979) to show the accuracy of the present method. The medium is isotropic ($\beta = 1.0$ and $\kappa = 0.0$) and the interface is in perfect condition ($k^* = 10^4$, same as the value in Chapter 8). Figure 9.3 illustrates the normalized ground surface displacement with the normalized horizontal coordinate x/R_1 for two embedment ratios $h/R_1 = 1.5$ and 5.0, and two

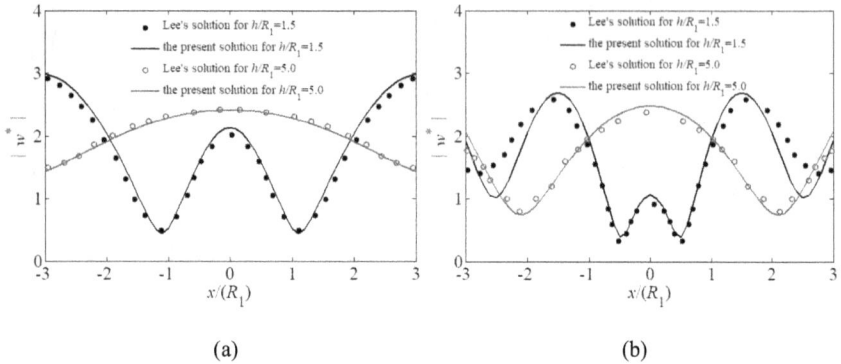

Figure 9.3 Verification of normalized displacement with different depths by the present solution with Lee's solution (1979). (a) $\eta = 0.5$; (b) $\eta = 1.0$.

cases of the dimensionless frequency $\eta = 0.1$ and 1.0. Even though the results obtained by Lee and Trifunac (1979) could not be read well with enough accuracy from the published figures, the present solutions are close to their solutions.

9.5.1 Case I: Medium Anisotropy Influence

For an illustration of the medium anisotropy influence, four sets of anisotropy characteristics $\beta = 1.0$ and $\kappa = 0$, $\beta = 0.5$ and $\kappa = 0$, $\beta = 0.5$ and $\kappa = 0.2$, and $\beta = 1.5$ and $\kappa = 0$ are employed. Four values of dimensionless frequency $\eta = 0.1, 0.5, 1.0$, and 1.5 (wave frequencies of 1.30 Hz, 6.46 Hz, 12.99 Hz, and 19.48 Hz), two incidence angles $\alpha = 0$ and $\pi/2$, and the perfect interface ($k^* = 10^4$) between the tunnel lining and its surrounding medium are considered. The geometric and material parameters of the system in Figure 9.1 are chosen as $h = 1.5R_1$, $R_1 = 5.0$ m, $\mu_1 = 3.26 \times 10^7$ Pa, $\rho_1 = 1932$ kg/m³, $\mu_2 = 0.35\mu_1$, and $\rho_2 = 2200$ kg/m³.

Figures 9.4 and 9.5 illustrate the ground surface displacement versus the normalized horizontal coordinate x/R_1 for four sets of anisotropy characteristics and two incidence angles of $\alpha = \pi/2$ and 0. In the absence of the tunnel, the normalized ground displacement in the uniform half-space is equal to 2.0 (Lee & Trifunac, 1979). In the presence of the tunnel, the normalized ground displacement departs significantly from the value of 2.0 due to the wave scattering and diffraction around the tunnel. The departure is dependent on the wave frequency. In the case of $\eta = 0.1$, the ground displacement shows a smooth variation pattern versus the dimensionless horizontal coordinate x/R_1 for the vertically and horizontally incident SH waves (Figures 9.4a and

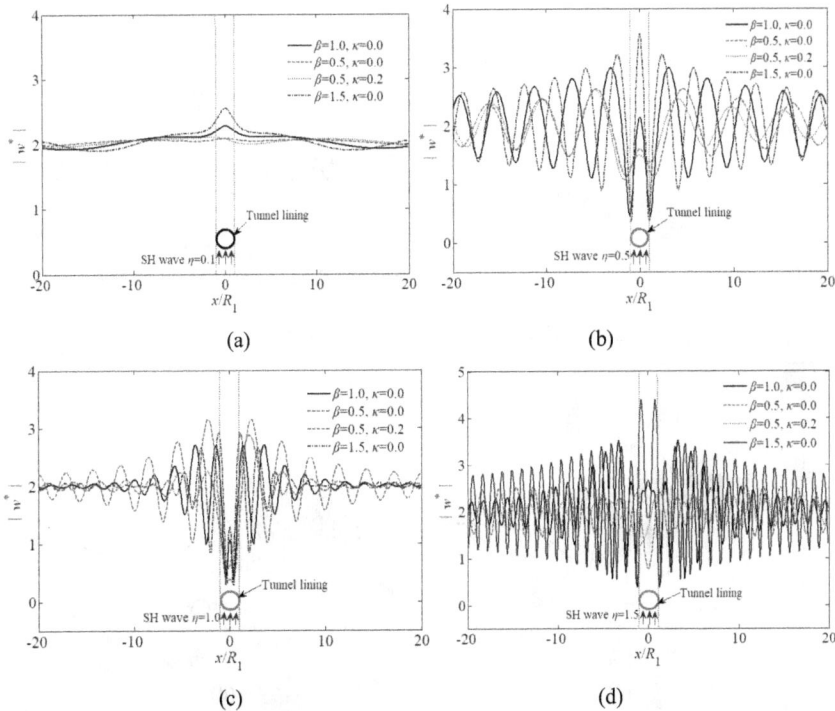

Figure 9.4 Displacement on the ground surface versus the normalized horizontal coordinate x/R_1 for four sets of anisotropy characteristics and incidence angle of $\alpha = \pi/2$. (a) $\eta = 0.1$; (b) $\eta = 0.5$; (c) $\eta = 1.0$; (d) $\eta = 1.5$.

9.5a). In the cases of $\eta = 0.5$ (Figures 9.4b and 9.5b), 1.0 (Figures 9.4c and 9.5c), and 1.5 (Figures 9.4d and 9.5d), the ground displacement fluctuates more and more rapidly. Such fluctuation phenomenon is related to the incidence wave frequency.

Figures 9.4 and 9.5 also show that the different parameters κ and β result in different the ground surface displacement distributions. Under the excitation of a vertical incident SH wave, the normalized ground surface displacements are symmetrical along the tunnel vertical axis when the medium is isotropic (in the case of $\beta = 1.0$ and $\kappa = 0$ with the black solid line in Figure 9.4) or orthotropic (in the cases of $\beta = 0.5$ and $\kappa = 0$ with the green dashed line, and $\beta = 1.5$ and $\kappa = 0$ with the blue dashed-dotted line in Figure 9.4). In the case of $\beta = 0.5$ and $\kappa = 0.2$ (red dotted line in Figure 9.4), the normalized ground surface displacement becomes asymmetrical because of the different propagation characteristics of the wave along with different directions due

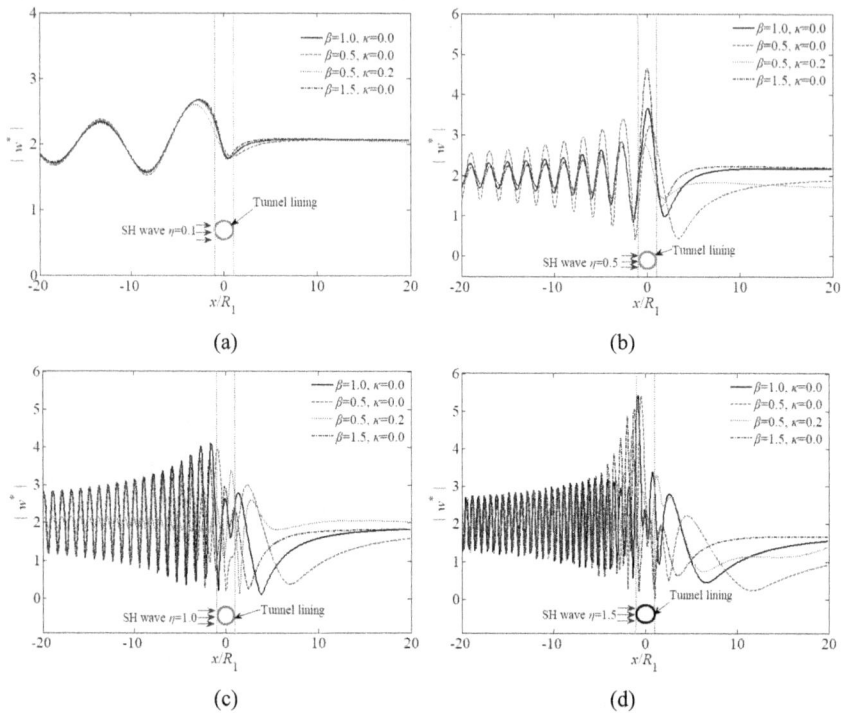

Figure 9.5 Displacement on the ground surface versus the normalized horizontal coordinate x/R_1 for four sets of anisotropy characteristics and incidence angle of $\alpha = 0$. (a) $\eta = 0.1$; (b) $\eta = 0.5$; (c) $\eta = 1.0$; (d) $\eta = 1.5$.

to the medium anisotropy. Thus, the anisotropy characteristics of the rock/soil medium could influence the ground surface displacement remarkably. When the seismic wave is excited vertically, the ground surface displacements right above the tunnel in the cases of $\beta = 0.5$ (the green dashed line and the red dotted line in Figure 9.4a) are smaller than those in the isotropic case ($\beta = 1.0$, $\kappa = 0$ with the black solid line in Figure 9.4a), while in the case of $\beta = 1.5$ (blue dash-dotted line in Figure 9.4a) right above the tunnel ground surface displacements turn out to be larger than those in the isotropic case. Nonetheless, the opposite condition is true (Figure 9.5a) under the excitation of a horizontal incident wave. The medium anisotropy influence is also dependent on the incidence angle.

Figures 9.6 and 9.7 show the medium DSCFs on tangential stresses at the interface with the polar coordinate θ for the four anisotropic medium cases and the two incidence angles. When the medium is isotropic ($\beta = 1.0$, $\kappa = 0$ denoted with a black solid line) or orthotropic ($\beta = 0.5$ and $\kappa = 0$ denoted with a green dashed line, and $\beta = 1.5$ and $\kappa = 0$ denoted with blue dashed-dotted

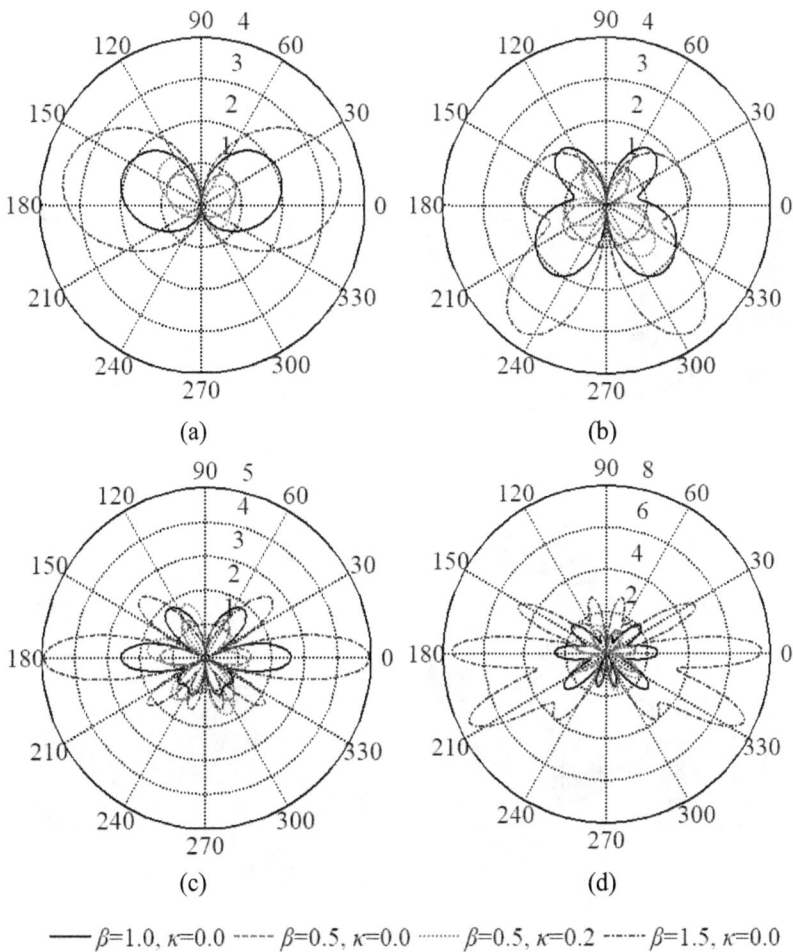

Figure 9.6 DSCFs of the medium at the interface versus the polar coordinate θ for four sets of anisotropy characteristics and incidence angle of $\alpha = \pi/2$. (a) $\eta = 0.1$; (b) $\eta = 0.5$; (c) $\eta = 1.0$; (d) $\eta = 1.5$.

line), the DSCFs on tangential stresses are symmetrical along the incidence direction regardless of the horizontal or vertical incident wave. However, the DSCFs become unsymmetrical in the case of $\beta = 0.5$ and $\kappa = 0.2$. This is also because of the different propagation characteristics along with different directions due to the medium anisotropy. Therefore, the dynamic stress distribution of the medium-lining system relates to the anisotropy characteristics of the rock/soil medium.

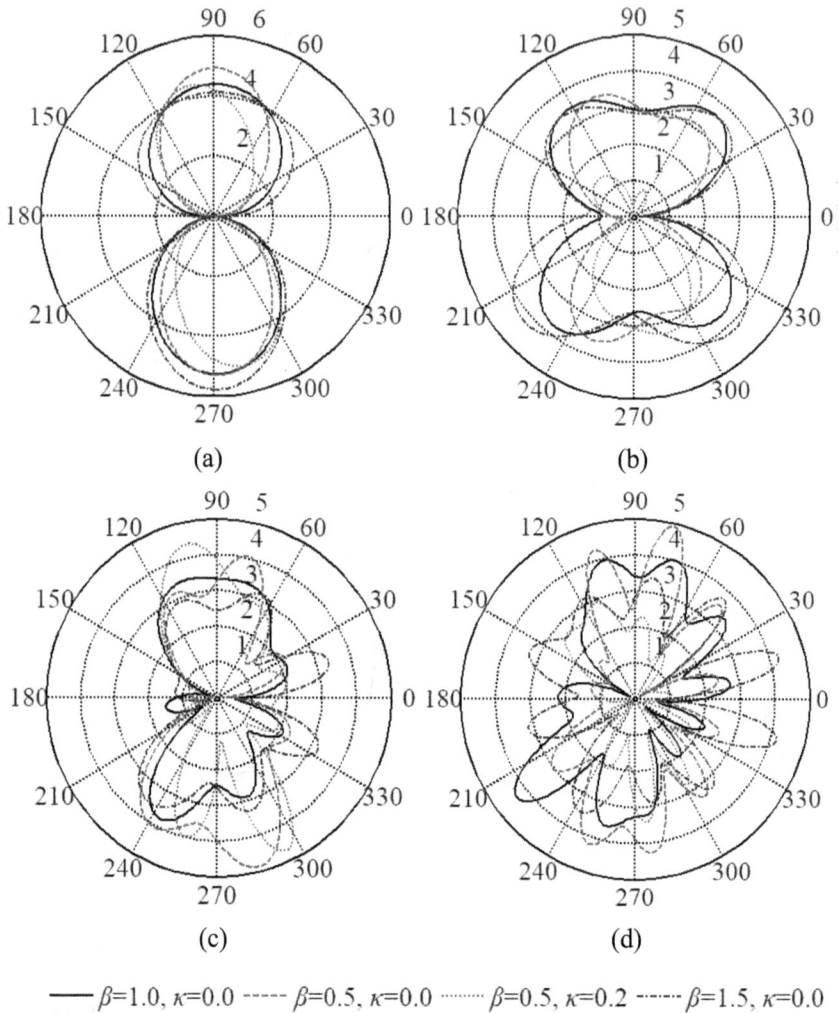

(a)

(b)

(c)

(d)

——— β=1.0, κ=0.0 ----- β=0.5, κ=0.0 ········· β=0.5, κ=0.2 --·--· β=1.5, κ=0.0

Figure 9.7 DSCFs of the medium at the interface versus the polar coordinate θ for four sets of anisotropy characteristics and incidence angle of α = 0. (a) η = 0.1; (b) η = 0.5; (c) η = 1.0; (d) η = 1.5.

Under the excitation of a vertical incident wave, as shown in Figure 9.6, the DSCFs on the tangential stress in the anisotropic cases with β = 0.5 (red dotted line and green dashed line) are smaller than those in the isotropic case (black solid line). However, the DSCFs on the tangential stresses in the aniso-tropic cases with β = 1.5 (blue dash-dotted line) are relatively larger than those (black solid line) in the isotropic case. Under the excitation of a hori-zontal incident wave, as shown in Figure 9.7, the DSCFs on the tangential stress in the anisotropic cases with $\beta = 0.5$ (red dotted line and green dashed line) are smaller than those in the isotropic case (black solid line) on the lining sidewall, while the opposite is true on the crown and floor. On the contrary, the DSCFs on the tangential stress in the anisotropic cases with $\beta = 1.5$ (blue dash-dotted line) are larger than those in the isotropic case (dashed-dotted line) on the lining sidewall, while the opposite is true on the crown and floor. The incidence angle is also one important factor affecting the influence of the medium characteristics.

Figures 9.8 and 9.9 present the DSCFs on the tangential stresses along the lining's inner side with the polar coordinate θ for the four cases of anisotropic

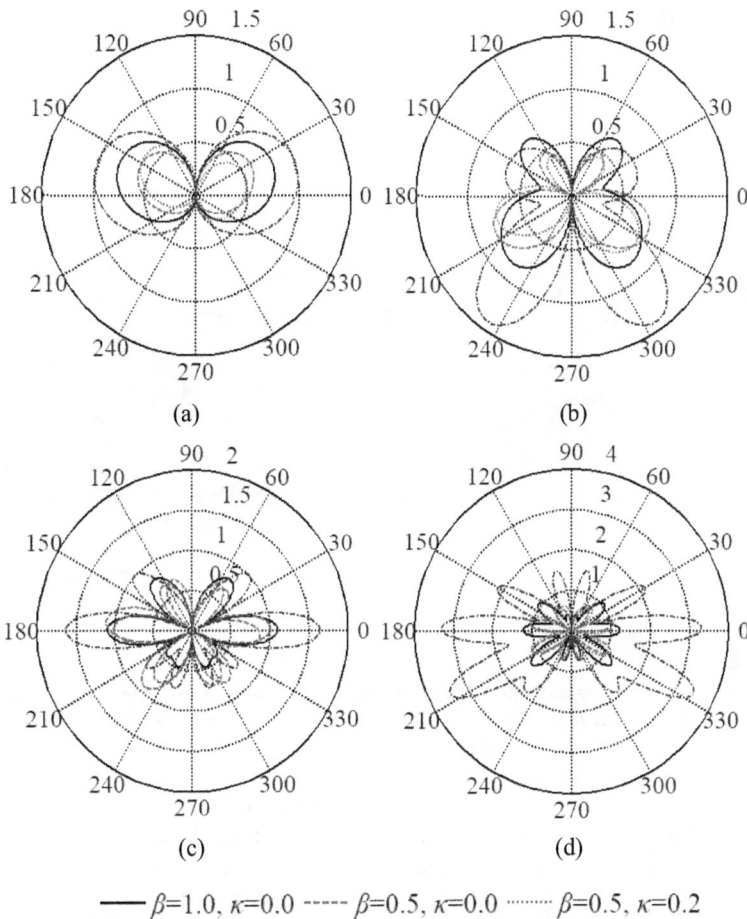

Figure 9.8 DSCFs along the inner side of the tunnel lining versus the polar coordinate θ for four sets of anisotropy characteristics and incidence angle of $\alpha = \pi/2$. (a) $\eta = 0.1$; (b) $\eta = 0.5$; (c) $\eta = 1.0$; (d) $\eta = 1.5$.

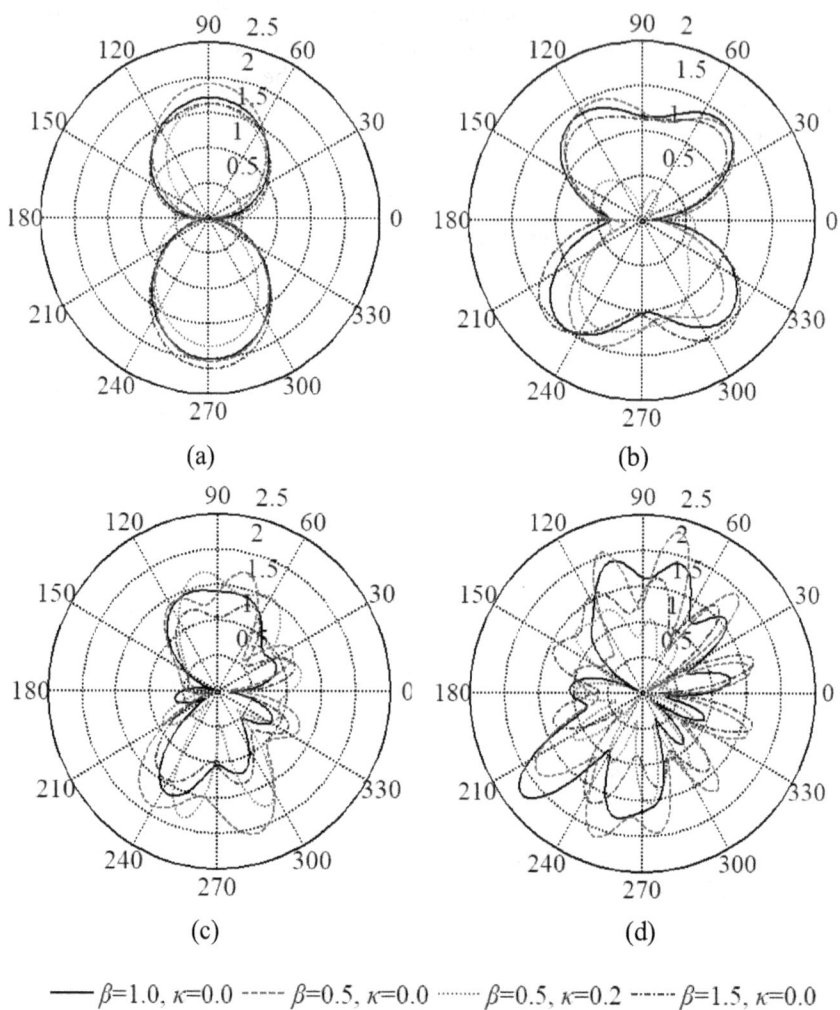

$$\text{──} \ \beta{=}1.0, \ \kappa{=}0.0 \quad \text{-----} \ \beta{=}0.5, \ \kappa{=}0.0 \quad \text{·······} \ \beta{=}0.5, \ \kappa{=}0.2 \quad \text{-·--·-} \ \beta{=}1.5, \ \kappa{=}0.0$$

Figure 9.9 DSCFs along the inner side of the tunnel lining versus the polar coordinate θ for four sets of anisotropy characteristics and incidence angle of $\alpha = 0$. (a) $\eta = 0.1$; (b) $\eta = 0.5$; (c) $\eta = 1.0$; (d) $\eta = 1.5$.

medium and the two incidence angles. A similar dynamic response of the medium-lining system could be observed regarding the influence of aniso-tropic medium and incidence angle.

9.5.2 Case II: Imperfect Interface Influence

For an illustration of the imperfect interface influence, four values of the interface stiffness $k^* = 50.0, 5.0, 0.5,$ and 0.1 are employed. Two values of dimensionless frequency $\eta = 0.1$ and 1.0, two sets of anisotropy characteristic $\beta = 1.0$ and $\kappa = 0$, and $\beta = 1.5$ and $\kappa = 0$, and an excitation of vertical wave are considered. The geometric and material parameters of the system in Figure 9.1 are chosen as $h = 1.5R_1$, $R_1 = 5.0$ m, $\mu_1 = 3.26 \times 10^7$ Pa, $\rho_1 = 1932$ kg/m^3, $\mu_2 = 0.35\mu_1$, and $\rho_2 = 2200$ kg/m^3.

Figure 9.10 illustrates the normalized ground surface displacement for the four stiffness cases with the dimensionless horizontal coordinate x/R_1. When the incidence frequency is low ($\eta = 0.1$, Figures 9.10a and 9.10b), the interface stiffness causes no significant difference in the ground surface displacement with the four stiffness cases. The stiffness influence on the ground surface displacement has no obvious relation to the anisotropy characteristic in a relatively low incidence frequency ($\eta = 0.1$) case. When the incidence frequency is high ($\eta = 1.0$, Figures 9.10c and 9.10d), the ground surface displacement is larger and fluctuates more seriously. The difference between each ground surface displacement for the four stiffness cases is much more obvious than that in the case of $\eta = 0.1$. When the incidence frequency becomes higher, the

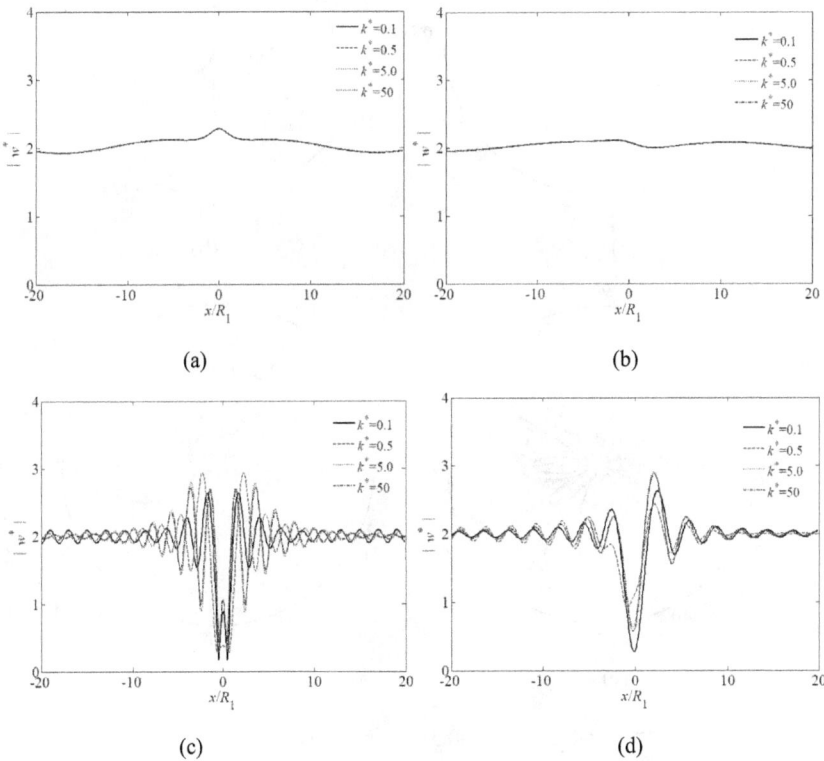

Figure 9.10 Displacements on the ground surface versus the dimensionless horizontal coordinate x/R_1 for four values of the interface stiffness. (a) $\eta = 0.1$, $\beta = 1.0$, $\kappa = 0$; (b) $\eta = 0.1$, $\beta = 0.5$, $\kappa = 0.2$; (c) $\eta = 1.0$, $\beta = 1.0$, $\kappa = 0$; (d) $\eta = 1.0$, $\beta = 0.5$, $\kappa = 0.2$.

imperfect interface influence also becomes obvious. Besides, $k^* = 0.5$ yields the minimum ground surface displacement right above the tunnel in the case of $\beta = 1.0$ and $\kappa = 0$ (red dashed line in Figure 9.10c); nonetheless, it yields the maximum displacement in the case of $\beta = 0.5$ and $\kappa = 0.2$ (red dashed line in Figure 9.10d). Thus, the medium anisotropy characteristic also makes a difference on the interface stiffness influence on the ground surface displacement. We can conclude that the imperfect interface influence on the ground response is not only related to the incidence frequency but also dependent on the medium anisotropy characteristics.

Figure 9.11 presents the DSCFs on the tangential stresses along the lining's inner side for the four stiffness cases with the polar coordinate θ. When the

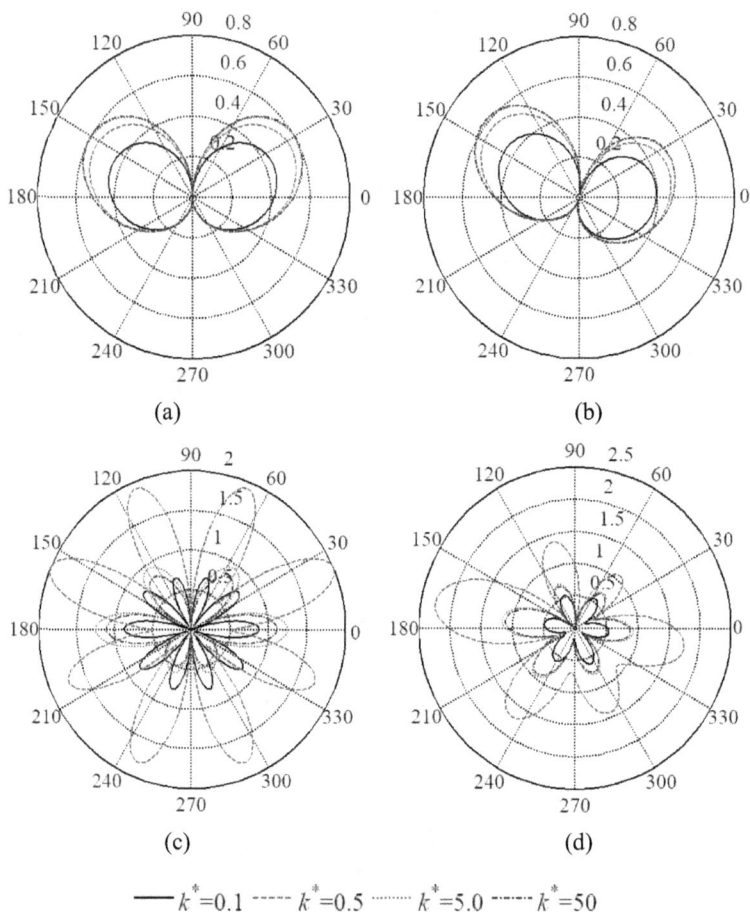

Figure 9.11 DSCFs along the inner side of the tunnel lining for four values of the interface stiffness versus the polar coordinate θ. (a) $\eta = 0.1$, $\beta = 1.0$, $\kappa = 0$; (b) $\eta = 0.1$, $\beta = 0.5$, $\kappa = 0.2$; (c) $\eta = 1.0$, $\beta = 1.0$, $\kappa = 0$; (d) $\eta = 1.0$, $\beta = 0.5$, $\kappa = 0.2$.

incidence frequency is low ($\eta = 0.1$, Figures 9.11a and 9.11b), the DSCFs on the tangential stress along the lining inner side increase, and the difference between the DSCFs for two adjacent stiffness values show a decreasing trend gradually as the stiffness increases. For example, the DSCF on the lining shoulder ($\theta = 30°$) in Figure 9.11a, is increased by 24.03% as the stiffness increases from 0.1 to 0.5, while the DSCF is increased by 0.71% as the stiffness increases from 5 to 50. With a sufficiently higher interface stiffness, the interface influence could be ignored.

When the incident wave's frequency is high ($\eta = 1.0$, Figures 9.11c and 9.11d), $k^* = 0.5$ yields the maximum DSCFs on the tangential stress regardless of the isotropic or anisotropic medium. The interface influence is slightly different from the case with lower incidence frequency ($\eta = 0.1$, Figures 9.11a and 9.11b). Compared to those in the lower frequency case, the DSCFs on the tangential stress in the lining are larger and distribute complicatedly with several peaks. The difference in the maximum DSCF between two adjacent stiffnesses turns out to be larger with a higher incidence frequency for the same rock mass anisotropy. For example, the increase rate of maximum DSCF as the stiffness increases from 0.1 to 0.5 in the case of $\eta = 0.1$, $\beta = 1.0$, $\kappa = 0$ is 32.9%, while it is 119.8% in the case of $\eta = 1.0$, $\beta = 1.0$, $\kappa = 0$. Besides, the DSCF in the anisotropic case ($\beta = 0.5$ and $\kappa = 0.2$, red dashed line in Figure 9.11d) distributes unsymmetrically with the incidence direction, instead of the symmetrical distribution in the isotropic case ($\beta = 1.0$ and $\kappa = 0.0$, red dashed line in Figure 9.11c). Thus, the seismic wave and medium anisotropic characteristics affect the interface's influence on the stress state of the underground structures.

The imperfect interface of the medium-lining system significantly affects the wave propagation characteristics and thereby influences the stresses and displacement distribution. As Son and Cording (2007) proposed, contact behavior around the tunnel lining is an essential factor during both the design and construction process of underground structures withstanding both static overburden and dynamic loading.

9.6 CONCLUSION

This chapter presents an analytical solution for the cross-sectional seismic performance of a shallow-buried tunnel in an anisotropic half-space based on the wave function expansion method and the multi-polar coordinate system. Case studies on the influences of medium anisotropy and imperfect interface as well as the incidence angle and wave frequency are discussed. Some important conclusions are as follows.

The medium anisotropy characteristics have a significant effect on the distribution of the ground surface displacements and stresses of the medium-lining system. Under the excitation of a vertical incident SH wave, the normalized ground surface displacements and the DSCFs on the tangential stress are symmetrical along the tunnel vertical axis when the medium is isotropic or orthotropic. When the medium is anisotropic, the normalized ground surface displacement and the DSCFs on the tangential stress becomes asymmetrical. The influence of the medium anisotropy characteristics depends on the incidence angle and incidence frequency. On the other hand, an imperfect interface of the medium-lining system exerts a significant influence on wave scattering. The influence of interface stiffness on the ground surface displacement is independent of the medium anisotropy characteristic when the incidence frequency is low ($\eta = 0.1$), while the opposite is correct when the incidence frequency is high ($\eta = 1.0$). Additionally, the influence of interface stiffness on DSCFs is related to both the incidence frequency and the medium anisotropy characteristic.

During both the design and construction process of underground structures withstanding both static overburden and dynamic loading, medium anisotropy characteristics and contact behavior of the medium-lining system are essential factors. Such factors should be considered to meet the stability and safety requirements of the designed structures as well as the robustness and reliability of the design scheme.

Chapter 10

Damage Assessment of Rock Tunnels under Earthquake Loading

10.1 SEISMIC DAMAGE CLASSIFICATION

Many instances with noticeable seismic damage were reported to indicate that the traditional view on seismic resistance of underground structures is no longer active. Various patterns of seismic damage to rock tunnels by earthquakes were reported as described in Chapter 1. Herein, some of the major patterns with significant characteristics including lining cracks, lining concrete spalling and collapse, pavement damage, and groundwater leakage are illustrated.

10.1.1 Lining Cracks

Lining cracks are the most frequently observed seismic damage after earthquakes. They can be ring crack, transversal crack, longitudinal crack, and inclined crack according to their spatial distribution state. Figure 10.1 illustrates the tunnel lining ring crack after earthquakes. Ring cracks with maximum dislocation of 8.0 mm in the Tawarayama tunnel were the most frequently observed (Zhang et al., 2018) (Figure 10.1a). Ring crack can be further classified into two types: transverse ring crack and inclined ring crack (Figure 10.1b). Damage of this pattern was also found in the Zipingpu Tunnel, the Longdongzi Tunnel, and the Longxi Tunnel built on the Chengdu-Wenchuan Line in China following the 2008 Wenchuan Earthquake (Li et al., 2012).

Figure 10.2 shows the tunnel lining longitudinal cracks after earthquakes. Longitudinal cracks can be further classified into three types: singular crack at the vault of the crown, symmetrical crack, and non-symmetric crack (Wang et al., 2001). Most of the longitudinal cracks were the former one in the Tawarayama tunnel (Figure 10.2a). The length of some longitudinal cracks exceeded the dimension of the lining section (about 10 m) (Figure 10.2b). In addition, a pair of symmetrical cracks were observed near the Nishihara Village side, and a few non-symmetric cracks occurred at the sidewall near the Minami Aso Village (Zhang et al., 2018, 2020). In the Chi-Chi earthquake, the No. 1 San-I railway tunnel, the New Chi-Chi tunnel on Highway No. 16, and the headrace

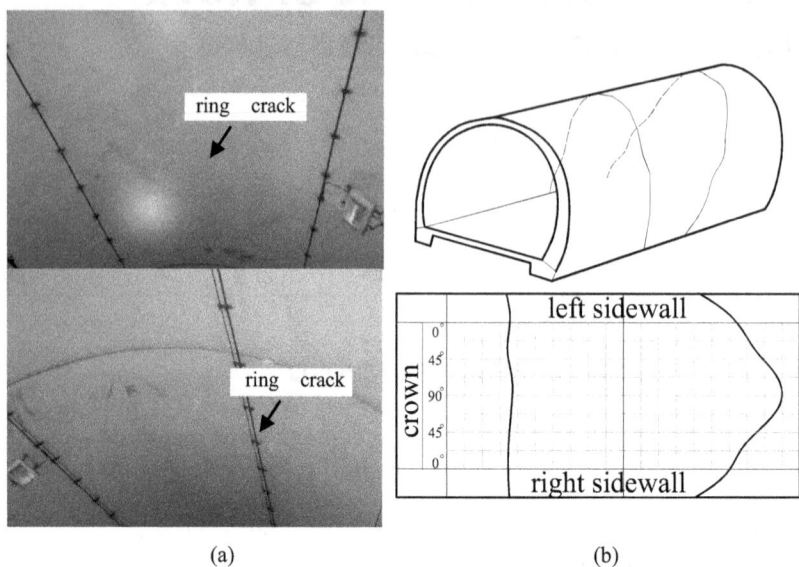

Figure 10.1   Tunnel lining ring crack after earthquakes. (a) Ring crack in the sections from S119 to S126 in the Tawarayama tunnel; (b) sketch and mapping result of ring crack.

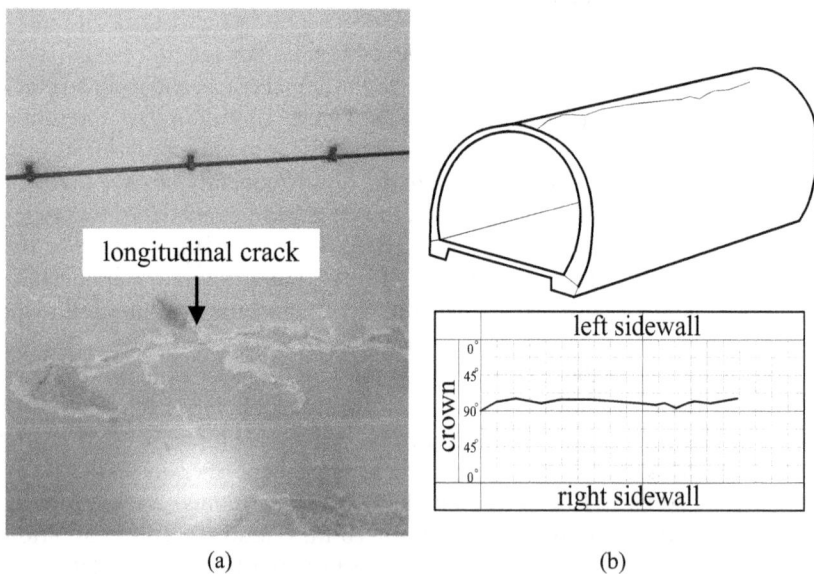

Figure 10.2   Tunnel lining longitudinal cracks after earthquakes. (a) Longitudinal crack in the Tawarayama tunnel; (b) sketch and mapping result of longitudinal crack.

tunnel of the New Tienlun power station are the most representative examples of this damage type (Wang et al., 2001). Longitudinal cracks also occurred in the Namutani Tunnel in the Great Kanto earthquake in Japan (Gong, 2007).

Transverse cracks that developed perpendicular to the tunnel axis direction and inclined cracks with an inclination angle to the tunnel axis direction mainly develop at the hance and sidewall, dominated by shearing type and tension-shearing type. Some ring cracks develop from propagation and interaction of transverse and inclined cracks. Both transverse and inclined cracks of tunnel lining after earthquakes are illustrated in Figure 10.3.

10.1.2 Spalling and Collapse of Lining Concrete

Figure 10.4 shows spalling and collapse of lining concrete after earthquakes. Concrete lining spalling at the sidewall often developed along with lining cracks, especially inclined cracks (Figure 10.4a). Besides, secondary concrete lining collapse is also a serious seismic damage occurred. In the 1995 Chi-Chi earthquake (Wang et al., 2001), the 2008 Wen-Chuan earthquake (Li et al., 2012; Yu et al., 2016b, 2016c), and the 2016 Kumamoto earthquake (Zhang et al., 2018), lining spalling and collapse were representative seismic damages in numerous mountain tunnels, such as the No. 1 San-I railway

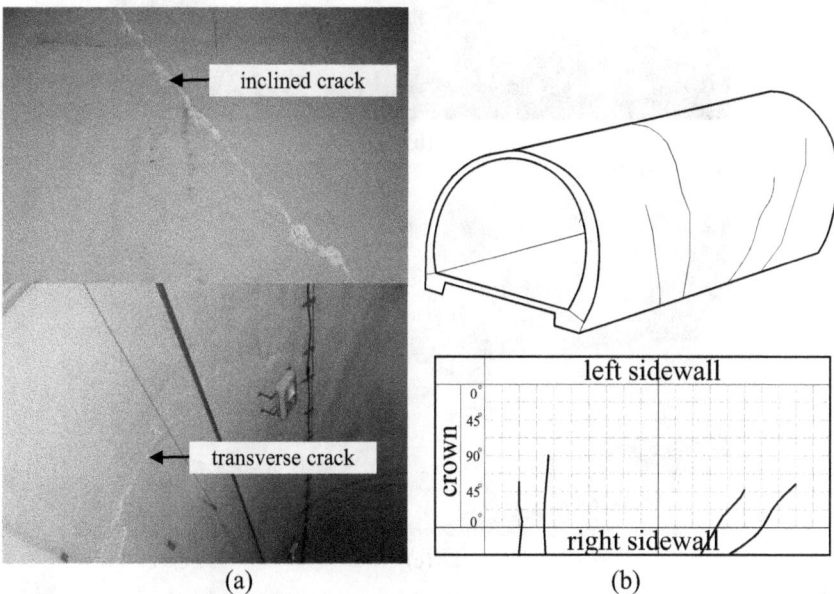

Figure 10.3 Transverse and inclined cracks of tunnel lining after earthquakes. (a) Transverse and inclined crack in the Tawarayama tunnel; (b) sketch and mapping result of transverse and inclined crack.

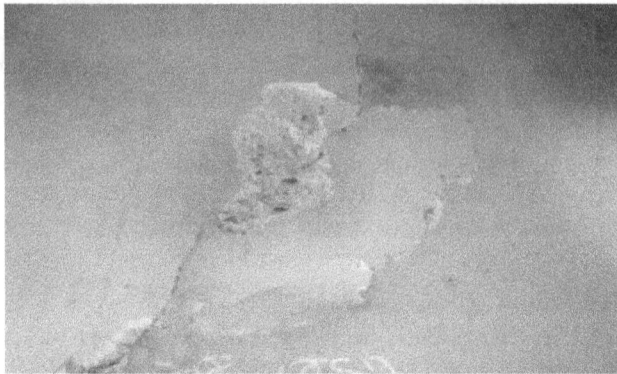

(a)

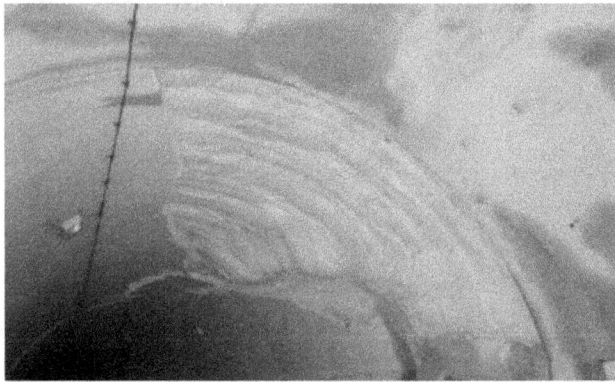

(b)

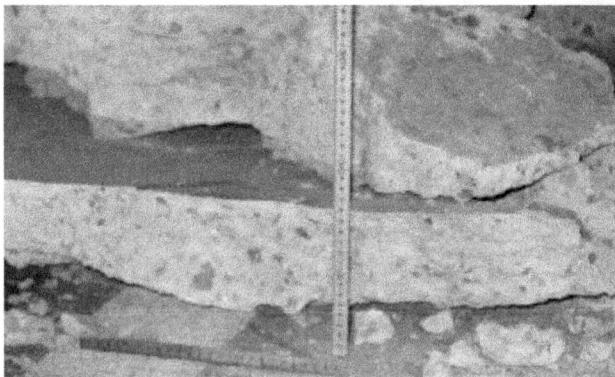

(c)

Figure 10.4 Spalling and collapse of lining concrete after earthquakes. (a) Concrete lining spalling along with inclined crack; (b) large area vault collapse of the crown at sections of S166 and S167; (c) concrete lining fallings.

tunnel, the Loingxi tunnel, and the Tawarayama tunnel (sections of S166 and S167 as shown in Figures 10.4b and 10.4c).

10.1.3 Pavement Damage

Figure 10.5 shows pavement damage after earthquakes. Damage of this pattern can be further classified into three types: transverse fracture and dislocation (Figure 10.5a), maintaining roadway uplift (Figure 10.5b), and invert uplift. In the Tawarayama tunnel, uplift and cracking of pavement are frequently observed at the portal with maximum of 55 cm (Zhang et al., 2018).

10.1.4 Groundwater Leakage

Figure 10.6 shows the groundwater leakage of tunnel lining after earthquakes. The pattern of groundwater leakage can be leakage through construction

Figure 10.5 Pavement damage after earthquakes. (a) Transverse cracking of pavement; (b) maintaining roadway uplift at the left side of the maintaining roadway.

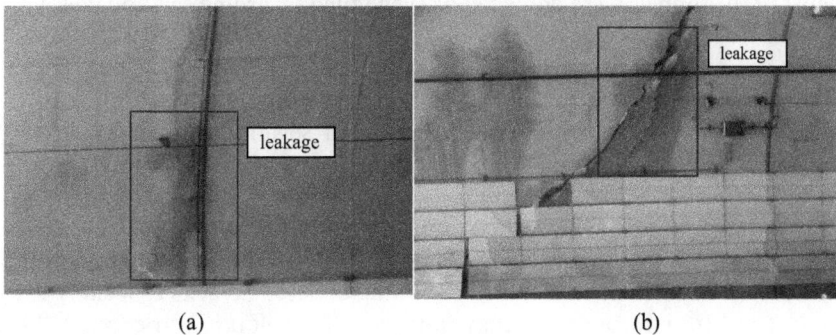

Figure 10.6 Groundwater leakage of tunnel lining after earthquakes. (a) Leakage in construction joint; (b) leakage in lining concrete with lining crack.

joints (Figure 10.6a) and leakage in tunnel linings (Figure 10.6b). The latter one often occurs in the concrete lining with cracks or concrete spalling. Large areas of leakage were found in the No. 2 and No. 3 lines of the Guanyin tunnel and the Old Guguan tunnel after the 1999 Chi-Chi earthquake (Chen et al., 2012). Seventeen groundwater leakages were counted totally in the Tawarayama tunnel after the 2016 Kumamoto earthquake (Zhang et al., 2018). Three types of groundwater inrush including soakage, dropping, and pouring in construction joint were observed in six tunnels after the 2008 Wenchuan earthquake (Wang et al., 2009b). Leakage may occur where groundwater is abundant. And the seismic damages to concrete lining and pavement due to earthquakes discussed above are preconditions for this damage pattern.

10.2 INFLUENTIAL FACTORS

A series of factors affecting the tunnel seismic response during earthquakes are summarized as follows: earthquake parameters, tunnel burial depth, rock mass quality, fault zone, and slope at portal.

10.2.1 Earthquake Parameters

Magnitude, focal depth, and epicentral distance are three key influential factors for earthquake parameters. Seismic waves with higher magnitude and shallower focal depth will cause tunnels to be under greater seismic forces if they are closer to the epicenter (Chen et al., 2012). In the 2016 Kumamoto earthquake, the Tawarayama tunnel is about 22.4 km away from the mainshock epicenter at a focal depth of about 12 km with a magnitude of 7.3 (Mj) (Zhang et al., 2018, 2020). The investigation results are in accordance with previous results after analyzing 192 cases of underground structures influenced by 78 earthquakes (Sharma & Judd, 1991; Chen et al., 2012). Most of the tunnels were damaged or even failed regarding earthquakes with magnitude 7 and above. The percentage of the damaged tunnel was up to 71% when the epicentral distance is less than 25 km. The percentage is near 75% when the epicentral distance is less than 50 km.

Additionally, the propagation direction of seismic waves may significantly affect the tunnel's seismic response and the resulting damages. A general axial deformation mode and mechanism of tunnels under seismic waves are depicted in Figure 10.7. When axial compressional/tensile stresses along the tunnel extension direction exceed the corresponding compressional/tensile strength of the lining concrete, cracks especially ring cracks and transverse cracks may initiate and develop (Chen et al., 2011). The response displacement method was also taken to illustrate that ring crack occurs on the lining when the actual strain in the lining concrete exceeds its ultimate strain (Yu et al., 2013). Numerical results in Chapter 7 could also provide further insight for seismic wave propagation direction

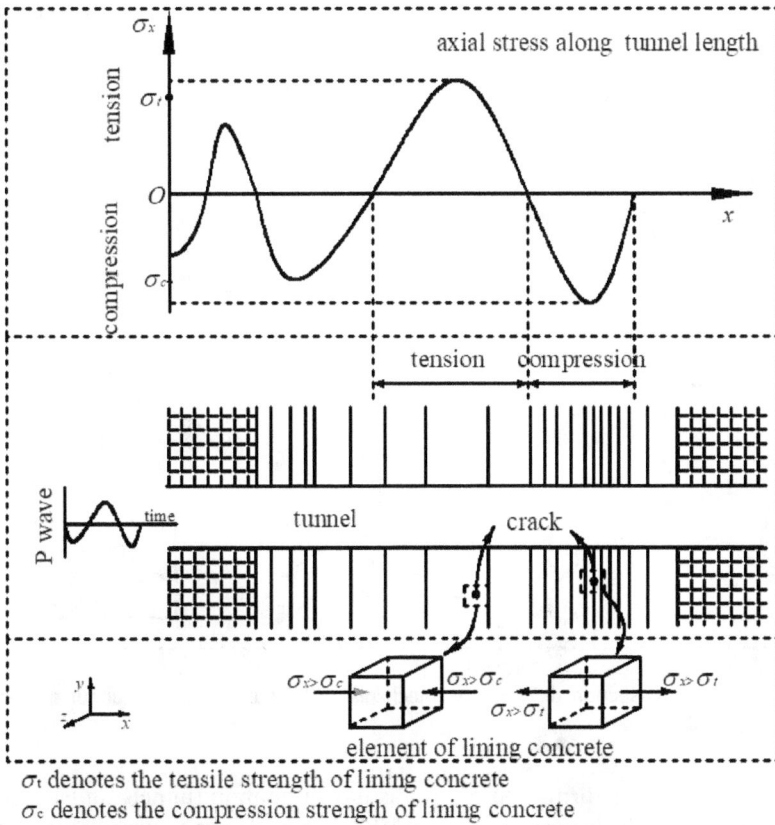

σₜ denotes the tensile strength of lining concrete
σ꜀ denotes the compression strength of lining concrete

Figure 10.7 Axial deformation mode and mechanism of tunnels under seismic waves.

influence. Axial stresses dominate in the case of P wave propagation with a shallow incidence angle of 15° (nearly parallel to the axis of tunnel lining). Ring/transverse cracks might occur due to the dominant axial deformation with tension and compression. Circumferential stresses become dominant in the case of a slightly larger incidence angle of 45°. Longitudinal cracks along the tunnel axis are likely to occur due to the superposition of axial deformation and curvature deformation. Besides, shear wave at 45° incidence angle to the tunnel axis is verified to have a significant effect on the initiation and propagation of longitudinal cracks by the finite element method (FEM) (Chen & Yu, 2006). Dislocation of the Futagawa fault zone obliquely crossing the axis of the Tawarayama tunnel with an estimated maximum of 2.2 m (in Mashiki) after the 2016 Kumamoto earthquake accelerated the tunnel axial deformation (Zhang et al., 2018). The lining cracks and invert failure due to compression/tension in the Tawarayama tunnel are the best examples of these types of failures (Zhang et al., 2020).

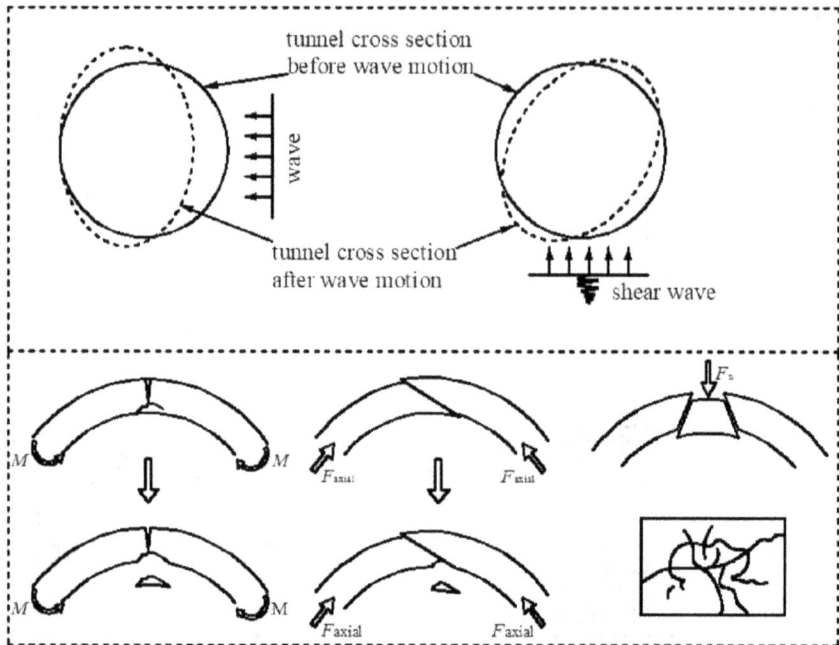

Figure 10.8 Cross-section deformation mechanism of mountain tunnels under a seismic shear wave.

Cross-section deformation mechanism of mountain tunnels under a seismic shear wave is depicted in Figure 10.8. When a seismic wave propagates normal or nearly normal to the tunnel axis, the shape of the tunnel cross-section would be distorted, resulting in the development of ovaling deformation or compression deformation of the tunnel cross-section. Moment and axial force then vary along with these deformations. Once local stress or moment surpasses corresponding strength, lining spalling may occur with the interaction of squeezing, and even highly excessive compression may cause a collapse of the concrete lining. In the 2016 Kumamoto earthquake, seismic waves propagated obliquely crossing the tunnel axis, therefore, the component of seismic wave normal to the tunnel axis could lead to a such seismic response for the tunnel cross-section (Zhang et al., 2018, 2020).

10.2.2 Tunnel Burial Depth

Tunnel burial depth is considered as one major factor for tunnel seismic damages. Chen et al. (2012) and Sharma et al. (1991) regard the burial depth of 50 m as the limit for the damage probability level for an underground structure after earthquakes concerning the results of site investigation, while the

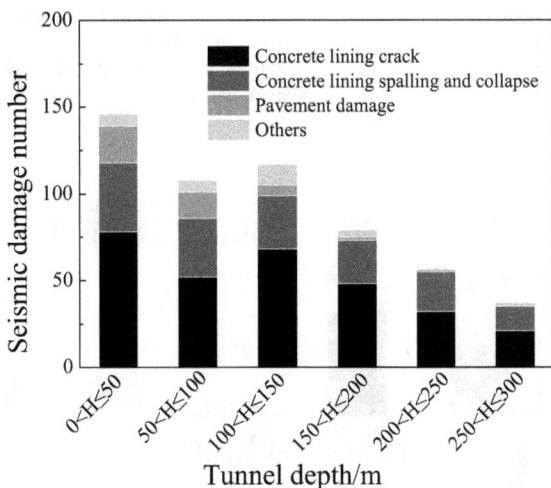

Figure 10.9 Relationship between the Tawarayama tunnel depth and resulting seismic damages due to the 2016 Kumamoto earthquake.

Source: Zhang et al., 2020.

burial depth threshold value in the Tawarayama tunnel after the 2016 Kumamoto earthquake is 150 m (Figure 10.9). The difference in the depth limit is related to rock mass quality and other geological conditions (Li, 2012).

10.2.3 Rock Mass Quality

Rock mass quality has a great impact on the tunnel's seismic damage. When tunnels run through the rock mass with poor quality, a higher degree of seismic damages can be noted due to poor mechanical properties, large damping and limited ability to constrain the lining structure of loose deposits, and broken and low-strength rock masses (Li et al., 2012). Figure 10.10 shows the distribution of rock type and seismic damage frequency in the Tawarayama tunnel after the 2016 Kumamoto earthquake (Zhang et al., 2020). When tunnel sections run through the talus and old talus deposits, seismic damages occur more frequently, especially at the portal. For the Andesite in the deeper mined part, tunnel structures in the crushed one are more vulnerable to seismic damages than those in the dense one due to the poor mechanical properties of the crushed Andesite. Besides, when grounds with different rock grades meet around the tunnel, seismic damages suffered by tunnel structure during an earthquake usually occur due to the ground squeeze in soft ground or ground relative deformation at the intersection of different grounds. Seismic damage at change zones between different grade rock masses is shown in Figure 10.11 (Zhang et al., 2018, 2020).

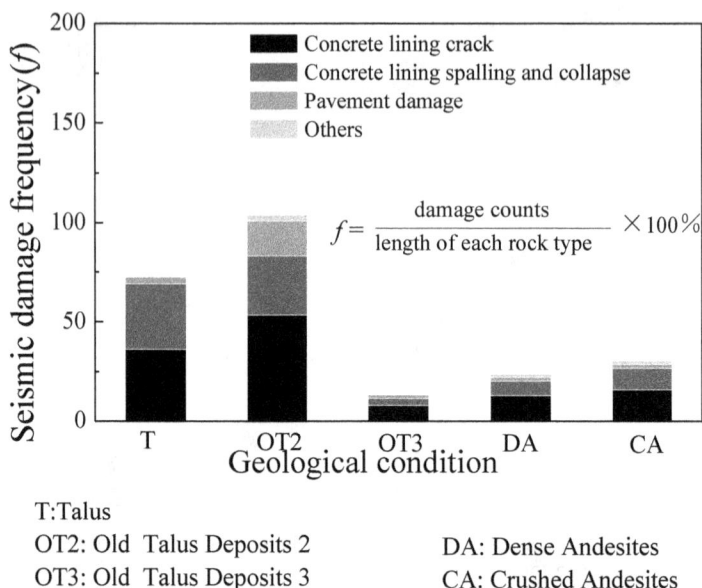

Figure 10.10 Distribution of rock type and seismic damage frequency in the Tawarayama tunnel subjected to the 2016 Kumamoto earthquake.

Source: Zhang et al., 2020.

10.2.4 Fault Zone

Tunnels through fault or shear areas or into a large plastic area are more likely to show collapse after earthquakes. One fault was detected to cross the axis of the tunnel during the site—investigation of the Tawarayama tunnel after the 2016 Kumamoto earthquake, which is shown in Figure 10.12 (the black dashed line). The serious collapse of the secondary lining (S167) occurred due to the large shear movement of fault as illustrated by Area C in Figure 10.12. Besides, such fault geological condition often influences the contact status between the tunnel lining and its surrounding rock mass. The stress and displacement behaviors of tunnel lining strongly after earthquakes depend on the contact status (Son & Cording, 2007). In the Tawarayama tunnel, some gravels at the waterproof for the lining spans crossing the afore-mentioned fault were observed. Therefore, the contact between the primary and secondary lining was imperfect, which intensified the dynamic response of the lining structure during earthquakes as discussed in Chapters 7–9.

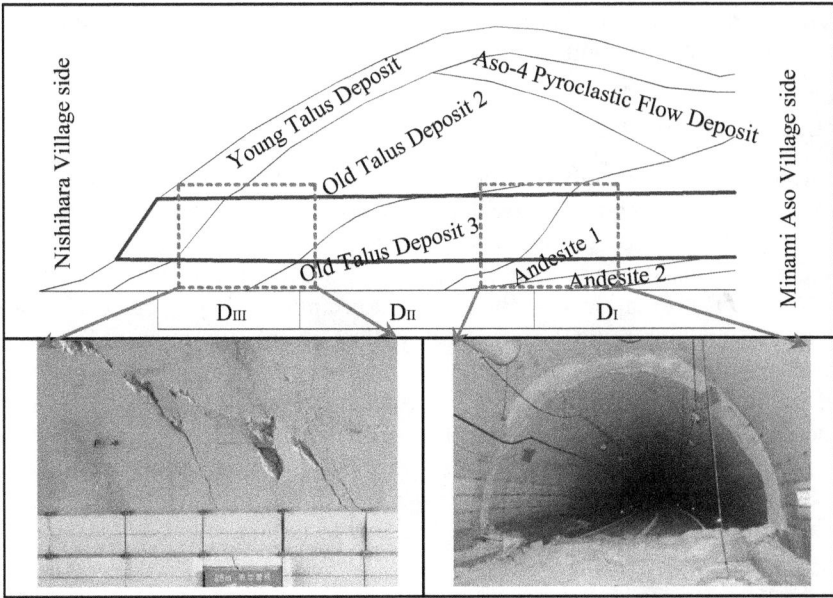

Figure 10.11 Seismic damage at change zones between different grade rock masses.

Source: Zhang et al., 2018, 2020.

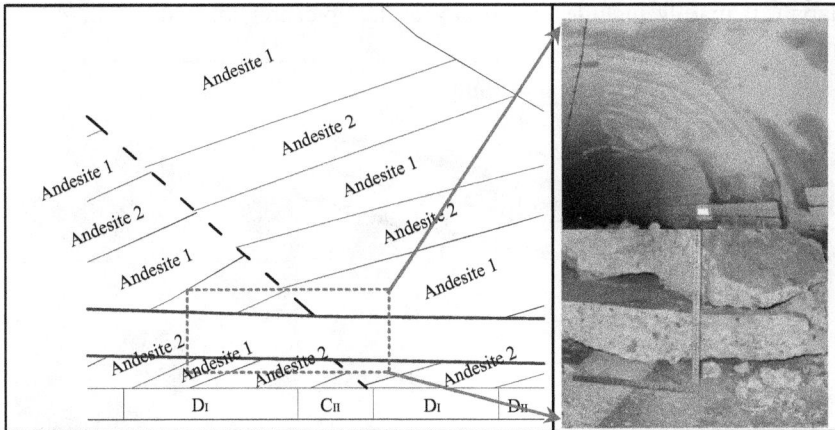

Figure 10.12 Lining collapse at the fault fracture zone.

Therefore, the contact condition between the tunnel lining and its surrounding rock is one significant factor that cannot be ignored for the seismic performance of underground structures during earthquakes.

10.2.5 Slope at Portal

Portal of tunnels are often buried underneath special geological condition such as slopes. Landslides are likely to occur due to the unsymmetrical loading of the slope during earthquakes. As a result, the lining structure underneath this ground might be forced to move toward the direction of the landslide. Deformation due to compression and bending occurred, which even induces longitudinal cracks and spalling of concrete lining. Moreover, the dislocation of the tunnel lining due to the movement could result in uplift of the maintaining roadway. Figure 10.13 shows the slope deformation at the portal and the resulting tunnel seismic damages (Zhang et al., 2020). Unsymmetrical loading due to the slope above the tunnel moved the lining span towards the south direction (lower side of the portal slope) by 10 cm. The tunnel lining was eventually sheared off to failure.

10.3 CASE STUDY

10.3.1 The 2016 Kumamoto Earthquake

The 2016 Kumamoto earthquake is a series of seismic events that occurred beneath the Kumamoto City of Kumamoto Prefecture in Kyushu Region, Japan. It mainly includes a foreshock (the epicenter located at 32.742 N,

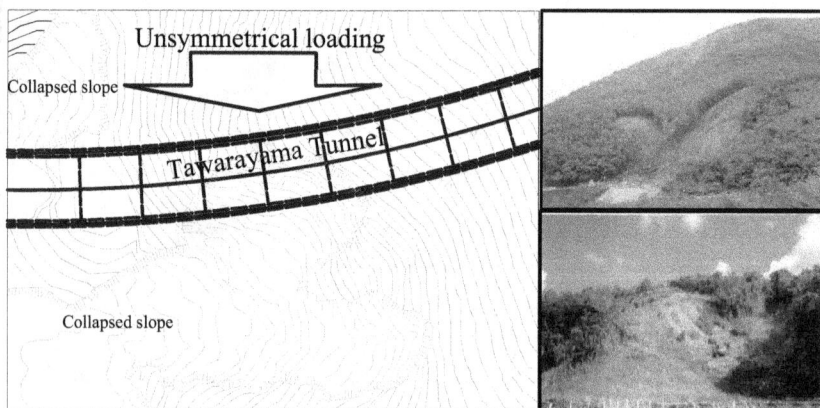

Figure 10.13 Slope deformation at the portal and the resulting tunnel seismic damages.
Source: Zhang et al., 2020.

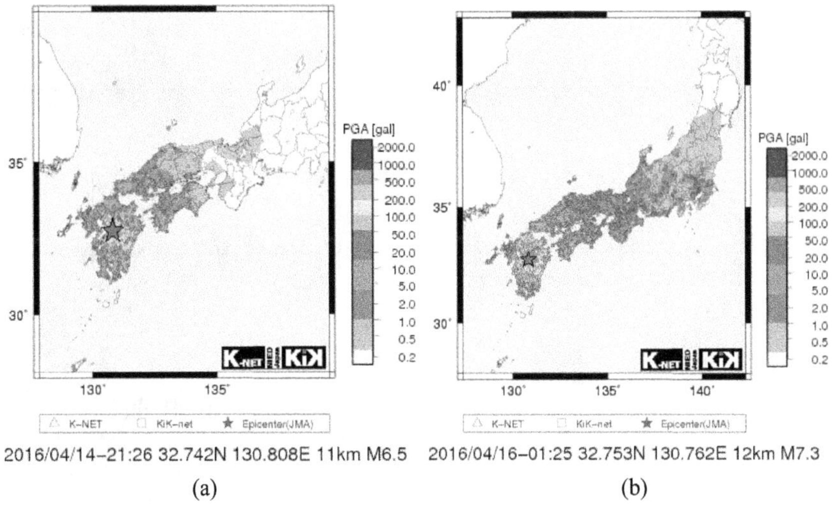

2016/04/14–21:26 32.742N 130.808E 11km M6.5   2016/04/16–01:25 32.753N 130.762E 12km M7.3

(a)   (b)

Figure 10.14 Peak acceleration distribution of the Kumamoto earthquake in Japan. (a) Peak acceleration contour map of foreshock on April 14, 2016; (b) peak acceleration contour map of mainshock on April 16, 2016.

Source: Data from K-NET, the National Research Institute for Earth Science and Disaster Prevention of Japan.

130.808 E) with a magnitude of 6.5 (Mj) at 21:26 JST on April 14, 2016, at a depth of about 11 km, and a mainshock with a magnitude of 7.3 (Mj) (the epicenter located at 32.753 N, 130.762 E) at 01:25 JST on April 16, 2016, at a depth of about 12 km (Asian Disaster Reduction Center, 2016). Figure 10.14 illustrates the distribution of peak acceleration of both foreshock and mainshock according to the National Research Institute for Earth Science and Disaster Prevention of Japan. The earthquake caused significant damage and disruption to lifeline systems, buildings, bridges, and transportation structures in Kumamoto Prefecture, especially in the areas near the fault (e.g., Mashiki Town, Nishihara Village, and Minami Aso Village). A total of 8550 buildings, mostly in Kumamoto City, were seriously damaged or collapsed. Fifty people lost their lives, mostly because of landslides or the collapse of buildings (Cabinet Office of Japan, 2016).

The acceleration waves measured at the Nakamatsu observation site during the Mj7.3 Kumamoto earthquake on April 16, 2016, are depicted in Figure 10.15. The Nakamatsu observation site is located at the northeast of the epicenter with a distance of 32.3 km. Acceleration of the mainshock from the time of 15 seconds to 30 seconds at the Nakamatsu observation site is shown in Figure 10.15b. The seismic records consist of three components along the NS (North-South), EW (East-West), and UD (Up-Down) directions. The NS

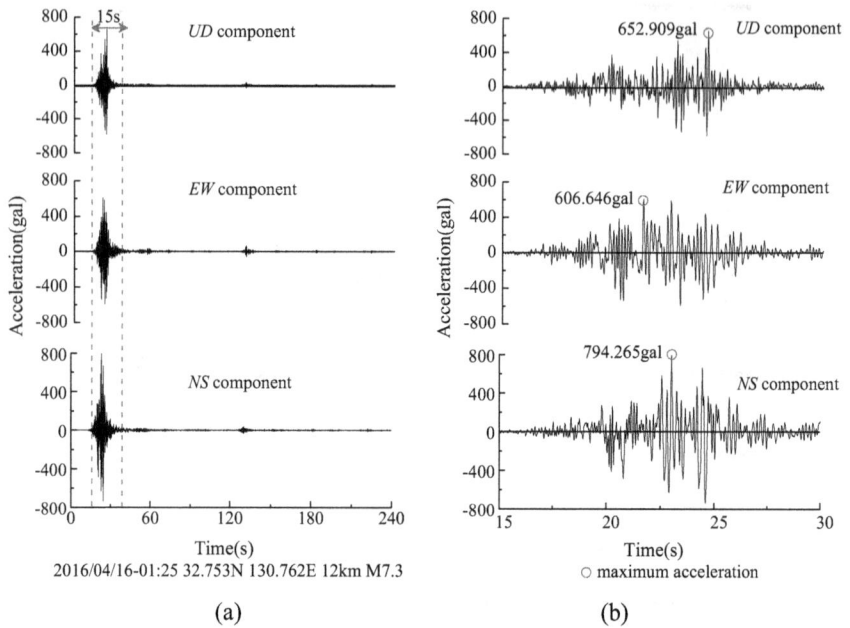

Figure 10.15 Acceleration waves measured at the Nakamatsu observation site during the Mj7.3 Kumamoto earthquake on April 16, 2016. (a) Acceleration waves of the NS, EW, and UD components; (b) acceleration wave of the NS, EW, and UD components from the time of 15 s to 30 s.

Source: Data from GSI, the Geospatial Information Authority of Japan.

and EW waves represent the horizontal motions of the ground surface, and the UD wave represents the vertical motion of the ground. Maximum values of the NS, EW, and UD accelerations are 794.3 gal (cm/s^2), 606.6 gal (cm/s^2), and 652.9 gal (cm/s^2), respectively. Table 10.1 lists measured maximum ground acceleration at various observation sites induced by the mainshock. Since the EW and NS components have larger seismic amplitude, damage to rock foundations and buildings caused by horizontal motion is much more severe than those caused by vertical motion. Besides, for the mainshock, at the Kumamoto GEONET station (32.8421 N, 130.7648 E), 0.75 m horizontal deformation in the ENE direction and 0.20 m downward deformation were recorded, at the Choyo GEONET station (32.8707 N, 130.9962 E), 0.97 m horizontal deformation in the SW direction and 0.23 m upward deformation were recorded (Goda et al., 2016). Field investigation by Lin et al. (2016) also showed that the horizontal displacement caused by the seismic event along fault accounted for a larger proportion.

Table 10.1 Measured Maximum Ground Acceleration at Various Observation Sites Induced by the Mainshock

Site Code	Site Location	Intensity	Maximum Ground Acceleration (gal = cm/s²)			Epicenter Distance (km)
			NS	EW	UD	
9CF	Matsubase-machi, Uki-shi	6	492.8	342.6	313.9	14.2
EEB	Kasuga, Nishi-ku	6	606.0	551.6	405.3	7.5
EED	Nakamatsu, Minami Aso Village	6	794.5	606.8	653.1	32.3
9D2	Ōyano-machi, Kami-amakusa-shi	6	262.1	334.4	122.3	36.3
5E5	Hirayamashin-machi, Yatsushiro-shi	5	171.8	175.6	82.5	34.6
9D0	Ashikita, Ashikita-machi	5	138.6	124.9	41.4	56.9
EF0	Nishiaida Shimo-machi, Hitoyoshi-shi	5	111.7	102.0	50.4	61.2

10.3.2 Tawarayama Tunnel Project

The Tawarayama tunnel is located at a distance of about 22.4 km from the epicenter of the mainshock (Mj7.3) as shown in Figure 10.16. The total length of the tunnel is 2057 m with a horseshoe-shaped cross-section. Figure 7.2 in Chapter 7 shows the typical cross-section of the tunnel. The typical cross-section has a total width of 10.20 m and a maximum height of 7.97 m. Figure 10.17 presents the geological profile of the tunnel. Its maximum overburden is about 300 m. The Tawarayama tunnel runs through three different formations: the Quaternary Holocene, the Quaternary Pleistocene, and the Tertiary Pliocene. The portal area is excavated in talus and early stage talus deposits composed of welded tuff, gravel, silt, and clay. The tunnel is excavated in the Andesite lava. Based on the Japanese Technical Standard for Structure Design of Road Tunnel (JARA, 2003), rock mass along the tunnel (Figure 10.17) is organized into four classes: C_{II}, D_I, D_{II}, and D_{III}.

Excavation of the Tawarayama tunnel uses the New Austrian tunneling method (NATM). NATM is assumed to be much better than the traditional method based on the conditions after earthquakes. This is because the interaction between the surrounding rock and tunnel using NATM performs better than that using the traditional method (Chen et al., 2012). Support systems of the tunnel consist of primary support, waterproof layer, and secondary support. The primary support includes shotcrete (0.10 m, 0.15 m, 0.20 m, and 0.25 m for rock class C_{II}, D_I, D_{II}, and D_{III}, respectively) and rockbolt. For rock class D_I, D_{II}, and D_{III}, the rockbolts are distributed on a

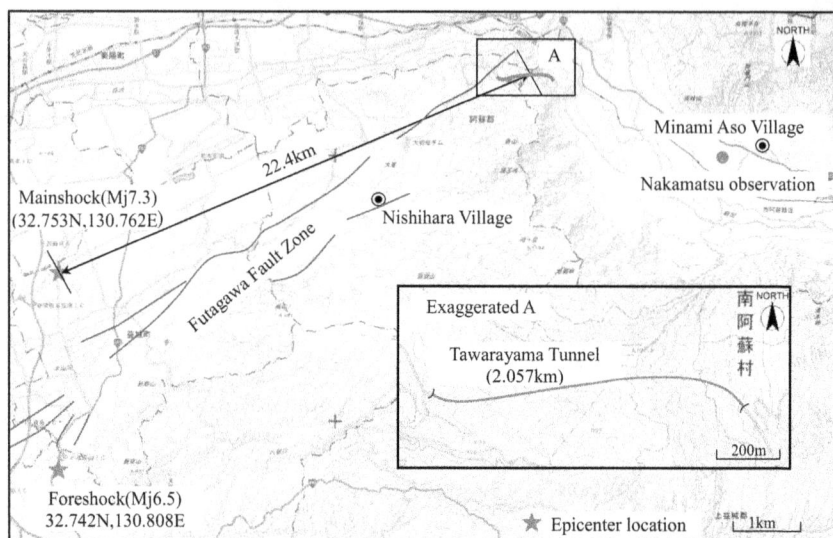

Figure 10.16 Location of the Tawarayama tunnel and epicenter of the Kumamoto earthquake.

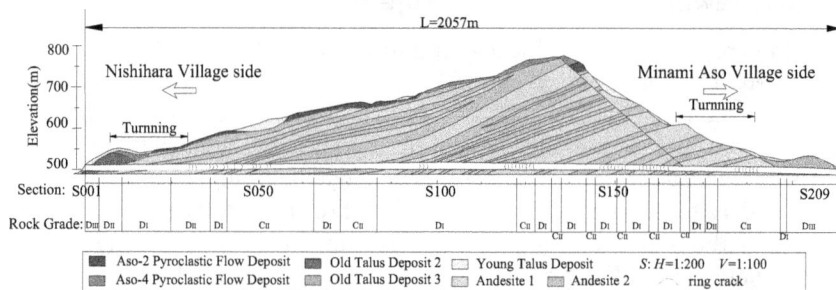

Figure 10.17 Geological profile of the Tawarayama tunnel.

grid of 1.2 m × 1.0 m and have a length of 4.0 m. For rock class C_{II}, the rock-bolts are distributed on a grid of 1.5 m × 1.2 m and have a length of 3.0 m. Besides, for rock class D_{III}, forepoling is conducted, especially at the portals. The rockbolts are spaced on a grid of 0.60 m × 1.0 m with a length of 3.0 m. The secondary lining is reinforced concrete with a thickness of 0.30 m.

10.3.3 Seismic Damages Statistics

After the 2016 Kumamoto earthquake, our research group visited Kuma-moto City to conduct field investigations on the seismic damages of the underground structures. Preliminary information on tunnel damage was gathered by performing quick visual inspections. With the help of the Kumamoto River and National Highway Office, the Kyushu Regional Development Bureau, and the Ministry of Land, Infrastructure, Transport and Tourism of Japan, detailed surveys were performed by using lining crack mapping, photo recording, and measuring characteristics of major cracks (including width, depth, and relative displacement direction). Numerous patterns of seismic damage to the Tawarayama tunnel were observed as follows: lining cracks, construction joint damage, groundwater leakage, spalling and collapse of concrete lining, and pavement damage. Figure 10.18 presents the distribution of seismic damages to the Tawarayama tunnel by the 2016 Kumamoto earthquake.

Among the seismic damages, the most frequently observed damage is the lining crack with an estimated probability of 66.5%. It can be classified into ring crack, transverse crack, longitudinal crack, and inclined crack. The construction joint damage, the concrete lining spalling/collapse, and the groundwater leakage are less and less common. The statistic and proportion of seismic damages to the Tawarayama tunnel by the 2016 Kumamoto earthquake are shown in Figure 10.19. The lining cracks, especially ring cracks, mainly occurred in the sections of S004–S005, S028–S053, S067–S071, S094–S097, S119–S126, S146–S168, and S184–S190 (S stands for

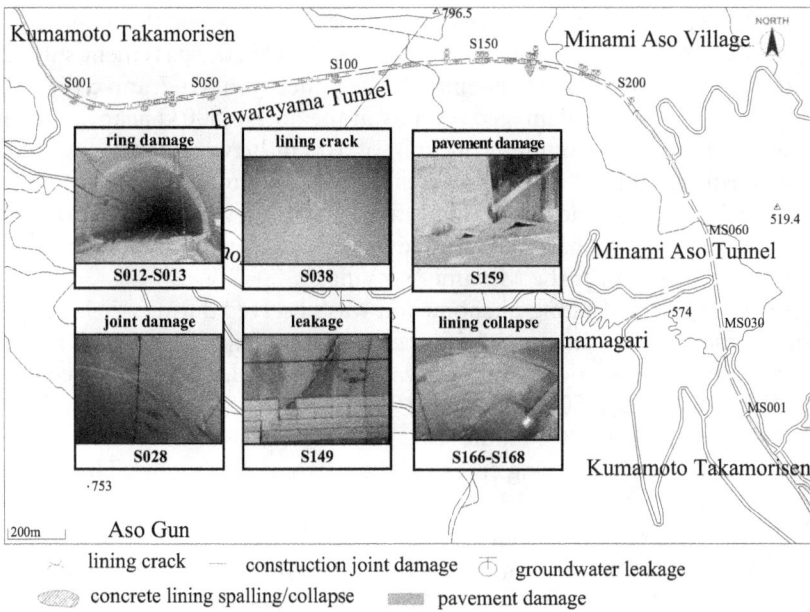

Figure 10.18 Distribution of seismic damages to the Tawarayama tunnel.

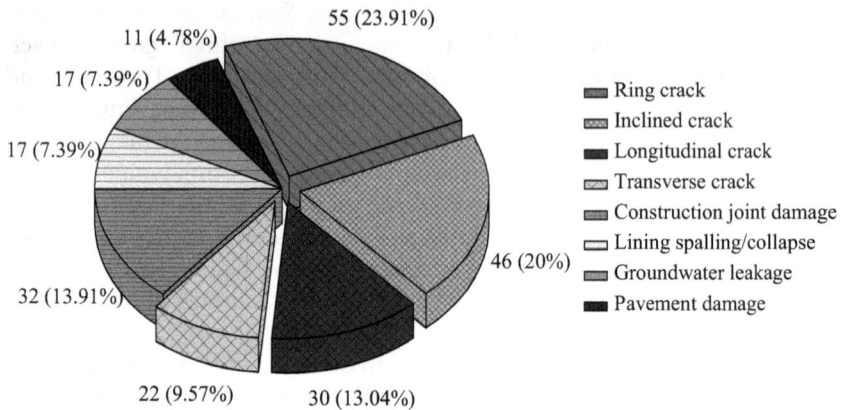

55 (23.91%)

11 (4.78%)

17 (7.39%)

17 (7.39%)

32 (13.91%)

22 (9.57%)   30 (13.04%)

46 (20%)

■ Ring crack
▨ Inclined crack
■ Longitudinal crack
▨ Transverse crack
▤ Construction joint damage
▭ Lining spalling/collapse
▤ Groundwater leakage
■ Pavement damage

Figure 10.19 Statistics and proportion of seismic damages to the Tawarayama tunnel by the 2016 Kumamoto earthquake.

Section), with a total length of 783.2 m (38.1% of the Tawarayama tunnel with a total length of 2057 m). Figure 10.20 shows a sketch of the seismic damages in these sections of the Tawarayama tunnel.

At the portal (S001) near the Nishihara Village, pavement in the sections from S001 to S030 was observed to uplift and crack continuously. And most of the pavement uplift developed on the left side along the maintaining roadway. Besides, in the sections from S158 to S168, the pavement suffered cracking with a maximum opening of 10 cm between S167 and S168. The construction joint was damaged severely at the portal (S001) near the Nishihara village with a maximum opening of 10 cm between S001 and S002. And construction joint opening decreased with the tunnel extending to the Minami Aso Village side. From the section of S098, severe construction joint damage was seldom observed. Groundwater leakage mainly occurred along with lining cracks near the Minami Aso Village.

A special and interesting phenomenon was observed during site investigation for the Tawarayama tunnel 55 ring cracks were distributed with regular spacing in 23.4% sections of the Tawarayama tunnel. They mainly concentrated in S030–S053, S095–S097, S119–S126, S146–S153, S157–S165, and S184–190 (Figure 10.20). Figure 10.21 illustrates the estimated spacing of the regularly distributed ring cracks in the Tawarayama tunnel. The spacing was estimated to be about 10.0 m.

The general seismic propagation direction during the strong earthquake is oblique to the tunnel axis, as discussed in section 10.3.1. Seismic waves parallel or obliquely crossing the tunnel axis cause longitudinal motion of the tunnel involving axial tension, compression, and bending deformation. Thus, once the concrete's corresponding (compression or tensile) strength is

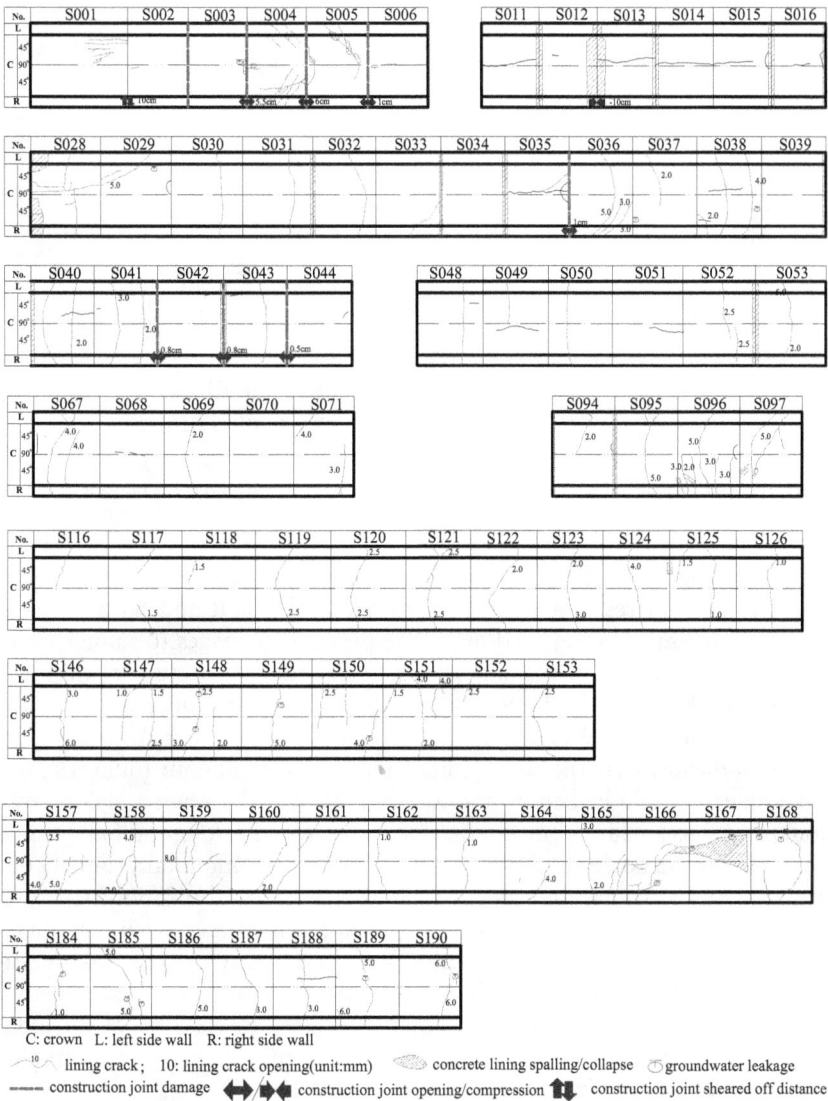

Figure 10.20 Sketch of the seismic damages in these sections of the Tawarayama tunnel.

reached, a ring crack is expected. This coincides with the ring cracks observed in the Tawarayama tunnel. Meanwhile, longitudinal motion is aggravated by the geological conditions to form this special phenomenon. According to the in-situ geological investigation, dense Andesite and crushed Andesite along the Tawarayama tunnel appear in tilt alternately with space between 10 m

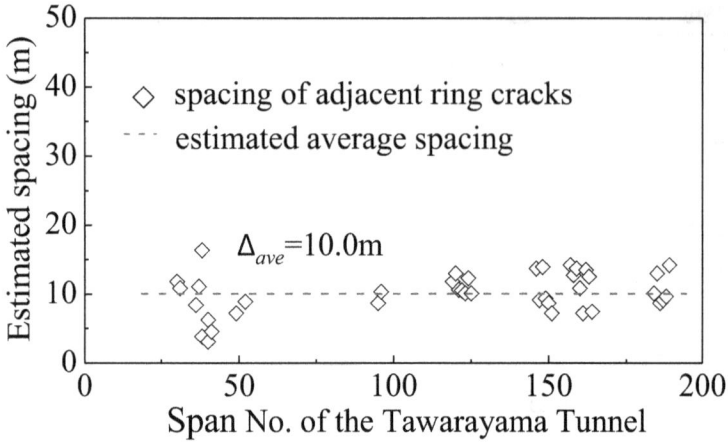

Figure 10.21 Estimated spacing of continuous ring cracks in the Tawarayama tunnel.

and 20 m (Figure 10.17). Because of different wave velocities, wave dispersion and ground resistance in soft and hard rock, soft and hard grounds behave differently during earthquakes. So seismic damages to tunnel structures normally occur in the soft ground or at the intersection of different rock grades by ground relative displacement or ground squeeze where soft and hard grounds meet (Yu et al., 2016b, 2016c).

This indicates that the longitudinal motion of mountain tunnels under earthquakes should be better monitored in the seismic design and construction procedure. Mitigation countermeasures for the longitudinal seismic response can be taken into consideration for further mountain tunnel construction and remediation process, such as ring shock absorption structure. It can absorb longitudinal seismic energy while maintaining the intact horseshoe shape of the tunnel cross-section with full ability to undertake vertical pressure from the surrounding rock and other external forces. Further detailed studies on the theoretical and engineering mechanism of the axial regularly distributed ring damages in mountain tunnels by the earthquake, and corresponding mitigation countermeasures such as ring shock absorption structure should be investigated.

10.4 CONCLUSION

The chapter summarizes the seismic performance of rock tunnels under earthquake loading. Influential factors including earthquake parameters, burial depth of tunnel, surrounding rock condition, fault zone, and slope at portal are discussed. Meanwhile, the Tawarayama tunnel in Kumamoto, Japan, subject to the 2016 Kumamoto earthquake is investigated as one illustration for the seismic damage assessment. A special and interesting phenomenon was observed: 55 ring cracks were distributed with an estimated average spacing of 10.0 m in 23.4% sections of the Tawarayama tunnel. This results from the interaction between seismic waves and special geological conditions around the Tawarayama tunnel with the presence of alternating layers of dense Andesite and crushed Andesite with spacing ranging from 10 m to 20 m.

Chapter 11

Quick Ground Deformation–Based Seismic Damage Assessment Method

11.1 BASIC PRINCIPLE

Ground shaking and deformation induced by earthquakes may generate tremendous forces on long and/or large structures, such as rock engineering structures (Aydan, 2017). It can yield relevant information about the evolution of the temporal and spatial distribution of the ground deformation. Therefore, it is essential to understand the causes, triggering factors, and mechanisms to delineate the most affected areas and achieve accurate assessment and mitigation of natural and anthropogenic hazards. Recent earthquakes, such as the 1995 Kobe earthquake, the 1999 Chi-Chi earthquake, the 2008 Wenchuan earthquake, and the 2016 Kumamoto earthquake have significantly influenced underground structures. Therefore, more and more attention has been paid to underground structures' seismic deformation and failure mechanism (e.g., Hashash et al., 2001; Wang et al., 2001; Yashiro et al., 2007; Wang et al., 2009b; Zhang et al., 2018). In Chapter 10, efforts were spent to study the underground structures' seismic damage assessment, including damage classification, influential factor analysis, and aseismic recommendation.

From the viewpoint of the geological condition, seismic damages are classified into three forms including damage by earthquake-induced ground failure, damage from fault displacement, and damage from ground shaking or vibration (Dowding & Rozen, 1978). However, there is a limited quantitative effort to classify the correlation between ground shaking and deformation and seismic damage to underground structures. The current study presents a quick ground deformation–based seismic damage assessment method aiming at detecting the ground deformation characteristics and studying its correlation with seismic damage to tunnels during earthquakes. The method includes data acquisition for the ground surface, 3-D ground deformation (horizontal and vertical displacement vectors) detection, and seismic damage investigation, as shown in Figure 11.1.

DOI: 10.1201/9781003401599-11

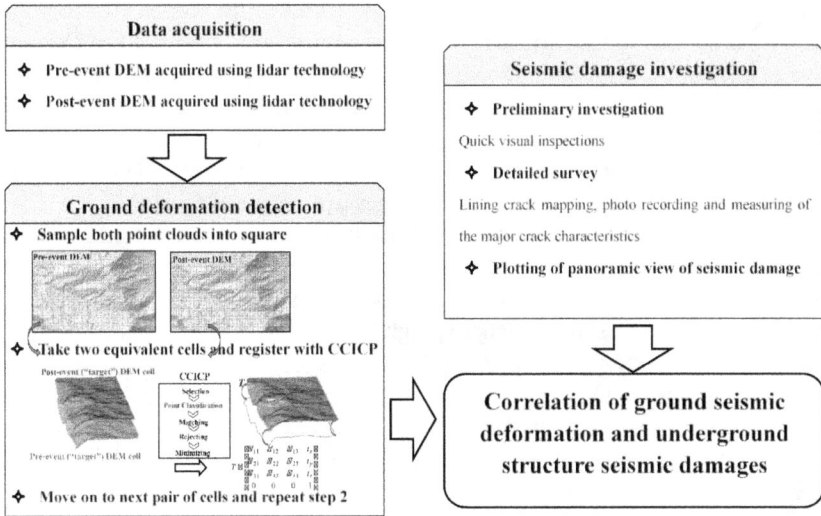

Figure 11.1 Schematic diagram of the quick ground deformation-based seismic damage assessment method.

11.1.1 Data Acquisition for Ground Surface

Inland earthquakes with a magnitude greater than 7 can cause ground shaking and deformation (Park et al., 2018). Record of the RTK-GNSS (real-time kinematic–global navigation satellite system) time series showed that strong shaking and deformation caused by earthquakes could be observed at the stations within 100 km of the epicenter. Clear permanent deformation was observed at nearby stations (Kawamoto et al., 2016a). Remarkable horizontal displacements were also observed by satellite synthetic aperture radar (SAR) analysis around the depressional (graben-like) rupture area (Nakano et al., 2016). The first step of the ground deformation observation is to acquire the ground deformation data. Currently, a series of methods have been applied. Expect for the methods mentioned above, various satellites and airborne remote sensing technologies by government agencies and aerial survey companies in Japan were applied to conduct surveys. Asia Air Survey Co., Ltd. (2016) conducted airborne light detection and ranging (LiDAR) survey. Geospatial Information Authority of Japan (2016) conducted a high-resolution vertical and oblique aerial photography survey. These surveys can be used to detect ground deformation and other hazards caused by an earthquake (Yamazaki & Liu, 2016; Kawamoto et al., 2016a, 2016b; Moya et al., 2017). Each method has its advantages and disadvantages. Taking GNSS as an example, the results of the GNSS can estimate the ground displacements, while its spatial resolution is low. The nationwide GNSS Earth Observation Network (GEONET) operated by the Geospatial Information Authority of Japan (GSI) has one station with a coverage interval of 20–30 km (Yamagiwa et al., 2006). It is difficult to obtain the spatial distribution of the ground deformation for a narrow area. Taking SAR as another example, it can provide a better spatial resolution, whereas it requires a pair of images with the same viewing condition to calculate the displacement along the line of sight (LOS) of radar. More pairs of SAR images are required to obtain the 2.5-D or 3-D deformation field, which might not be realistic (Moya et al., 2017). As a result, LiDAR is adopted in the current chapter to obtain both the pre- and post-event DEM (Digital Elevation Model) datasets with the advantages of high accuracy, high sample density, high penetrative abilities, and minimum human dependence.

The airborne LiDAR system is composed of an inertial navigation system (INS), and a laser scanner (Shih et al., 2008). The system sends pulses of laser light toward the ground and records the return time for calculating the distance between the sensor and the ground surface (Lilles& et al., 2007; Moya et al., 2017). Currently, it is the most detailed and accurate method for creating digital elevation models. One of its major advantages compared to photogrammetry is the ability to filter out reflections from vegetation due to its high penetrative abilities. The model can represent the actual situation of the ground surfaces such as rivers, paths, and cultural heritage sites concealed by trees (Sturzenegger & Stead, 2009; Slob, 2010; Gigli et al., 2011; Riquelme et al., 2014; Riquelme et al., 2015). For rock mechanics, the system has been widely applied to rock characterization, slope stability detection, as well as ground deformation estimation (Muller & Harding, 2007; Nissen et al., 2014; Moya et al., 2017). Usually, the application of LiDAR technology is limited because only post-event LiDAR data are available in most cases.

11.1.2 Ground Deformation Detection

The CCICP algorithm was originally developed to accurately register the MMS (mobile mapping system) point cloud because the point clouds often differ when the same areas are scanned several times by MMS (Takai et al., 2013). In the current chapter, the pre-event DEM is marked with S as a source point cloud and the post-event DEM is marked with T as a target point cloud. Firstly, each set of points sampled from the two-point clouds is classified into linear points, planar points, and scatter points (Figure 11.2) depending on the results of the PCA (principal component analysis) method used by Demantke et al. (2011). Using this method, local distributions of the points are evaluated by the eigenvalues λ_1, λ_2, and λ_3 of the variance-covariance matrix, which are calculated from the position of each point and its neighbors. Linear-planar and scatter-planar correspondences are rejected

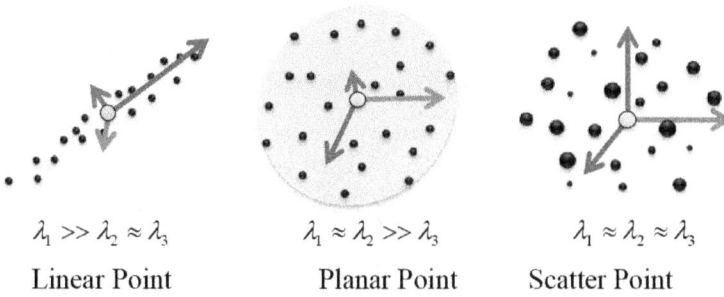

$\lambda_1 \gg \lambda_2 \approx \lambda_3$ $\lambda_1 \approx \lambda_2 \gg \lambda_3$ $\lambda_1 \approx \lambda_2 \approx \lambda_3$

Linear Point   Planar Point   Scatter Point

Figure 11.2 Point classification by PCA.

Source: After Oda et al., 2016.

as incorrect correspondences for accurate registration. After the above preprocessing, execute the calculation for minimizing point-to-plane and point-to-point distance. We applied point-to-plane distance minimization to planar-planar correspondences and point-to-point distance minimization to the other correspondences (Low, 2004; Takai et al., 2013; Oda et al., 2016).

Point-to-point distance D_{PT_PT} and point-to-plane distance D_{PT_PL} are derived as follows (Oda et al., 2016):

$$D_{pt\_pt} = \sqrt{\left|(\mathbf{T}\cdot\mathbf{p}_s - \mathbf{p}_t)\right|^2} \tag{11.1}$$

$$D_{PT\_PL} = \sqrt{\left|(\mathbf{T}\cdot\mathbf{p}_s - \mathbf{p}_t)\cdot\mathbf{n}_t\right|^2} \tag{11.2}$$

where T is a transformation matrix in the homogeneous coordinate system, p_s is a point in the source point cloud, p_t is the matching point in the target point cloud, and n_t is the normal vector of point p_t calculated by PCA. The transformation matrix T can be described by Equation (11.3).

$$\mathbf{T} = \begin{pmatrix} \alpha_{11} & \alpha_{12} & \alpha_{13} & t_x \\ \alpha_{21} & \alpha_{22} & \alpha_{23} & t_y \\ \alpha_{31} & \alpha_{32} & \alpha_{33} & t_z \\ 0 & 0 & 0 & 1 \end{pmatrix} \tag{11.3}$$

where α_{ij} (i,j = 1, 2, 3) are components of a rotation matrix and t_x, t_y, and t_z are the translation components. The algorithm computes 3-D displacements and rotations by iteratively minimizing the sum of the square difference of corresponding coordinates of points between local subsets of the pre-event ("source") and post-event ("target") data as illustrated in Figure 11.1.

11.1.3 Seismic Damage Investigation

Details on seismic damage investigation refer to Chapter 10. The ground deformation was spatially compared with seismic damages to tunnels to explore whether the seismic damages of underground structures are related to the ground deformation.

11.2 METHODOLOGY VALIDATION

A resulting comparison of the CCICP algorithm and the existing method is conducted to show the accuracy of the current method in the chapter. Mukoyama et al. (2017) detected the ground deformation using the 3D-GIV (geomorphic image velocity) method for Mt. Tawarayama after the 2016 Kumamoto earthquake. Figure 11.3 shows the deformation distribution detected using the 3D-GIV method. Westward vectors are dominant on the west side of the Tawarayama tunnel although southwestward vectors account for a large portion of the whole area. The horizontal displacement (combination of the East-West (WE), and North-South (NS) components) was also calculated by the CCICP algorithm as illustrated in Figure 11.4. The result of the CCICP method shows a good agreement with that of the 3D-GIV method.

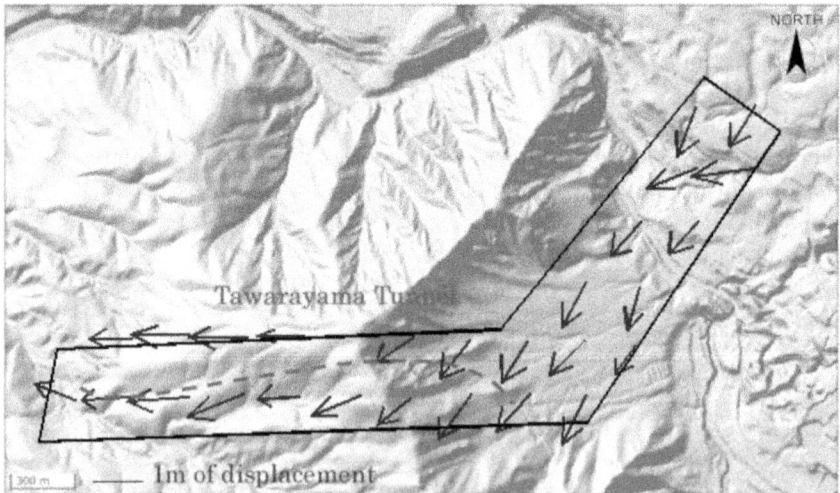

Figure 11.3 Horizontal displacement vectors by previous work (blue arrow denotes the horizontal displacement vector).

Source: Modified after Mukoyama et al., 2017.

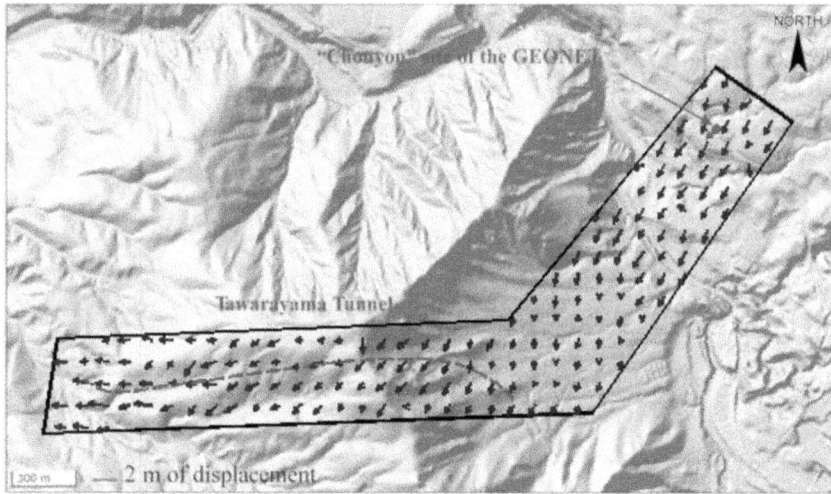

Figure 11.4 Horizontal displacement vectors detected by the CCICP method (blue arrow denotes the horizontal displacement vector).

An additional reference site station named Chouyou (location: 32.8707 N, 130.9962 E; code: EL04930274901) was investigated as well. It is applied to verify the accuracy of the detection method for ground deformation. During the 2016 Kumamoto earthquake, the real-time GEONET analysis (REGARD) system successfully observed and recorded the co-seismic displacement of each GEONET station (Kawamoto et al., 2016a, 2016b). Figure 11.5 illustrates the observed (red arrow) and detected (blue arrow) displacements of the horizontal component at Chouyou station. The Chouyou station was observed to move in 0.97 m horizontally in the southwest direction. Using the present method, the horizontal displacement around the Chouyou station was detected to be 0.97 m (an average value of the 6 points around the Chouyou station in Figure 11.5) in the southwest direction. The horizontal displacement detected by the CCICP method is consistent with that observed by the REGARD system. The consistency of the observed and detected deformation illustrates the accuracy of the CCICP method.

11.3 CASE APPLICATION—TAKING THE TAWARAYAMA TUNNEL AS ONE EXAMPLE

The Tawarayama tunnel with a total length of 2057 m was excavated under Mt. Tawarayama on the Takamori Line of Kumamoto Prefectural Route 28, as illustrated in Figure 11.6a. The height of Mt. Tawarayama is 1095 m. Most of the ground surface is covered with dense forests (Figure 11.6b). The

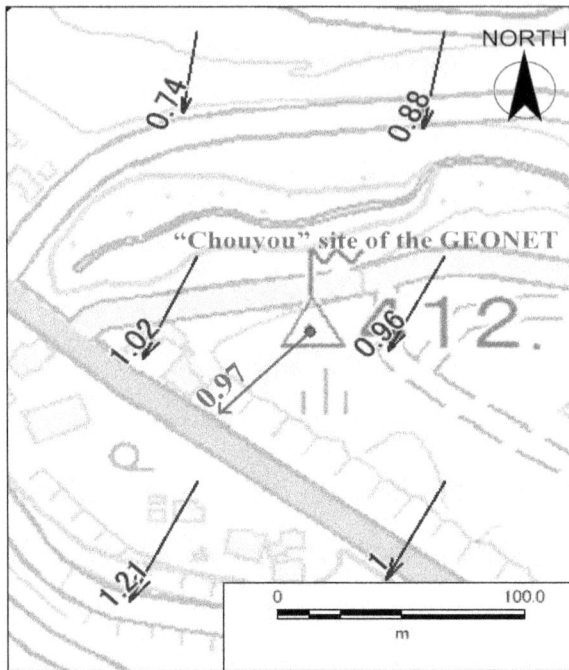

Figure 11.5 Enlarged map from Figure 11.4 at the site station of Chouyou (blue arrow denotes the detected horizontal deformation vector using the CCICP method, the red arrow denotes the observed horizontal deformation vector) (unit: meters).

tunnel is located about 22.4 km away from the epicenter of the mainshock (Mj7.3) and was destroyed severely by the earthquake. An overview of the seismic damages has been presented in Chapter 10. The study area was set to the ground of Mt. Tawarayama above the Tawarayama tunnel.

Airborne LiDAR acquired a pair of DEM datasets before and after the 2016 Kumamoto earthquake with the ability to filter out reflections of forests. The DEM dataset before the earthquake was surveyed from January to February 2013 and has a resolution of 1.0 m. It is provided by the Geospatial Information Authority of Japan (GSI). The DEM dataset after the earthquake was surveyed from April to July 2016 and has a resolution of 0.5 m. It is provided by the Forestry Agency of Japan. Since the results of the detected deformation depend on the lower resolution, the resolution was unified into 1.0 m. The precision of the detected deformation can be up to 0.1 m with a resolution of 1.0 m of the DEM datasets. Therefore, the horizontal deformation with a magnitude larger than 0.1 m is deemed effective. For the sake of brevity, we would call the DEM datasets acquired in 2013

(a)   (b)

(c)   (d)

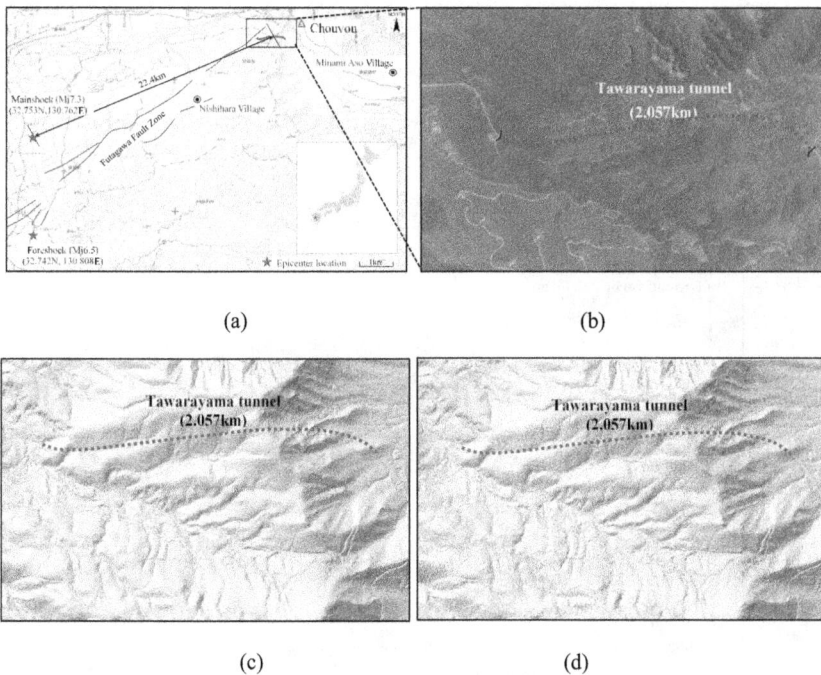

Figure 11.6 Location of the Tawarayama tunnel and the epicenter of the 2016 Kumamoto earthquake (a); post-event earth image of the neighborhood of the Tawarayama tunnel captured on May 31, 2016 (Geospatial Information Authority of Japan) (b); (c) pre- and (d) post-event airborne LiDAR DEMs.

and 2016 pre-event DEM and post-event DEM, respectively. Figure 11.6c and 11.6d show the pre- and post-event DEM of the study area.

To illustrate the correlation between the ground deformation and seismic damage to the tunnel, a spatial combination of the ground deformation and seismic damage to the Tawarayama tunnel was performed, as shown in Figure 11.7. The ground deformation includes NS, WE, and vertical components. Detection lines in the south-north direction were set at an interval of 100 m along the axis of the tunnel to illustrate the ground deformation and were numbered from L_1 to L_20 (from the Nishihara Village side to the Minami Aso Village side), as illustrated in Figure 11.7.

At the ground near the west entrance of the tunnel (L_1 to L_6 in Figure 11.7a, Nishihara Village side), strong horizontal deformation occurred with significantly large displacement. The corresponding spans of the Tawarayama tunnel are in the range of S001–S050. The maximum and average values of the horizontal displacements are 1.75 m (L_2) and 1.15 m, respectively. The strong deformation resulted from the activities of the Futagawa

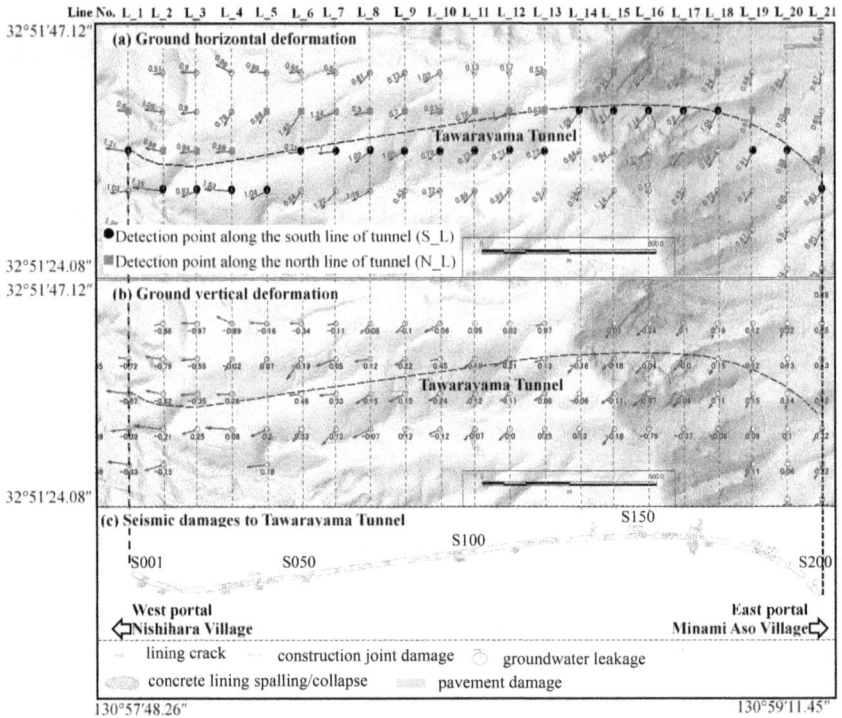

Figure 11.7 Ground (a) horizontal and (b) vertical displacement combined with (c) seismic damages to the Tawarayama tunnel subjected to the 2016 Kumamoto earthquake (blue arrow denotes horizontal displacement vector). For the ground vertical displacement, positive value denotes uplift of the ground, and vice versa.

fault zone under the earthquake. It locates on the western side of the tunnel. The fault strikes northeast-southwest while the axis of the tunnel strikes west-east (Figure 11.6a). The strike of the fault crosses obliquely with that of the tunnel axis. The large dislocation of the fault with a maximum displacement of 2.2 m (in Mashiki Machi) contributed to the strong deformation. Site investigation on the tunnel showed that the spans near the Nishihara Village side (S001–S050 in Figure 11.7c) were seriously damaged. The seismic damages included lining cracks, concrete spalling, construction joint damage, and pavement failure. As a result, there is a possibility that the ground's horizontal deformation and shaking are related to the seismic response of the tunnel. In contrast with the horizontal deformation, vertical displacement revealed a small, smoothly varying pattern of uplift and subsidence (Figure 11.7b). The ground at the lines L_1, L_2, and L_3 subsided with a maximum displacement of 0.62 m and an average displacement of 0.40 m, while the ground at the other lines (L_4 to L_21) uplifted with a maximum

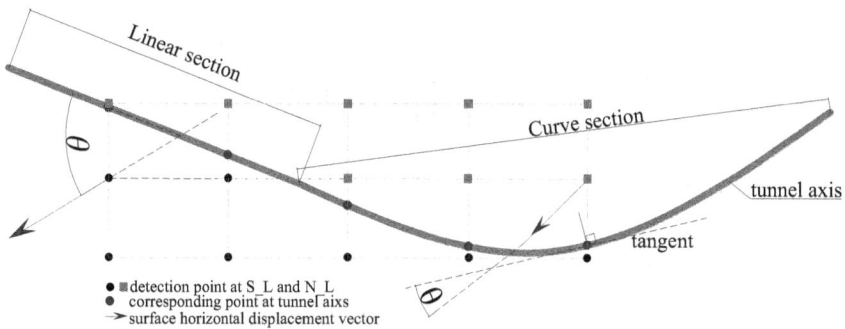

Figure 11.8 Definition of the angle θ between ground horizontal deformation direction and tunnel axis.

displacement of 0.46 m and an average displacement of 0.16 m. No special significance for the seismic response of the tunnel was observed.

To highlight the correlation between the ground horizontal deformation and the seismic response of the tunnel, two detection lines in the east-west direction, named north line (N_L) and south line (S_L), were set along the axis of the Tawarayama tunnel. Both of the detection lines are illustrated in Figure 11.7a. The N_L detection point is marked with a red solid square, and the S_L detection point is marked with a black solid circle. In addition, we also defined the acute angle ($\leq 90°$) between the horizontal displacement vector and the axis of the tunnel as angle θ. Figure 11.8 shows the definition of the angle θ. Figure 11.9 presents two components (NS and WE components) of the horizontal displacement on both of the detection lines and the angle θ along the tunnel axis.

At the lines in the range of L_1 to L_6 (Nishihara Village side), EW displacement (Figure 11.9a) was larger than NS displacement (Figure 11.9b). This contributes to the fact that the angle θ was less than 10° (Figure 11.9c) in most cases. This kind of deformation pattern indicates that ground in the range of L_1 to L_6 moved westwards, approximately parallel to the axis of the tunnel. Generally speaking, seismic wave nearly parallel to the tunnel axis results in such kinds of axial deformations, as axial tension and compression. Therefore, once the compression or tensile strength of the concrete is reached, the damage takes place. The axial deformation mode and mechanism of mountain tunnel under seismic waves are illustrated in Figure 10.7 of Chapter 10. Under the ground in this range, the tunnel section in the spans S001–S050 suffered from widespread damages, as illustrated in Figure 10.1 of Chapter 10. Most of the construction joint damages and pavement failures that occurred in these spans were tensile/compressive failures with large displacement. The construction joint was compressed to fail

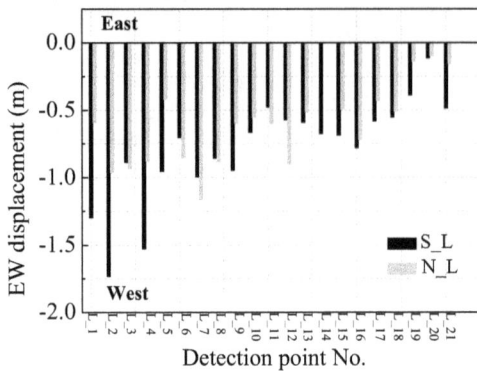

(a)

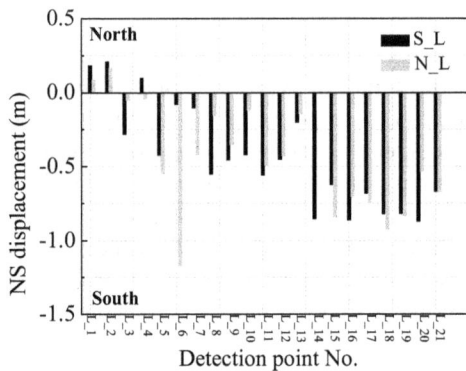

(b)

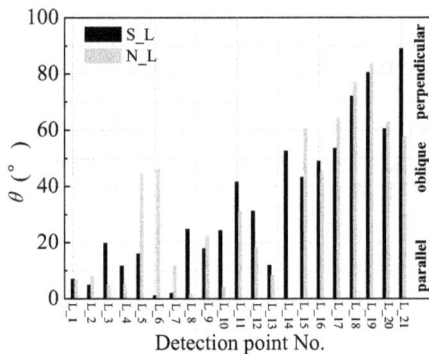

(c)

Figure 11.9 Three components of 3-D displacement on both detection lines and the variation of angle θ along the tunnel axis. (a) EW horizontal displacement; (b) NS horizontal displacement; (c) variation of angle θ along the tunnel axis.

(a)

(b)

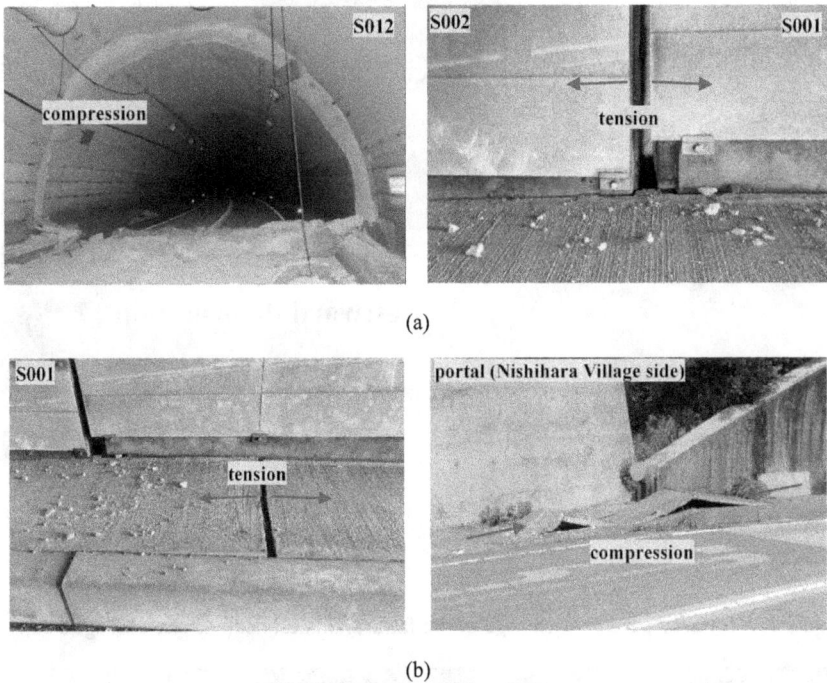

Figure 11.10 Site investigation of seismic damages to the Tawarayama tunnel. (a) Construction joint; (b) pavement damage.

circumferentially between S012 and S013 and opened due to tension between S001 and S002 (Figure 11.10a). The pavement in the span of S001 and at the portal deformed in tension and compression patterns, respectively (Figure 11.10b). Furthermore, the tunnel portal (Nishihara Village side) moved westwards by 10 cm. The portal pavement was compressed horizontally to uplift by 15 cm (Figure 11.11). These facts coincide with the observation of the ground westward deformation of Mt. Tawarayama.

At the lines L_1 and L_2, the ground moved northward by 0.19 m (average value at the S_L) and 0.13 m (average value at the N_L) (Figure 11.9b), while the ground at the other lines moved southward. The change in the direction of the NS displacement contributed to ground dislocation. Between the spans of S001 and S002, the tunnel lining dislocated in the north-south direction at the construction joint, as shown in Figure 11.12. The aforementioned results show that strong horizontal ground deformation can reflect the seismic performance of the tunnel to some extent.

The horizontal deformation at the lines L_7 to L_21 (Minami Aso Village side) decreased compared with that in the western part (L_1 to L_6). The

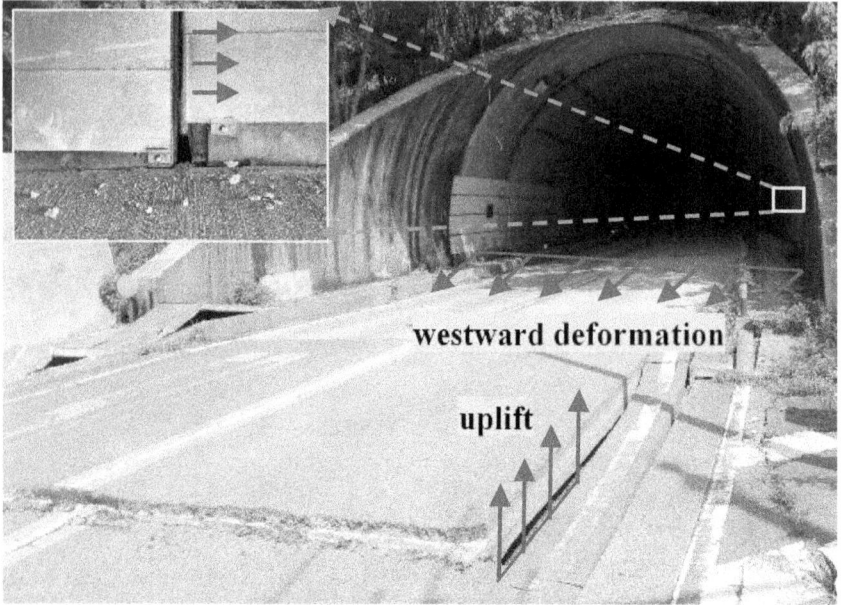

Figure 11.11 Portal pavement damage of the Tawarayama tunnel (Nishihara Village side).

Figure 11.12 Dislocation of the construction joint between S001 and S002.

maximum and average values of the horizontal displacements are 1.24 m (L_7) and 0.87 m, respectively. However, seismic damages to the tunnel still occurred frequently in the spans (S50–S200) under the ground in the range of L_7 to L_21. The previous study noted that geological conditions, such as rock mass quality and fault zone as well as tunnel lining conditions have a significant influence on the seismic response of the tunnel (Zhang et al., 2018). From the viewpoint of the ground deformation, EW displacement decreased gradually with the increase of distance from the Futagawa fault segment (Figure 11.9a), but NS displacement increased, even larger than EW displacement (Figure 11.9b). The variation of the two components of the horizontal deformation contributes to the fact that the angle θ became larger, as illustrated in Figure 11.9c, indicating that the ground turned to move southwestward. The direction of the ground deformation obliquely crosses the axis of the tunnel. That is to say, the tunnel is subjected to seismic waves obliquely crossing the axis of the tunnel, providing further validation of the assumption of seismic wave propagation for the regular distribution of ring cracks discussed in Chapter 10. This consistency also provides further verification of the assumption of the effect of seismic wave propagation on other seismic damages. Therefore, the direction of the ground deformation can be taken as an intuitive index to reveal the seismic wave propagation direction for a better understanding of the seismic response of the underground structures.

11.4 CONCLUSION

The chapter presents a quick ground deformation-based seismic damage assessment method. The method includes data acquisition using the LiDAR system, 3-D ground deformation (horizontal and vertical displacement vectors) detection using the Combination and Classification ICP (CCICP) algorithm, and investigation of seismic damages to mountain tunnels. The method is applied in the Tawarayama tunnel after the 2016 Kumamoto earthquake as one case study to illustrate its application. The detected 3-D deformation field agreed well with the permanent displacements, which were detected using the 3D-GIV (Geomorphic Image Velocity) method and were observed at the site station of Chouyou (location: 32.8707 N, 130.9962 E; code: EL04930274901).

Case application results show that strong horizontal deformation of the ground can reflect seismic damage to the tunnel to some extent. Furthermore, the direction of the ground deformation can be taken as an intuitive index to reveal the seismic wave propagation direction for a better understanding of the seismic response of the underground structures. Ground deformation was observed to be parallel to or obliquely crossing the axis of the tunnel, which contributes to axial compression and tension deformation of the concrete lining and pavement. It provides a further explanation for the phenomenon that the ring cracks are distributed regularly along the axis of the tunnel. On the other hand, the seismic response of the underground structure does not have special significance for vertical ground deformation.

Chapter 12

Performance-Based Design for Seismic Restoration and Aseismic Design

12.1 RESTORATION DESIGN

After an earthquake, the damaged tunnel should be restored as soon as possible for later usage. Discussion on the influential factors is to provide a fundamental suggestion for restoration and aseismic design for mountain tunnels directly and eventually to improve the performance-based seismic design of the tunnel. Kunita et al. (1994) presented a restoration work of the Kinoura tunnel by the Noto Peninsular Offshore earthquake. Most literature regarding the seismic effect of earthquakes on mountain tunnels focused on the aseismic design method (e.g., St. John & Zahrah, 1987; Sharma & Judd, 1991; Hashash et al., 2001; Wang et al., 2001; Yashiro et al., 2007; Chen et al., 2012; Li, 2012; Maugeri & Soccodato, 2014; Shen et al., 2014; Isago & Kusaka, 2018; Zhang et al., 2018). Here, the case of the restoration work for the Tawarayama tunnel during the 2016 Kumamoto earthquake is presented in detail with both aspects of the restoration design criterion and restoration method.

12.1.1 Restoration Design Criterion

According to the classification of the soundness in the Road Tunnel Maintenance Handbook (JARA, 2015) as listed in Table 12.1, the soundness of tunnel lining can be classified into four categories: (1) soundness I with sound status; (2) soundness II with the necessity for preventive maintenance; (3) soundness III with the necessity for early measure; and (4) soundness IV with emergency measure. For convenience, the four degrees are referred to as I, II, III, and IV in sequence. The former chapters have documented the earthquake-induced damages to the Tawarayama tunnel. Diagnosis results show that 54 spans among the 209 spans in the Tawarayama tunnel are with the soundness of I, 66 spans with the soundness of II, 31 spans with the soundness of III, and 58 spans with the soundness of IV (Isago & Kusaka, 2018). For the spans with soundness from II to IV, repair and reinforcement are

Table 12.1 Soundness Classification for Road Tunnel

Classification		Definition
I	Sound status	There is no problem with the function of the road tunnel
II	With the necessity for preventive maintenance	There is no hindrance to the function of the road tunnel, but it is desirable to take measures from the viewpoint of preventive maintenance
III	With the necessity for an early measure	There is a possibility that the function of the road tunnel may be impaired, and it is necessary to take early measures
IV	With emergency measure	There is a situation in which the function of the road tunnel has been hindered or the hindrance is likely to occur, and emergency measures should be taken

Source: JARA, 2015.

necessary. The preliminary classification provides a reference for a further selection of detailed restoration measures. It can avoid unnecessary work for the condition without the necessity for repair.

Since restoration measures depend on the status of damage, a design criterion based on the site investigation of the Tawarayama tunnel is established to make an accurate judgment and determine reasonable restoration measures. This new criterion is developed concerning the restoration design criteria for the Touya tunnel (Suzuki et al., 2001). The restoration criterion for the Touya tunnel was developed for the design of restoration measures against the eruption of Mount Usu, yet the effect of the earthquake was not fully taken into consideration.

However, spalling of lining concrete in spans of S4 and S5 and portal sections occurred in the Tawarayama tunnel due to strong ground shaking. Moreover, the collapse of lining concrete was observed in spans S166 and S167. Therefore, the deformation and failure of tunnel lining due to strong ground shaking should not be ignored in the design criterion for tunnel restoration after the earthquake. Figure 12.1 shows the flowchart of the newly proposed restoration design criterion for the Tawarayama tunnel. Similar to the design criterion for the Touya tunnel, crack width, crack distribution, and geological condition are considered in the new design criterion. Besides, two aspects—the spalling/collapse of lining due to ground shaking and underground water leakage—are introduced into the present criterion.

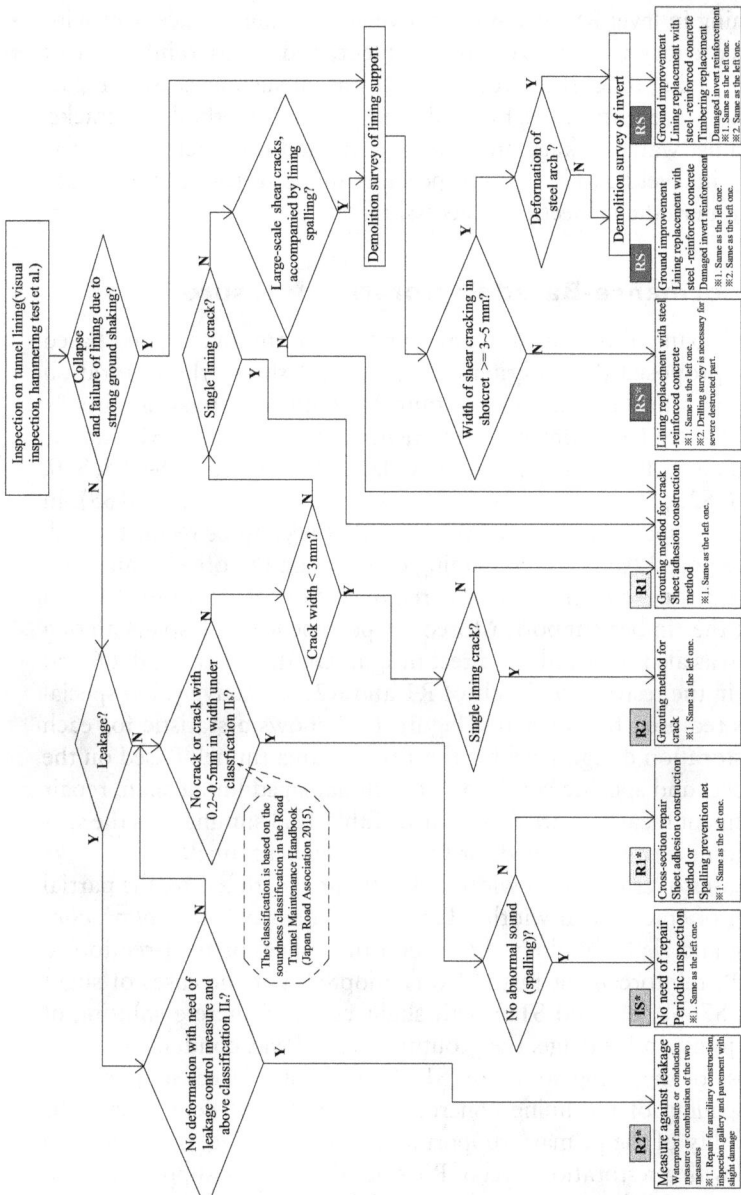

Figure 12.1 Flow of the newly proposed restoration criterion for the Tawarayama tunnel.

Source: Kumamoto River and National Highway Office, Kyushu Regional Development Bureau, Ministry of Land, Infrastructure, Transport and Tourism, 2017.

*Note:

R1 refers to the restoration level with need of repair 1; R2 refers to the restoration level with need of repair 2;
RS refers to the restoration level with need of reconstruction; IS refers to the restoration level with need of inspection.

Based on the newly proposed design criterion, the restoration level can be classified into four categories: reconstruction (referred to as RS), repair 1 (referred to as R1), repair 2 (referred to as R2), and inspection (referred to as IS). For lining in level RS, ground improvement, lining replacement with steel-reinforced concrete, timbering replacement, and invert reinforcement are necessary. Lining in level R1 requires restoration such as section repair method, spalling prevention net method, and grouting method for cracks. Measures against water leakage and grouting method for cracks are necessary for lining in level R2. There is no special required restoration for lining in level IS, but periodic inspection is necessary.

12.1.2 Performance-Based Restoration Measure

There are 207 spans of the Tawarayama tunnel that are investigated based on the newly proposed design criterion. Table 12.2 shows the restoration level at each span in the Tawarayama tunnel. As site investigation already discussed in Chapter 10, reconstructions are needed in the portal (S1) and the spans with lining spalling/collapse and other large deformations (S4, S5, S10, S12, S13, S28, S29, S96, S97, S158, S159, S165, S166, S167, and S168). In total, there are 16 spans among the 207 spans that need to be reconstructed. Except for the portal (span S1), the lining concrete on the other spans with the requirement for reconstruction was removed to provide a direct visual inspection of the timber support. Moreover, pavement in 11 spans among the 16 spans was also removed for direct insight. On the other hand, 66 and 65 spans are in the restoration levels of R1 and R2, respectively. No special restoration is required for 60 spans. Figure 12.2 shows a statistic for each span with restoration design level for the Tawarayama tunnel. Based on the restoration level and specific condition of each damaged lining span, repair and reinforcement measures are determined. Table 12.3 summarizes the specific restoration measures adopted for the Tawarayama tunnel.

For three cases of cracks with width ≥ 3 mm and length ≥ 5 m, the partial concentration of cracks with width ≥ 0.3mm and density ≥ 0.2 m/m^2, concrete splitting and slight spalling, the carbon fiber sheet or fiber-reinforced polymer (FRP) reinforcement method was adopted. For the cases of slight damage (i.e., S7, S8, S9, and S11) with slight cracks, concrete splitting of construction joint, and leakage, the grouting method was conducted.

The collapse of the lining structure (S167), buckling of steel support, and widespread spalling of the lining concrete (S12, and S28) deteriorated the bearing capability of the primary support and concrete lining. In spans from S165 to S167 with a restoration level of RS, since the primary support and lining cannot function well with a very low capability, reconstruction is needed. Figure 12.3 shows the reconstruction design for the lining from spans S165 to S167. Due to 10 cm deformation in both the transversal and longitudinal directions, the steel arch buckled severely as shown in Figure 12.4 and replacement

Table 12.2 Restoration Level at Each Span in the Tawarayama Tunnel according to the Newly Proposed Restoration Design Criterion

Span	1	2	3	4	5	6	7	8	9	10	11	12	13	14	15	16	17	18	19	20	21	22	23	24	25	26	27	28
Level	RS	R1	R1	RS	RS	R1	R2	R2	R2	RS	R2	RS	RS	R2	R2	R2	R1	R2	R2	IS	R2	R2	R2	R1	R1	R2	R1	RS
Span	29	30	31	32	33	34	35	36	37	38	39	40	41	42	43	44	45	46	47	48	49	50	51	52	53	54	55	56
Level	RS	R1	R1	R1	R2	R2	R2	R1	R1	R1	R1	R1	R1	R2	R2	R2	R2	IS	IS	IS	R2	R2	IS	R1	R1	IS	R2	R2
Span	57	58	59	60	61	62	63	64	65	66	67	68	69	70	71	72	73	74	75	76	77	78	79	80	81	82	83	84
Method	IS	IS	IS	IS	R1	R1	R2	R2	R1	R1	R1	R2	R1	R1	R1	R2	R2	R2	R2	R1	R1	R2	R2	IS	R2	R2	R2	R2
Span	85	86	87	88	89	90	91	92	93	94	95	96	97	98	99	100	101	102	103	104	105	106	107	108	109	110	111	112
Level	R2	R2	R2	R2	IS	IS	IS	R2	R2	R1	R1	R2	R1	IS	IS	R2	R1	R2	R2	IS	IS	IS	IS	IS	IS	IS	IS	R2
Span	113	114	115	116	117	118	119	120	121	122	123	124	125	126	127	128	129	130	131	132	133	134	135	136	137	138	139	140
Level	R1	R2	IS	R1	R2	R1	R1	R1	R1	R1	R1	R1	R1	R1	IS	IS	IS	R1	IS	IS	IS	R2	R2	R1	IS	IS	R2	R1
Span	141	142	143	144	145	146	147	148	149	150	151	152	153	154	155	156	157	158	159	160	161	162	163	164	165	166	167	168
Level	IS	IS	IS	R2	R2	R1	R1	R1	R1	IS	IS	IS	IS	IS	IS	IS	IS	RS	RS	R1	R1	R1	R1	R1	RS	RS	RS	RS
Span	169	170	171	172	173	174	175	176	177	178	179	180	181	182	183	184	185	186	187	188	189	190	191	192	193	194	195	196
Level	IS	IS	IS	IS	IS	IS	IS	IS	IS	IS	IS	IS	IS	IS	IS	IS	R1	R1	R1	R1	R1	R1	R1	IS	RS	RS	RS	IS
Span	197	198	199	200	201	202	203	204	205	206	207																	
Level	IS	IS	IS	IS	IS	R2	R1	R2	R1	R2	IS																	

Source: Kumamoto River and National Highway Office, Kyushu Regional Development Bureau, Ministry of Land, Infrastructure, Transport and Tourism, 2017.

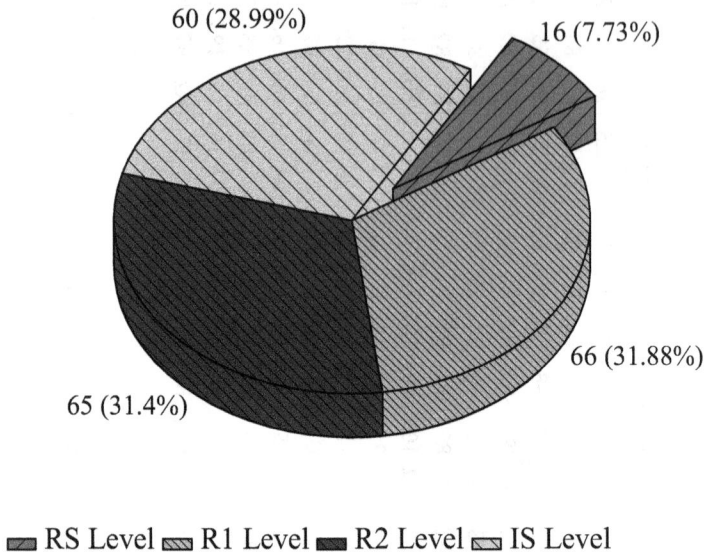

Figure 12.2 Span percentage of each mitigation countermeasure for the Tawarayama tunnel.

of the primary support is necessary. The rockbolt (solid circle in Figure 12.3) and steel arch (gray area in Figure 12.3) were replaced and the support pattern was changed from DI to DIIIa. On the other hand, the removal of the failed steel support might affect the stability of the surrounding lining and ground and even induce secondary damage. Therefore, before removing the steel arch support, the injection type forepoling as an auxiliary method was conducted around the buckled areas to reinforce their stability.

Whether to reconstruct the primary support such as steel support, shotcrete, and rock bolt should be decided based on their damage degree. Since no obvious damage was observed for the other 13 spans in restoration level RS, the steel arch support was left as it was. Only the failed shotcrete was replaced and the number of rock-bolt was increased. For the reconstructed lining, concrete was reinforced with a reinforcing steel bar of φ19mm@200mm as the primary reinforcement and φ16mm@300mm as the secondary reinforcement. In addition, the invert concrete of 6 spans (S12 and S13, S165 to S168) with large deformation due to compression was also replaced by steel-reinforced concrete. Figure 12.5 shows the tunnel condition after reconstruction.

Concerning the portal of the Tawarayama tunnel (at the Nishihara Village side), unsymmetrical loading due to slope induced the deformation and even

Table 12.3 Summary of Restoration Measures Adopted for the Tawarayama Tunnel

Tunnel Section			Damage Condition	Restoration Method and Soundness
The exterior of portal section	Electric room	Before entrance	The ground collapse of the electric room	Demolition and reconstruction of the electric room; Ground re-embankment
	Unsymmetrical terrain	S1–S6	Continuous and longitudinal cracks due to tension along the left shoulder of the tunnel, tunnel lining deformation Partial collapse of the slope right above the tunnel	Soil removal work to reduce the influence of unsymmetrical terrainv
Entrance	S1		Movement deformation, shear cracking of lining shoulder	Reconstruction of the lining at the entrance
Tunnel interior	Remarkable damage	S4, S5	Collapse, movement deformation, shear crack, spalling, exposure of reinforcing steel, etc.	Reconstruction method Auxiliary construction method (FRP injection rock-bolt for sewing effect)
		S10, S12, S13	Collapse, movement deformation, compressive damage	Reconstruction with reinforcement steel
		S96, S97, S166, S167, S168	Collapse, movement deformation, shear crack, spalling, compressive damage	Reconstruction with reinforcement steel Auxiliary construction method (FRP injection rock-bolt for ground improvement and sewing effect)
		S28, S29, S158, S159, S165	Collapse, movement deformation, shear crack, spalling, compressive damage	Reconstruction with reinforcement steel
	Inclined crack	S2, S3, S6, S36, S37, S121 etc.	Inclined crack due to tension with width of 3 mm or more, length of 5 m or more Splitting of concrete along the crack and slight spalling Partial concentration of crack with width of 0.3 mm or more and density of 0.2 m/m² or more and concrete splitting	Sheet adhesion construction method with carbon fiber ※Repair I (RI)

(Continued)

Table 12.3 Continued

Tunnel Section		Damage Condition	Restoration Method and Soundness
Transverse crack	S38, S41, S61, S62, S67, S95 etc.	Transverse crack due to tension with width of 3 mm or more, length of 5 m or more Splitting of concrete along the crack and slight spalling Partial concentration of crack with width of 0.3 mm or more and density of 0.2 m/m² or more and concrete splitting	Sheet adhesion construction method with carbon fiber ※ Repair 1 (R1)
Slight damage	S7, S8, S9, S11 etc.	Slight cracks, concrete splitting of construction joint and leakage	Grouting method for crack, measures against leakage and splitting ※ Repair 2 (R2)
Slight damage without the need for repair	S38, S41, S61, S62 etc.	Slight damage or no damage	Periodic inspection
Exit	S207	No damage	Sound status
Auxiliary construction	Pavement, inspection gallery, drainage	Slight damage without the need for repair except for partial pavement heave and movement deformation	Replacement of pavement, reconstruction of the circular gutter, replacement of inspection gallery, etc.

Source: Kumamoto River and National Highway Office, Kyushu Regional Development Bureau, Ministry of Land, Infrastructure, Transport and Tourism, 2017.

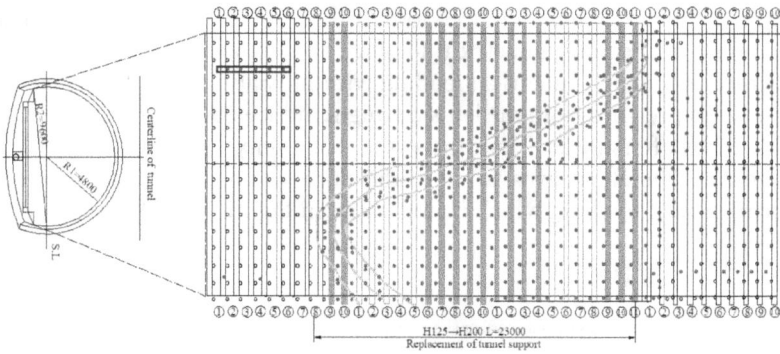

Figure 12.3 Reconstruction of the lining from spans S165 to S167 (the area with the light gray line is the fault zone behind the tunnel lining; the solid circle denotes the replaced rock-bolt; and gray areas denote the replaced steel arch).

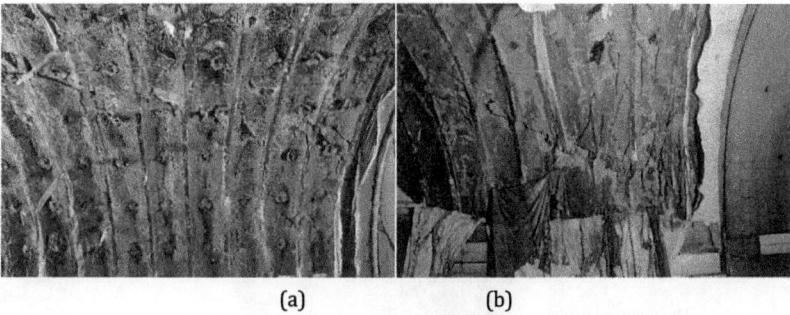

(a)   (b)

Figure 12.4 Damage condition of the steel support. (a) Right side of span S166; (b) left side of span S167.

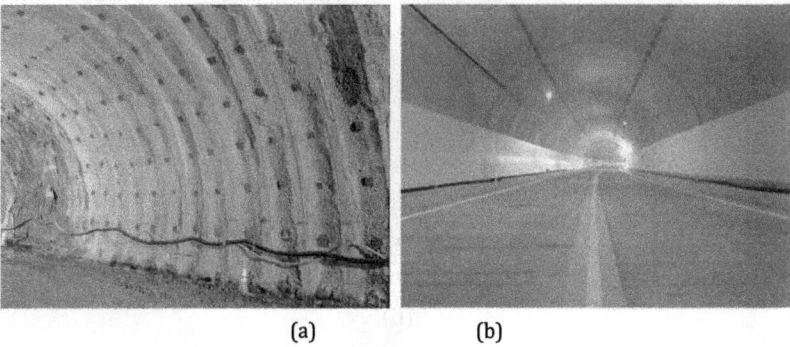

(a)   (b)

Figure 12.5 Tunnel condition after reconstruction. (a) Condition after the replacement of steel support at the left side in the spans of S166 and S167; (b) condition after reconstruction in the spans from S165 to S168.

failure of the lining structure. Even worse, heavy rainfall caused the collapse of part of the earthquake-induced loosened slope. The slope collapse aggravated the effect of unsymmetrical loading on the deformation of the lining structure. If the unsymmetrical loading without stability is always there, the continuous pressure from the left side might induce continuous deformation followed by aftershocks. Thus, there is a probability that secondary damage may occur. Therefore, excavation of the slope was conducted to relieve the unsymmetrical loading, which is illustrated in Figure 12.6. The earth at the

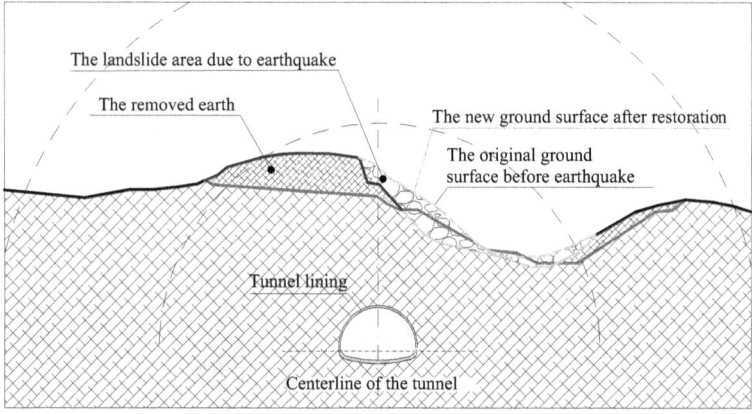

(a)

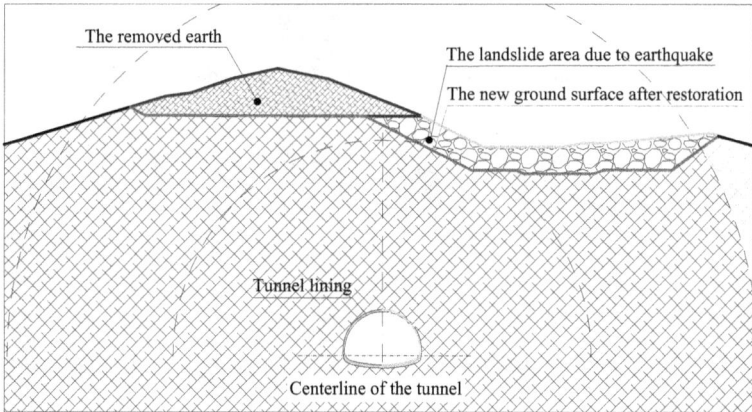

(b)

Figure 12.6 Excavation of the slope. (a) Cross-section of spans 4 and 5; (b) cross-section of spans 6 and 7.

upper side above the tunnel in Figure 12.6 was removed. In addition, some loosened soil due to the earthquake was also removed against secondary damage to the slope. After removal, a new ground surface in Figure 12.6 was formed to relieve the unsymmetrical loading effect.

12.2 ASEISMIC DESIGN

12.2.1 Overview of Recommendations

Following the analysis of the seismic damages and corresponding influencing factors to the mountain tunnels under earthquakes, some recommendations for future tunnel planning are given as follows:

1. The tunnel should be placed far away from slope faces if possible. If not, it is important to simultaneously consider the slope stability evaluation and integrated design of the non-buried tunnel section and tunnel portal structure.
2. It is advised to avoid tunnels running across active fault zones or weak surrounding rocks if possible. When crossing faults, reinforced countermeasures should be taken into consideration for the concrete lining.
3. Longitudinal motion of mountain tunnels under earthquake should be paid much more attention to in the aseismic design and construction procedures. Mitigation countermeasures for the longitudinal seismic response can be taken into consideration for further mountain tunnel construction and remediation process, such as ring shock absorption structure.

For further understanding of theoretical and engineering seismic response and mechanism of mountain tunnels under earthquake, efforts should be taken on such aspects as (1) longitudinal response of mountain tunnel under an earthquake in three dimensions; (2) effect of ground surface motion under a seismic wave on the performance of mountain tunnel; and (3) corresponding mitigation countermeasures with consideration of longitudinal shock absorption.

12.2.2 Specific Longitudinal Aseismic Design

According to the study on the seismic response of mountain tunnels with both site investigation and analytical seismic analysis discussed in Chapters 7–11, a longitudinal laminated shear energy dissipation (LSED) structure is conceptually prosed concerning the longitudinal performance of tunnel under earthquake (Jiang & Zhang, 2019).

1. Structure of the proposed LSED structure and its shock absorption principle

The LSED structure consists of two parts: one energy absorption body (A in Figure 12.7) and two connecting parts (B in Figure 12.7). The elastic rubber layer (A1 in Figure 12.7) with deformation capability and steel plate (A2 in Figure 12.7) with high rigidity are superimposed alternately to form

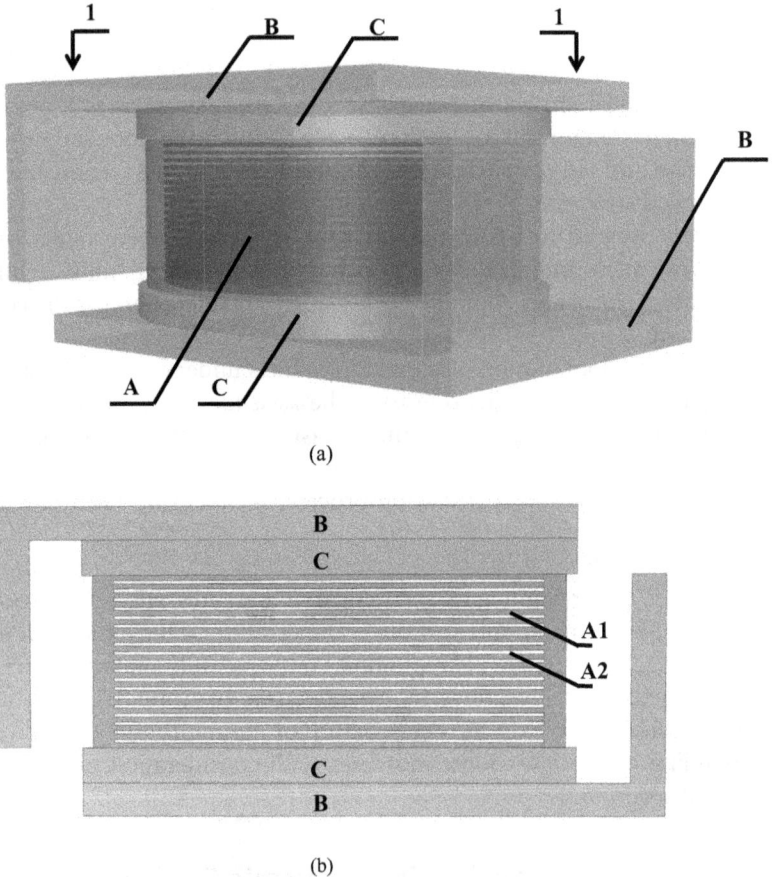

(a)

(b)

Figure 12.7 Diagram of the proposed LSED structure (a) LSED structure; (b) LSED section in 1–1. (Symbol A denotes energy absorption body, symbol A1 denotes elastic rubber layer, symbol A2 denotes steel plate, symbol B denotes L-type steel-reinforced concrete slab, and symbol C denotes flange plate.)

the energy absorption body. The connecting part is made from an L-type steel-reinforced concrete slab. The energy absorption body is connected to the L-type connecting part through the flange plate (C in Figure 12.7). Horizontal equivalent stiffness is a key parameter for evaluating its anti-shear characteristics. The stiffness could be influenced by rubber shear modulus, rubber layer thickness, rubber layer number, and steel plate dimension. The influence weight ratio of the four influential factors is 22(rubber shear modulus): 29(rubber layer thickness): 21(rubber layer number): 28(steel plate dimension) using the coefficient of variation based on a series of compression-shear experimental studies. The ratio could provide a reference for the LSED structure design for the application.

Since the thickness of the thin rubber layer A1 is less than that of the steel plate A2, the energy absorption body keeps nearly rigid in the laminated direction (vertical direction) due to the rigidity of the steel plate. However, the energy absorption body keeps the deformation capability of rubber in the horizontal direction thanks to the deformation capability of the elastic rubber layer. Figure 12.8 shows the deformation contour of the LESD structure under compression-shear loading. When the LSED structure is compressed in the laminated direction, there is no obvious deformation due to its vertical rigidity. Once horizontal force (tension or compression) is applied in the horizontal direction, it can deform outwards or inwards because of its horizontal deformability.

2. Application method of the longitudinal shock absorber in mountain tunnels

Figure 12.9 shows the installation design of the LSED structure in a mountain tunnel. The structure can be directly connected with tunnel lining with annual distribution in a cross-section of the tunnel lining at a certain angle interval. Bolt connection between the connection part and tunnel lining can be used to combine the LSED structure and tunnel lining into a whole. The laminated direction of the structure should coincide with the radial direction of the tunnel cross-section and be perpendicular to the longitudinal direction.

In the radial direction of the tunnel cross-section, the LSED structure keeps nearly rigid so that there is no obvious effect on the bearing capability of radial stress (F3 in Figure 12.8b) resulting from the surrounding rock mass. With the LSED structure, the tunnel lining has a certain flexibility in the longitudinal direction. When longitudinal seismic performance is much more obvious in the tunnel subjected to an earthquake or other dynamic loadings, dislocation of the rubber layer along the longitudinal direction contributes to the bearing capacity of the phase-derived stress and allows relative compressive or tensile deformation to a particular extent. The aseismic effect

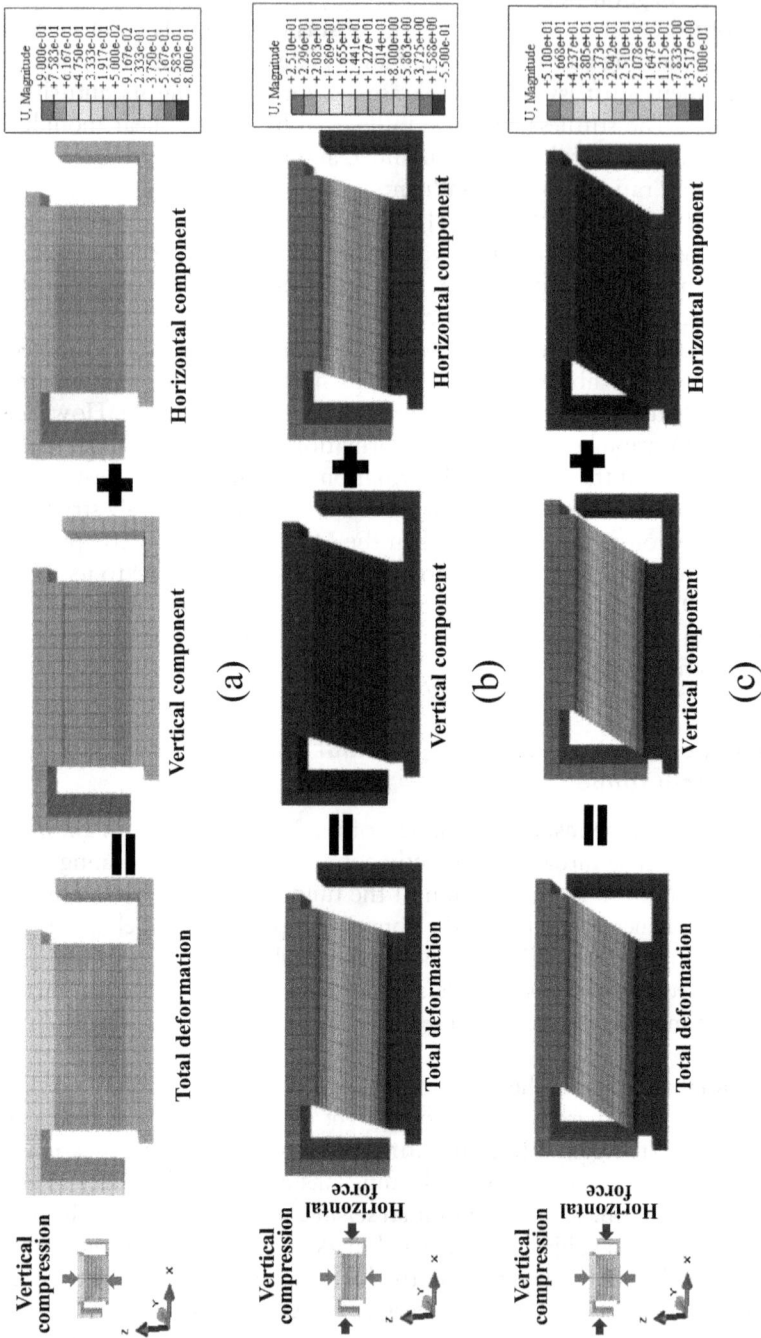

Figure 12.8 Deformation contour of LSED structure under (a) compression only, (b) compression-shear with 50% shear strain, and (c) compression-shear with 100% shear strain (units: mm).

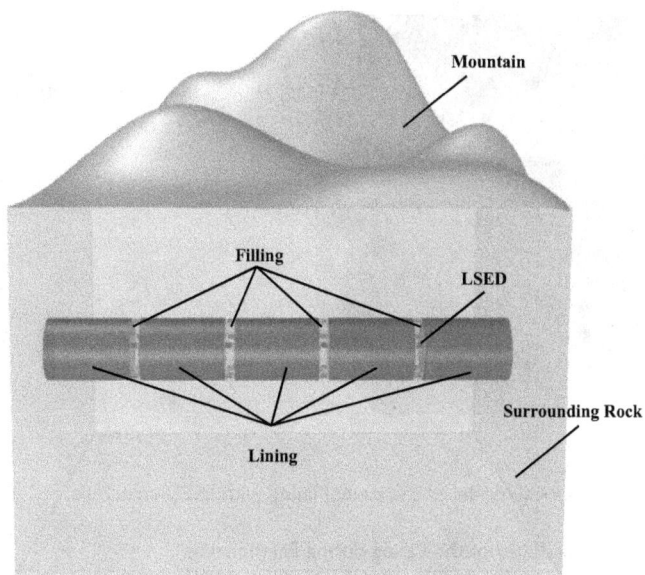

(a)

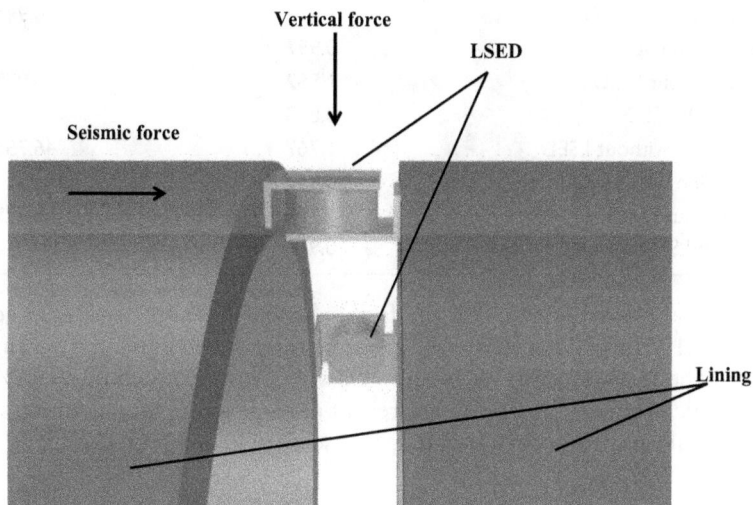

(b)

Figure 12.9 Installation design of the LSED structure in a mountain tunnel. (a) 3D stereo-gram; (b) schematic of LSED partial installation.

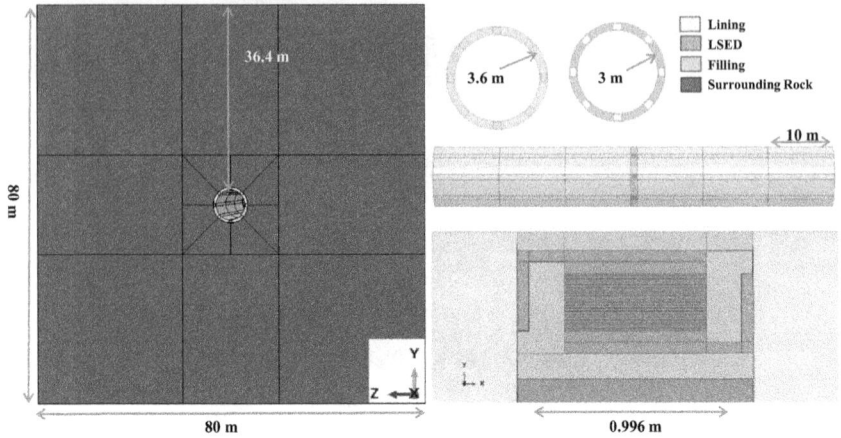

Figure 12.10 Numerical model of the tunnel lining with LSED structure.

Table 12.4 Aseismic Ratio of the Lining during Earthquakes

	Lining Axial Stress/MPa	Aseismic Ratio
Crown without LSED	1.955	46.09%
Crown with LSED	1.054	
Spandrel without LSED	1.879	46.94%
Spandrel with LSED	0.997	
Haunch without LSED	1.562	45.07%
Haunch with LESD	0.858	
Springing line without LSED	1.767	46.75%
Springing line with LSED	0.941	
Invert without LSED	1.798	46.72%
Invert with LSED	0.958	

is numerically conducted using the finite element method, as illustrated in Figure 12.10. Table 12.4 shows the aseismic ratio of the lining during earthquakes. The aseismic ratio for the lining axial stress can reach up to 45%. It plays a role in shock absorption to prevent damage to the tunnel lining, such as the tensile and compressive crack, as discussed in Chapter 10.

12.3 CONCLUSION

The chapter provides a fundamental suggestion for performance-based design for seismic restoration and the aseismic design of tunnels. The case of the restoration work for the Tawarayama tunnel by the 2016 Kumamoto earthquake is presented in detail with both aspects of the restoration design criterion and restoration method. The restoration design criterion for the Tawarayama tunnel is developed with concerning restoration design criteria for the Touya tunnel. Similar to the design criterion for the Touya tunnel, crack width, crack distribution, and geological conditions are considered in the new design criterion. Besides, two aspects—the spalling/collapse of lining due to ground shaking and underground water leakage—are introduced into the present criterion.

Some recommendations for future tunnel planning are provided and a tunnel longitudinal LSED structure is designed for mountain tunnels. In the radial direction of the tunnel cross-section, the LSED structure keeps nearly rigid so that there is no obvious effect on the bearing capability of radial stress resulting from the surrounding rock. With the LSED structure, the tunnel lining has certain flexibility in the longitudinal direction. When longitudinal seismic performance is much more obvious in the tunnel subjected to an earthquake or other dynamic loadings, dislocation of the rubber layer along the longitudinal direction contributes to the bearing capacity of the phase-derived stress and allows relative compressive or tensile deformation to a particular extent. It plays a role in shock absorption to prevent damage to the tunnel lining.

Appendices

APPENDIX A

Displacements and stresses in the isotropic rock medium ($m = 1$) and the isotropic concrete lining ($m = 2$) based on the three-dimensional elastodynamics analysis are given as follows.

$$u_{r(m)} = \frac{\partial \varphi_m^{\text{total}}}{\partial r} + \frac{1}{r} \frac{\partial \psi_m^{\text{total}}}{\partial \theta} + \frac{1}{k_{s(m)}} \frac{\partial^2 \chi_m^{\text{total}}}{\partial r \partial z} \tag{A.1}$$

$$u_{\theta(m)} = \frac{1}{r} \frac{\partial \varphi_m^{\text{total}}}{\partial \theta} - \frac{\partial \psi_m^{\text{total}}}{\partial r} + \frac{1}{k_{s(m)} r} \frac{\partial^2 \chi_m^{\text{total}}}{\partial \theta \partial z} \tag{A.2}$$

$$u_{z(m)} = \frac{\partial \varphi_m^{\text{total}}}{\partial z} - \frac{1}{k_{S(m)}} \left[\frac{1}{r} \frac{\partial}{\partial r} \left(r \frac{\partial \chi_m^{\text{total}}}{\partial r} \right) + \frac{1}{r^2} \frac{\partial^2 \chi_m^{\text{total}}}{\partial \theta^2} \right] \tag{A.3}$$

$$\sigma_{rr(m)} = \lambda \nabla^2 \varphi_m^{\text{total}} + 2\mu \left[\frac{\partial^2 \varphi_m^{\text{total}}}{\partial r^2} + \frac{\partial}{\partial r} \left(\frac{1}{r} \frac{\partial \psi_m^{\text{total}}}{\partial \theta} \right) + \frac{1}{k_{s(m)}} \frac{\partial^3 \chi_m^{\text{total}}}{\partial r^2 \partial z} \right] \tag{A.4}$$

$$\sigma_{\theta\theta(m)} = \lambda \nabla^2 \varphi_m^{\text{total}} + 2\mu \begin{bmatrix} \frac{1}{r} \left(\frac{\partial \varphi_m^{\text{total}}}{\partial r} + \frac{1}{r} \frac{\partial^2 \varphi_m^{\text{total}}}{\partial \theta^2} \right) + \frac{1}{r} \left(\frac{1}{r} \frac{\partial \psi_m^{\text{total}}}{\partial \theta} - \frac{\partial^2 \psi_m^{\text{total}}}{\partial r \partial \theta} \right) + \\ \frac{1}{k_{s(m)} r} \left(\frac{\partial^2 \chi_m^{\text{total}}}{\partial r \partial z} + \frac{1}{r} \frac{\partial^3 \chi_m^{\text{total}}}{\partial \theta^2 \partial z} \right) \end{bmatrix} \tag{A.5}$$

$$\sigma_{zz(m)} = \lambda \nabla^2 \varphi_m^{\text{total}} + 2\mu \left[\frac{\partial^2 \varphi_m^{\text{total}}}{\partial z^2} - \frac{1}{k_{s(m)}} \frac{\partial}{\partial z} \left(\nabla^2 \chi_m^{\text{total}} - \frac{\partial^2 \chi_m^{\text{total}}}{\partial z^2} \right) \right] \tag{A.6}$$

$$\sigma_{r\theta(m)} = \mu \left\{ \begin{matrix} 2 \left(\frac{1}{r} \frac{\partial^2 \varphi_m^{\text{total}}}{\partial \theta \partial r} - \frac{1}{r^2} \frac{\partial \varphi_m^{\text{total}}}{\partial \theta} \right) + \left[\frac{1}{r^2} \frac{\partial^2 \psi_m^{\text{total}}}{\partial \theta^2} - r \frac{\partial}{\partial r} \left(\frac{1}{r} \frac{\partial \psi_m^{\text{total}}}{\partial \theta} \right) \right] \\ + \frac{2}{k_{s(m)}} \left[\frac{1}{r} \frac{\partial^3 \chi_m^{\text{total}}}{\partial r \partial \theta \partial z} - \frac{1}{r^2} \frac{\partial^2 \chi_m^{\text{total}}}{\partial \theta \partial z} \right] \end{matrix} \right\} \tag{A.7}$$

$$\sigma_{rz(m)} = \mu \left\{ 2\frac{\partial^2 \varphi_m^{total}}{\partial r \partial z} + \frac{1}{r}\frac{\partial^2 \psi_m^{total}}{\partial \theta \partial z} + \frac{2}{k_{s(m)}}\left[2\frac{\partial^3 \chi_m^{total}}{\partial r \partial z^2} - \frac{\partial}{\partial r}\left(\nabla^2 \chi_m^{total}\right)\right]\right\} \quad (A.8)$$

$$\sigma_{\theta z(m)} = \mu \left\{ \frac{2}{r}\frac{\partial^2 \varphi_m^{total}}{\partial \theta \partial z} - \frac{1}{r}\frac{\partial^2 \psi_m^{total}}{\partial r \partial z} + \frac{1}{k_{s(m)}}\left[\frac{2}{r}\frac{\partial^3 \chi_m^{total}}{\partial \theta \partial z^2} - \frac{1}{r}\frac{\partial}{\partial \theta}\left(\nabla^2 \chi_m^{total}\right)\right]\right\} \quad (A.9)$$

Appendix B

Coefficients for expressions of stress and displacement in tunnel lining and surrounding rock mass in Chapter 7 are as follows.

$$X_{11}^{(b)} = \left(n^2 + n - \frac{k_s^2 r^2}{2} + \gamma^2 r^2 \right) \zeta_n^{(b)}(\alpha r) - \alpha r \zeta_{n-1}^{(b)}(\alpha r) \tag{B.1}$$

$$X_{12}^{(b)} = n \left[-(n+1)\zeta_n^{(b)}(\beta r) + \beta r \zeta_{n-1}^{(b)}(\beta r) \right] \tag{B.2}$$

$$X_{13}^{(b)} = \frac{i\gamma}{k_s} \left[\left(n^2 + n - \beta^2 r^2 \right) \zeta_n^{(b)}(\beta r) - \beta r \zeta_{n-1}^{(b)}(\beta r) \right] \tag{B.3}$$

$$X_{21}^{(b)} = -\left(n^2 + n + \frac{k_s^2 r^2}{2} - k_p^2 r^2 \right) \zeta_n^{(b)}(\alpha r) + \alpha r \zeta_{n-1}^{(b)}(\alpha r) \tag{B.4}$$

$$X_{22}^{(b)} = n \left[(n+1)\zeta_n^{(b)}(\beta r) - \beta r \zeta_{n-1}^{(b)}(\beta r) \right] \tag{B.5}$$

$$X_{23}^{(b)} = \frac{i\gamma}{k_s} \left[-\left(n^2 + n \right) \zeta_n^{(b)}(\beta r) + \beta r \zeta_{n-1}^{(b)}(\beta r) \right] \tag{B.6}$$

$$X_{31}^{(b)} = \left(\alpha^2 r^2 - \frac{k_s^2 r^2}{2} \right) \zeta_n^{(b)}(\alpha r) \tag{B.7}$$

$$X_{32}^{(b)} = 0 \tag{B.8}$$

$$X_{33}^{(b)} = \frac{i\gamma}{k_s} \beta^2 r^2 \zeta_n^{(b)}(\beta r) \tag{B.9}$$

$$X_{41}^{(b)} = -n \left[-(n+1)\zeta_n^{(b)}(\alpha r) + \alpha r \zeta_{n-1}^{(m)}(\alpha r) \right] \tag{B.10}$$

$$X_{42}^{(b)} = -\left(n^2 + n - \frac{\beta^2 r^2}{2} \right) \zeta_n^{(b)}(\beta r) + \beta r \zeta_{n-1}^{(m)}(\beta r) \tag{B.11}$$

$$X_{43}^{(b)} = -n\frac{i\gamma}{k_s}\left[-(n+1)\zeta_n^{(b)}(\beta r) + \beta r\zeta_{n-1}^{(m)}(\beta r)\right] \tag{B.12}$$

$$X_{51}^{(b)} = i\gamma r\left[-n\zeta_n^{(b)}(\beta r) + \alpha r\zeta_{n-1}^{(m)}(\alpha r)\right] \tag{B.13}$$

$$X_{52}^{(b)} = n\frac{i\gamma r}{2}\zeta_n^{(b)}(\beta r) \tag{B.14}$$

$$X_{53}^{(b)} = \frac{1}{2rk_s}\left(\beta^2 r^2 - \gamma^2 r^2\right)\left[-n\zeta_n^{(b)}(\beta r) + \beta r\zeta_{n-1}^{(m)}(\beta r)\right] \tag{B.15}$$

$$X_{61}^{(b)} = -ni\gamma r\zeta_n^{(b)}(\alpha r) \tag{B.16}$$

$$X_{62}^{(b)} = \frac{i\gamma r}{2}\left[n\zeta_n^{(b)}(\beta r) - \beta r\zeta_{n-1}^{(m)}(\beta r)\right] \tag{B.17}$$

$$X_{62}^{(b)} = -n\frac{1}{2rk_s}\left(\beta^2 r^2 - \gamma^2 r^2\right)\zeta_n^{(b)}(\beta r) \tag{B.18}$$

$$X_{71}^{(b)} = \alpha r\zeta_{n-1}^{(b)}(\alpha r) - n\zeta_n^{(m)}(\alpha r) \tag{B.19}$$

$$X_{72}^{(b)} = n\zeta_n^{(b)}(\beta r) \tag{B.20}$$

$$X_{73}^{(b)} = \frac{i\gamma}{k_s}\left[\beta r\zeta_{n-1}^{(b)}(\beta r) - n\zeta_n^{(m)}(\beta r)\right] \tag{B.21}$$

$$X_{81}^{(b)} = -n\zeta_n^{(b)}(\alpha r) \tag{B.22}$$

$$X_{82}^{(b)} = -\left[\beta r\zeta_{n-1}^{(b)}(\beta r) - n\zeta_n^{(m)}(\beta r)\right] \tag{B.23}$$

$$X_{83}^{(b)} = -n\frac{i\gamma}{k_c}\zeta_n^{(b)}(\beta r) \tag{B.24}$$

$$X_{91}^{(b)} = i\gamma r\zeta_n^{(b)}(\alpha r) \tag{B.25}$$

$$X_{92}^{(b)} = 0 \tag{B.26}$$

$$X_{93}^{(b)} = \frac{\beta^2 r}{k_s}\zeta_n^{(m)}(\beta r) \tag{B.27}$$

where $\zeta_n^{(1)}(\cdot) = J_n(\cdot)$, $\zeta_n^{(3)}(\cdot) = H_n^{(1)}(\cdot)$, and $\zeta_n^{(4)}(\cdot) = H_n^{(2)}(\cdot)$.

Appendix C

Coefficients K_n^{mj} and Q^j in Equation (8.28) in Chapter 8 are as follows.

$$K_n^{11} = H_n^{(1)}\left(k_1\left|\chi(\varsigma_1)\right|\right)\left[\frac{\chi(\varsigma_1)}{\left|\chi(\varsigma_1)\right|}\right]^n$$

$$+ \frac{C_{55}k_1}{4k}\left\{\begin{array}{l}\left[(a+ic)\sum_{n=-\infty}^{\infty}A_nH_{n-1}^{(1)}\left(k_1\left|\chi(\varsigma_1)\right|\right)\left[\frac{\chi(\varsigma_1)}{\left|\chi(\varsigma_1)\right|}\right]^{n-1} - \\ (b-ic)\sum_{n=-\infty}^{\infty}A_nH_{n+1}^{(1)}\left(k_1\left|\chi(\varsigma_1)\right|\right)\left[\frac{\chi(\varsigma_1)}{\left|\chi(\varsigma_1)\right|}\right]^{n+1}\end{array}\right]\frac{\varsigma}{R}\frac{\Omega'(\varsigma_1)}{\left|\Omega'(\varsigma_1)\right|} + \\ \left[(b+ic)\sum_{n=-\infty}^{\infty}A_nH_{n-1}^{(1)}\left(k_1\left|\chi(\varsigma_1)\right|\right)\left[\frac{\chi(\varsigma_1)}{\left|\chi(\varsigma_1)\right|}\right]^{n-1} - \\ (a-ic)\sum_{n=-\infty}^{\infty}A_nH_{n+1}^{(1)}\left(k_1\left|\chi(\varsigma_1)\right|\right)\left[\frac{\chi(\varsigma_1)}{\left|\chi(\varsigma_1)\right|}\right]^{n+1}\end{array}\right]\frac{\bar{\varsigma}}{R}\frac{\overline{\Omega'(\varsigma_1)}}{\left|\Omega'(\varsigma_1)\right|}\right\} \quad \text{(C.1)}$$

$$K_n^{12} = -H_n^{(1)}\left(k_2\left|\Omega(\varsigma_1)\right|\right)\left[\frac{\Omega(\varsigma_1)}{\left|\Omega(\varsigma_1)\right|}\right]^n \quad \text{(C.2)}$$

$$K_n^{13} = -H_n^{(2)}\left(k_2\left|\Omega(\varsigma_1)\right|\right)\left[\frac{\Omega(\varsigma_1)}{\#\Omega(\varsigma_1)}\right]^n \quad \text{(C.3)}$$

$$K_n^{21} =$$

$$\frac{C_{55}k_1}{4}\left\{\begin{bmatrix} (a+ic)\sum_{n=-\infty}^{\infty}A_nH_{n-1}^{(1)}\left(k_1|\chi(\zeta_1)|\right)\left[\frac{\chi(\zeta_1)}{|\chi(\zeta_1)|}\right]^{n-1} - \\ (b-ic)\sum_{n=-\infty}^{\infty}A_nH_{n+1}^{(1)}\left(k_1|\chi(\zeta_1)|\right)\left[\frac{\chi(\zeta_1)}{|\chi(\zeta_1)|}\right]^{n+1} \end{bmatrix}\frac{\zeta_1}{R_1}\frac{\Omega'(\zeta_1)}{|\Omega'(\zeta_1)|} + \\ \begin{bmatrix} (b+ic)\sum_{n=-\infty}^{\infty}A_nH_{n-1}^{(1)}\left(k_1|\chi(\zeta_1)|\right)\left[\frac{\chi(\zeta_1)}{|\chi(\zeta_1)|}\right]^{n-1} - \\ (a-ic)\sum_{n=-\infty}^{\infty}A_nH_{n+1}^{(1)}\left(k_1|\chi(\zeta_1)|\right)\left[\frac{\chi(\zeta_1)}{|\chi(\zeta_1)|}\right]^{n+1} \end{bmatrix}\frac{\overline{\zeta_1}}{R_1}\frac{\overline{\Omega'(\zeta_1)}}{|\Omega'(\zeta_1)|}\right\}$$

(C.4)

$$K_n^{22} = \frac{\mu_2 k_2}{2}\left\{ H_{n-1}^{(1)}\left(k_2|\Omega(\zeta_1)|\right)\left[\frac{\Omega(\zeta_1)}{|\Omega(\zeta_1)|}\right]^{n-1}\frac{\zeta_1}{R_1}\frac{\Omega'(\zeta_1)}{|\Omega'(\zeta_1)|} - H_{n+1}^{(1)}\left(k_2|\Omega(\zeta_1)|\right)\left[\frac{\Omega(\zeta_1)}{|\Omega(\zeta_1)|}\right]^{n+1}\frac{\overline{\zeta_1}}{R_1}\frac{\overline{\Omega'(\zeta_1)}}{|\Omega'(\zeta_1)|} \right\}$$

(C.5)

$$K_n^{23} = \frac{\mu_2 k_2}{2}\left\{ H_{n-1}^{(2)}\left(k_2|\Omega(\zeta_1)|\right)\left[\frac{\Omega(\zeta_1)}{|\Omega(\zeta_1)|}\right]^{n-1}\frac{\zeta_1}{R_1}\frac{\Omega'(\zeta_1)}{|\Omega'(\zeta_1)|} - H_{n+1}^{(2)}\left(k_2|\Omega(\zeta_1)|\right)\left[\frac{\Omega(\zeta_1)}{|\Omega(\zeta_1)|}\right]^{n+1}\frac{\overline{\zeta_1}}{R_1}\frac{\overline{\Omega'(\zeta_1)}}{|\Omega'(\zeta_1)|} \right\}$$

(C.6)

$$K_n^{31} = 0$$

(C.7)

$$K_n^{32} = \frac{\mu_2 k_2}{2}\left\{ H_{n-1}^{(1)}\left(k_2|\Omega(\zeta_2)|\right)\left[\frac{\Omega(\zeta_2)}{|\Omega(\zeta_2)|}\right]^{n-1}\frac{\zeta_2}{R_2}\frac{\Omega'(\zeta_2)}{|\Omega'(\zeta_2)|} - H_{n+1}^{(1)}\left(k_2|\Omega(\zeta_2)|\right)\left[\frac{\Omega(\zeta_2)}{|\Omega(\zeta_2)|}\right]^{n+1}\frac{\overline{\zeta_2}}{R_2}\frac{\overline{\Omega'(\zeta_2)}}{|\Omega'(\zeta_2)|} \right\}$$

(C.8)

$$K_n^{33} = \frac{\mu_2 k_2}{2} \left\{ \begin{array}{l} H_{n-1}^{(2)}\left(k_2 \left|\Omega(\zeta_2)\right|\right)\left[\frac{\Omega(\zeta_2)}{\left|\Omega(\zeta_2)\right|}\right]^{n-1} \frac{\zeta_2}{R_2} \frac{\Omega'(\zeta_2)}{\left|\Omega'(\zeta_2)\right|} - \\ H_{n+1}^{(2)}\left(k_2 \left|\Omega(\zeta_2)\right|\right)\left[\frac{\Omega(\zeta_2)}{\left|\Omega(\zeta_2)\right|}\right]^{n+1} \frac{\overline{\zeta_2}}{R_2} \frac{\overline{\Omega'(\zeta_2)}}{\left|\Omega'(\zeta_2)\right|} \end{array} \right\} \tag{C.9}$$

$$Q^1 = -w_0 \sum_{n=-\infty}^{\infty} i^n J_n\left(k_{in}\left|\Omega(\zeta_1)\right|\right)\left[\frac{\Omega(\zeta_1)}{\left|\Omega(\zeta_1)\right|}\right]^n e^{-in\alpha} +$$

$$\frac{C_{55} k_{in} w_0}{4k} \left\{ \begin{array}{l} \left[(1+\beta) \sum\limits_{n=-\infty}^{\infty} i^n J_{n-1}\left(k_{in}\left|\Omega(\zeta_1)\right|\right)\left[\frac{\Omega(\zeta_1)}{\left|\Omega(\zeta_1)\right|}\right]^{n-1} e^{-in\alpha} - \right. \\ \left. (1-\beta-2i\kappa) \sum\limits_{n=-\infty}^{\infty} i^n J_{n+1}\left(k_{in}\left|\Omega(\zeta_1)\right|\right)\left[\frac{\Omega(\zeta_1)}{\left|\Omega(\zeta_1)\right|}\right]^{n+1} e^{-in\alpha} \right] \frac{\zeta_1}{R_1} \frac{\Omega'(\zeta_1)}{\left|\Omega'(\zeta_1)\right|} + \\ \left[(1-\beta+2i\kappa) \sum\limits_{n=-\infty}^{\infty} i^n J_{n-1}\left(k_{in}\left|\Omega(\zeta_1)\right|\right)\left[\frac{\Omega(\zeta_1)}{\left|\Omega(\zeta_1)\right|}\right]^{n-1} e^{-in\alpha} - \right. \\ \left. (1+\beta) \sum\limits_{n=-\infty}^{\infty} i^n J_{n+1}\left(k_{in}\left|\Omega(\zeta_1)\right|\right)\left[\frac{\Omega(\zeta_1)}{\left|\Omega(\zeta_1)\right|}\right]^{n+1} e^{-in\alpha} \right] \frac{\overline{\zeta_1}}{R_1} \frac{\overline{\Omega'(\zeta_1)}}{\left|\Omega'(\zeta_1)\right|} \end{array} \right\} \tag{C.10}$$

$$Q^2 = -\frac{C_{55} k_{in} w_0}{4} \left[\begin{array}{l} (1+\beta) \sum\limits_{n=-\infty}^{\infty} i^n J_{n-1}\left(k_{in}\left|\Omega(\zeta_1)\right|\right)\left[\frac{\Omega(\zeta_1)}{\left|\Omega(\zeta_1)\right|}\right]^{n-1} e^{-in\alpha} - \\ (1-\beta-2i\kappa) \sum\limits_{n=-\infty}^{\infty} i^n J_{n+1}\left(k_{in}\left|\Omega(\zeta_1)\right|\right)\left[\frac{\Omega(\zeta_1)}{\left|\Omega(\zeta_1)\right|}\right]^{n+1} e^{-in\alpha} \end{array} \right] \frac{\zeta_1}{R_1} \frac{\Omega'(\zeta_1)}{\left|\Omega'(\zeta_1)\right|}$$

$$+\frac{C_{55} k_a w_0}{4} \left[\begin{array}{l} (1-\beta+2i\kappa) \sum\limits_{n=-\infty}^{\infty} i^n J_{n-1}\left(k_{in}\left|\Omega(\zeta_1)\right|\right)\left[\frac{\Omega(\zeta_1)}{\left|\Omega(\zeta_1)\right|}\right]^{n-1} e^{-in\alpha} - \\ (1+\beta) \sum\limits_{n=-\infty}^{\infty} i^n J_{n+1}\left(k_{in}\left|\Omega(\zeta_1)\right|\right)\left[\frac{\Omega(\zeta_1)}{\left|\Omega(\zeta_1)\right|}\right]^{n+1} e^{-in\alpha} \end{array} \right] \frac{\overline{\zeta_1}}{R_1} \frac{\overline{\Omega'(\zeta_1)}}{\left|\Omega'(\zeta_1)\right|} \tag{C.11}$$

$$Q^3 = 0 \tag{C.12}$$

where $\zeta_1 = R_1 e^{i\theta}$ and $\zeta_2 = R_2 e^{i\theta}$.

Appendix D

Coefficients K_n^{mj} and Q^j in Equation (9.22) in Chapter 9 are as follows:

$$K_n^{11} = H_n^{(1)}\left(k_1|\chi(z_1)|\right)\left[\frac{\chi(z_1)}{|\chi(z_1)|}\right]^n + H_n^{(1)}\left(k_1|\chi'(z_1)|\right)\left[\frac{\chi'(z_1)}{|\chi'(z_1)|}\right]^{-n}$$
$$-\frac{C_{55}k_1}{4k}\left\{\begin{array}{l}\left[(a+ic)(F_{n-1}-F'_{n+1})-(b-ic)(F_{n+1}-F'_{n-1})\right]e^{i\theta} + \\ \left[(b+ic)(F_{n-1}-F'_{n+1})-(a-ic)(F_{n+1}-F'_{n-1})\right]e^{-i\theta}\end{array}\right\} \quad \text{(D.1)}$$

$$K_n^{12} = -H_n^{(1)}\left(k_2|z_1|\right)\left[\frac{z_1}{|z_1|}\right]^n \quad \text{(D.2)}$$

$$K_n^{13} = -H_n^{(2)}\left(k_2|z_1|\right)\left[\frac{z_1}{|z_1|}\right]^n \quad \text{(D.3)}$$

$$K_n^{21} = \frac{C_{55}k_1}{4}\left\{\begin{array}{l}\left[(a+ic)(F_{n-1}-F'_{n+1})-(b-ic)(F_{n+1}-F'_{n-1})\right]e^{i\theta} + \\ \left[(b+ic)(F_{n-1}-F'_{n+1})-(a-ic)(F_{n+1}-F'_{n-1})\right]e^{-i\theta}\end{array}\right\} \quad \text{(D.4)}$$

$$K_n^{22} = -\frac{\mu_2 k_2}{2}\left\{H_{n-1}^{(1)}\left(k_2|z_1|\right)\left[\frac{z_1}{|z_1|}\right]^{n-1}e^{i\theta} - H_{n+1}^{(1)}\left(k_2|z_1|\right)\left[\frac{z_1}{|z_1|}\right]^{n+1}e^{-i\theta}\right\} \quad \text{(D.5)}$$

$$K_n^{23} = -\frac{\mu_2 k_2}{2}\left\{H_{n-1}^{(2)}\left(k_2|z_1|\right)\left[\frac{z_1}{|z_1|}\right]^{n-1}e^{i\theta} - H_{n+1}^{(2)}\left(k_2|z_1|\right)\left[\frac{z_1}{|z_1|}\right]^{n+1}e^{-i\theta}\right\} \quad \text{(D.6)}$$

$$K_n^{31} = 0 \quad \text{(D.7)}$$

$$K_n^{32} = \frac{\mu_2 k_2}{2} \left\{ H_{n-1}^{(1)}\left(k_2\left|z_2\right|\right)\left[\frac{z_2}{\left|z_2\right|}\right]^{n-1} e^{i\theta} - H_{n+1}^{(1)}\left(k_2\left|z_2\right|\right)\left[\frac{z_2}{\left|z_2\right|}\right]^{n+1} e^{-i\theta} \right\} \tag{D.8}$$

$$K_n^{33} = \frac{\mu_2 k_2}{2} \left\{ H_{n-1}^{(2)}\left(k_2\left|z_2\right|\right)\left[\frac{z_2}{\left|z_2\right|}\right]^{n-1} e^{i\theta} - H_{n+1}^{(2)}\left(k_2\left|z_2\right|\right)\left[\frac{z_2}{\left|z_2\right|}\right]^{n+1} e^{-i\theta} \right\} \tag{D.9}$$

$$Q^1 = w_0 e^{\frac{ik_{in}}{2}\left[(z_1+ih)\zeta_{in}+\left(z_1-ih\right)\zeta_{in}\right]} \left\{ \begin{array}{l} -1+\dfrac{ik_{in}}{4k}\left[\left(C_{55}+C_{44}\right)\zeta_{in}+\left(C_{55}-C_{44}-2iC_{45}\right)\overline{\zeta}_{in}\right]e^{i\theta} \\[2mm] +\dfrac{ik_{in}}{4k}\left[\left(C_{55}-C_{44}+2iC_{45}\right)\zeta_{in}+\left(C_{55}+C_{44}\right)\overline{\zeta}_{in}\right]e^{-i\theta} \end{array} \right\}$$

$$+w_0 e^{\frac{ik_r}{2}\left[(z_1+ih)\zeta_r+\left(z_1-ih\right)\zeta_r\right]} \left\{ \begin{array}{l} -1+\dfrac{ik_r}{4k}\left[\left(C_{55}+C_{44}\right)\zeta_r+\left(C_{55}-C_{44}-2iC_{45}\right)\overline{\zeta}_r\right]e^{i\theta} \\[2mm] +\dfrac{ik_r}{4k}\left[\left(C_{55}-C_{44}+2iC_{45}\right)\zeta_r+\left(C_{55}+C_{44}\right)\overline{\zeta}_r\right]e^{-i\theta} \end{array} \right\} \tag{D.10}$$

$$Q^2 = -\frac{ik_{in}}{4} w_0 e^{\frac{ik_{in}}{2}\left[(z_1+ih)\zeta_{in}+\left(z_1-ih\right)\zeta_{in}\right]} \left\{ \begin{array}{l} \left[\left(C_{55}+C_{44}\right)\zeta_{in}+\left(C_{55}-C_{44}-2iC_{45}\right)\overline{\zeta}_{in}\right]e^{i\theta} + \\[2mm] \left[\left(C_{55}-C_{44}+2iC_{45}\right)\zeta_{in}+\left(C_{55}+C_{44}\right)\overline{\zeta}_{in}\right]e^{-i\theta} \end{array} \right\}$$

$$-\frac{ik_r}{4} w_0 e^{\frac{ik_r}{2}\left[(z_1+ih)\zeta_r+\left(z_1-ih\right)\zeta_r\right]} \left\{ \begin{array}{l} \left[\left(C_{55}+C_{44}\right)\zeta_r+\left(C_{55}-C_{44}-2iC_{45}\right)\overline{\zeta}_r\right]e^{i\theta} + \\[2mm] \left[\left(C_{55}-C_{44}+2iC_{45}\right)\zeta_r+\left(C_{55}+C_{44}\right)\overline{\zeta}_r\right]e^{-i\theta} \end{array} \right\} \tag{D.11}$$

$$Q^3 = 0 \tag{D.12}$$

where $z_1 = R_1 e^{i\theta}$ and $z_2 = R_2 e^{i\theta}$.

References

Achenbach, J. D. (1973). *Wave propagation in elastic solids*. North-Holland Publisher.

AFPS/AFTES. (2001). *Guidelines on earthquake design and protection of underground structures*. Working Group of the French Association for Seismic Engineering (AFPS) and French Tunneling Association (AFTES) Version 1.

Allred, B., Daniels, J. J., & Ehsani, M. R. (2008). *Handbook of agricultural geophysics*. CRC Press.

Anderson, D. L., Minster, B., & Cole, D. (1974). The effect of oriented cracks on seismic velocities. *Journal of Geophysical Research, 79*(26), 4011–4015.

Arulanandan, K., Canclini, J., & Anandarjah, A. (1982). Simulation of earthquake motions in the centrifuge. *ASCE-Journal of the Geotechnical Division, 108*(5), 730–742.

Asakura, T., Shiba, Y., Matsuoka, S., Oya, T., & Yashiro, K. (2000). Damage to mountain tunnels by the earthquake and its mechanism. *Doboku Gakkai Ronbunshu, 659*, 27–38. (in Japanese)

Asadi, M., Rasouli, V., & Barla, G. (2012). A bonded particle model simulation of shear strength and asperity degradation for rough rock fractures. *Rock Mechanics and Rock Engineering, 45*, 649–675.

Asia Air Survey Co., Ltd. (2016). *The 2016 Kumamoto earthquake*. http://www.ajiko.co.jp/article/detail/ID5725UVGCD/. Accessed 14 June 2016.

Asian Disaster Reduction Center. (2016). *2016 Kumamoto earthquake survey report (preliminary)*. http://www.adrc.asia/publication ns/201604_KumamotoEQ/ADRC_2016KumamotoEQ_Repor t_1.pdf. Accessed 21 June 2016.

Aydan, O. (2017). *Rock dynamics*. CRC Press.

Bahaaddini, M., Hagan, P., Mitra, R., & Hebblewhite, B. (2014). Scale effect on the shear behavior of rock joints based on a numerical study. *Engineering Geology, 181*, 212–223.

Bahaaddini, M., Sharrock, G., & Hebblewhite, B. (2013). Numerical direct shear tests to model the shear behavior of rock joints. *Computers and Geotechnics, 51*, 101–115.

Bamant, Z. (2000). Size effect. *International Journal of Solids and Structures, 37*, 69–80.

Bandis, S. C. (1980). *Experimental studies of scale effects on shear strength and deformation of rock joints* [PhD Thesis]. University of Leeds.

Bandis, S. C., Barton, N. R., & Christianson, M. (1985). Application of a new numerical model of joint behaviour to rock mechanics problems. In *Proceedings of the International Symposium on Fundamentals of Rock Joints* (pp. 345–356). A.A. Balkema.

Bandis, S.C., Lumsden, A.C., & Barton, N.R. (1983). Fundamentals of rock joint deformation. *International Journal of Rock Mechanics and Mining Sciences & Geomechanics Abstracts, 20*(6), 249–268.

Bao, Z., Yuan, Y., & Yu, H. (2017). Multi-scale physical model of shield tunnels applied in shaking table test. *Soil Dynamics and Earthquake Engineering, 100,* 465–479.

Barpi, F., & Peila, D. (2012). Influence of the tunnel shape on shotcrete lining stresses. *Computer-Aided Civil and Infrastructure Engineering, 27*(4), 260–275.

Barton, N., & Choubey, V. (1977). The shear strength of rock joints in theory and practice. *Rock Mechanics, 10*(1), 1–54.

Barton, N.K., Bandis, S.C., & Baktar, K. (1985). Strength, deformation and conductivity coupling of rock joints. *International Journal of Rock Mechanics and Mining Sciences, 22*(3), 121–140.

Bayliss, A., & Turkel, E. (1980). Radiation boundary conditions for wave-like equations. *Communications on Pure and Applied Mathematics, 33*(6), 707–725.

Beskos, D.E. (1987). Boundary element methods in dynamic analysis. *Applied Mechanics Reviews, 40*(1), 1–23.

Bewick, R.P., & Kaiser, P.K. (2009). Influence of rock mass anisotropy on tunnel stability. In *ROCKENG09, Proceedings of the 3rd CANUS Rock Mechanics Symposium* (Paper No. 3995). Canadian Rock Mechanics Association.

Biggs, J.M. (1964). *Introduction to structural dynamics*. McGraw-Hill Companies.

Bolton, M.D., & Steedman, R.S. (1982). Centrifuge testing of microconcrete retaining walls subjected to base shaking. In: *Proceedings of the International Conference on Soil Dynamics and Earthquake Engineering* (pp. 311–330). University of Cambridge Engineering Department.

Brady, B. H. G., & Brown, E.T. (2004). *Rock mechanics for underground mining*. Kluwer Academic Publishers.

Brekhovskikh, L.M. (1960). *Waves in layered media*. Academic Press.

Cabinet Office of Japan. (2016). *Summary of damage situation in the Kumamoto earthquake sequence*. www.bousai.go.jp/updates/h280414jishin/index.html.

Cai, J.G., & Zhao, J. (2000). Effects of multiple parallel fractures on apparent attenuation of stress waves in rock masses. *International Journal of Rock Mechanics and Mining Sciences, 37,* 661–682.

Cai, M.F., He, M.C., & Liu, D.Y. (2002). *Rock mechanics and engineering*. Science Press. (in Chinese)

Červený, V., & Pšenčík, I. (2005). Plane waves in viscoelastic anisotropic media-I. Theory. *Geophysical Journal International, 161*(1), 197–212.

Chapman, C.H. (2004). *Fundamentals of seismic wave propagation*. Cambridge University Press.

Che, A., & Iwatate, T. (2002). Shaking table test and numerical simulation of seismic response of subway structures. In N. Jones, C.A. Brebbia & A.M. Rajendran (Eds.), *Structure under shock and impact VII* (pp. 367–376). WIT Press.

Chen, C., Wang, T., & Huang, T. (2011). Case study of earthquake-induced damage patterns of rock tunnel and associated reason. *Chinese Journal of Rock Mechanics and Geotechnical Engineering, 30*(1), 045–057. (in Chinese)

Chen, G. X., Wang, Z. H., Zuo, X., Du, X. L., & Han, X. J. (2010a). Development of laminar shear soil container for shaking table tests. *Chinese Journal of Geotechnical Engineering*, 32(1), 89–97. (in Chinese)

Chen, J. T., Shi, X., & Li, J. (2010b). Shaking table test of utility tunnel under non-uniform earthquake wave excitation. *Soil Dynamics and Earthquake Engineering*, 30(11), 1400–1416.

Chen, J. T., Yu, H. T., & Yuan, Y. (2018). Seismic response of shallow buried rock tunnel: Shaking table test and numerical simulation. *Proceedings of the 3rd International Conference on Rock Dynamics and Applications 3* (pp. 449–454). CRC Press.

Chen, W. Z., Song, W. P., & Zhao, W. S. (2017). Research progress of seismic analysis methods and performance evaluation in underground engineering. *Chinese Journal of Rock Mechanics and Engineering*, 36(2), 310–320. (in Chinese)

Chen, Z., Shi, C., & Li, T. (2012). Damage characteristics and influence factors of mountain tunnels under strong earthquakes. *Nature Hazards*, 61(2), 387–401.

Chen, Z. F., & Yu, Y. Z. (2006). A review on development of geotechnical dynamic centrifugal model test. *Chinese Journal of Rock Mechanics and Engineering*, 25(s2), 4026–4033. (in Chinese)

Chen, Z. G. (2012). Dynamic stress concentration around shallow cylindrical cavity by SH wave in anisotropically elastic half-space. *Rock and Soil Mechanics*, 33(3), 899–905. (in Chinese)

Chen, Z. G. (2015). Effect of shallow buried cavity on anti-plane motion of ground surface in anisotropic half-space. *Acta Seismologica Sinica*, 37(4), 617–628.

Cho, N., Martin, C., & Sego, D. (2007). A clumped particle model for rock. *International Journal of Rock Mechanics and Mining Sciences*, 44, 997–1010.

Clough, R. W., & Penzien, J. (1993). *Dynamics of structures* (2nd ed.). McGraw-Hill.

Coe, C. J., Prevost, J. H., & Scanlan, R. H. (1985). Dynamic stress wave reflections/attenuation: Earthquake simulation in centrifuge soil models. *Earthquake Engineering and Structural Dynamics*, 13, 109–128.

Coelho, P. A. L. F., Haigh, S. K., & Madabhushi, S. P. G. (2003). Boundary effects in dynamic centrifuge modelling of liquefaction in sand deposits. In: *Proceedings of the 16th ASCE Engineering Mechanics Conference* (pp. 1–12). American Society of Civil Engineers (ASCE).

Costa, P. A., Calçada, R., & Cardoso, A. S. (2012). Track-ground vibrations induced by railway traffic: In-situ measurements and validation of a 2.5 D FEM-BEM model. *Soil Dynamics and Earthquake Engineering*, 32(1), 111–128.

Coulomb, C. A. (1773). Essai sur une application des regles des maximis et minimis a quelquels problemesde statique relatifs, a la architecture. *Academie Royale Des Sciences*, 7, 343–382.

Crampin, S. (1984). Effective anisotropic elastic constants for wave propagation through cracked solids. *Geophysical Journal of the Royal Astronomical Society*, 76(1), 135–145.

Cundall, P. A. (1971). A computer model for simulating progressive, large-scale movement in blocky rock system. In *Proceedings of the International Symposium on Rock Mechanics* (pp. 129–136). International Society for Rock Mechanics.

Cundall, P. A., & Hart, R. D. (1984). *Analysis of block test No. 1—inelastic rock mass behavior: Phase 2-A characterization of joint behavior (final report)*.

Itasca Consulting Group Report, Rockwell Hanford Operations, Subcontract SA-957.

Cundall, P.A., & Lemos, J. (1990). Numerical simulation of fault instabilities with a continuously-yielding joint model. In: *Proceedings of 2nd International Symposium on Rockbursts and Seismicity in Mines*. South African Institute of Mining and Metallurgy.

Curran, J.H., & Leong, P.K. (1983). Influence of shear rate on rock joint strength. In: *Proceedings of the 5th Congress of the International Society for Rock Mechanics* (pp. 235–240). A.A. Balkema.

Da Cunha, A.P. (1993). *Scale effects in rock masses*. Balkema.

Datta, S.K., O'Leary, P.M., & Shah, A.H. (1985). Three-dimensional dynamic response of buried pipelines to incident longitudinal and shear waves. *Journal of Applied Mechanics, 52*(4), 919–926.

Davis, C.A., Lee, V.W., & Bardet, J.P. (2010). Transverse response of underground cavities and pipes to incident SV waves. *Earthquake Engineering & Structural Dynamics, 30*(3), 383–410.

Day, R.W. (2002). *Geotechnical earthquake engineering handbook*. McGraw-Hill.

Deere, D.U. (1963). Technical description of rock cores for engineering purposes. *Rock Mechanics and Engineering Geology, 1*, 18–22.

Degrande, G., Clouteau, D., Othman, R., Arnst, M., Chebli, H., Klein, R., & Janssens, B. (2006). A numerical model for ground-borne vibrations from underground railway traffic based on a periodic finite element-boundary element formulation. *Journal of Sound and Vibration, 293*(3–5), 645–666.

Demantke, J., Mallet, C., David, N., & Vallet, B. (2011). Dimensionality based scale selection in 3D LiDAR point cloud. *International Archives of the Photogrammetry, Remote Sensing and Spatial Information Sciences, 38*(Part 5), W12.

Dong, L.L., Yan, G.R., & Liao, H.J. (2000). The dynamic centrifuge model test in geotechnical engineering. *Chinese Journal of Rock Mechanics and Engineering, 19*(6), 789–793. (in Chinese)

Dowding, C.H., & Rozen, A. (1978). Damage to rock tunnels from earthquake shaking. *Journal of Geotechnical & Geoenvironmental Engineering, 104*(2), 175–191.

Dravinski, M., & Yu, C.W. (2011). Peak surface motion due to scattering of plane harmonic P, SV, or Rayleigh waves by a rough cavity embedded in an elastic half-space. *Journal of Seismology, 15*, 131–145.

Du, S.T. (1996). *Seismic wave mechanics*. Petroleum University Press (in Chinese)

Du, X.L. (2009). *Theories and methods of wave motion for engineering*. Science Press (in Chinese)

Duke, C.M., & Leeds, D.J. (1959). Effects of earthquakes on tunnels. In *Proceedings of 2nd Protective Construction Symposium* (pp. 303–191). O'Sullivan, Rand Corporation.

Earthquake geotechnical engineering design Greek Seismic Code – EAK2000 (2003). Ministry of public works, Athens, Greece (in Greek).

Elliott, G.M. (1982). *An investigation of a yield criterion for rock* [PhD thesis]. University of London.

Engquist, B., & Majda, A. (1977). Absorbing boundary conditions for numerical simulation of waves. *Proceedings of the National Academy of Sciences, 74*(5), 1765–1766.

Engquist, B., & Majda, A. (1979). Radiation boundary conditions for acoustic and elastic wave calculations. *Communications on Pure and Applied Mathematics, 32*(3), 313–357.

Ewing, W.M., Jardetzky, W.S., & Press, F. (1957). *Elastic waves in layered media.* McGraw-Hill.

Fang, X.Q., & Jin, H.X. (2016a). Visco-elastic imperfect bonding effect on dynamic response of a non-circular lined tunnel subjected to P and SV waves. *Soil Dynamics and Earthquake Engineering, 88*, 1–7.

Fang, X.Q., & Jin, H.X. (2017). Dynamic response of a non-circular lined tunnel with visco-elastic imperfect interface in the saturated poroelastic medium. *Computers and Geotechnics, 83*, 98–105.

Fang, X.Q., Jin, H.X., Liu, J.X., & Huang, M.J. (2016). Imperfect bonding effect on dynamic response of a non-circular lined tunnel subjected to shear waves. *Tunnelling and Underground Space Technology, 56*, 226–231.

Fang, X.Q., Jin, H.X., & Wang, B.L. (2015). Dynamic interaction of two circular lined tunnels with imperfect interfaces under cylindrical P-waves. *International Journal of Rock Mechanics and Mining Sciences, 79*, 172–182.

Fardin, N., Jing, L., & Stephansson, O. (2001). The scale dependence of rock joint surface roughness. *International Journal of Rock Mechanics and Mining Sciences, 38*, 659–669.

Farmer, I.W. (1983). *Engineering behaviour of rocks.* Chapman and Hall.

Fujii, N. (1991). Development of an electromagnetic centrifuge earthquake simulator. In *Proceedings of Centrifuge 91* (pp. 351–354). A.A. Balkema.

Galvín, P., François, S., Schevenels, M., Bongini, E., Degrande, G., & Lombaert, G. (2010). A 2.5 D coupled FE-BE model for the prediction of railway induced vibrations. *Soil Dynamics and Earthquake Engineering, 30*(12), 1500–1512.

Gao, F., Sun, C.X., Tan, X.K., Zhu, Y., Li, H. (2015). Shaking table tests for seismic response of tunnels with different depths. *Rock and Soil Mechanics, 36*(9), 2517–2522. (in Chinese)

Geospatial Information Authority of Japan. (2016). *The 2016 Kumamoto earthquake.* www.gsi.go.jp/BOUSAI/H27-kumamoto-earthquake-index.html.

Gigli, G., & Casagli, N. (2011). Semi-automatic extraction of rock mass structural data from high resolution LIDAR point clouds. *International Journal of Rock Mechanics and Mining Sciences, 48*(2), 187–198.

Goda, K., Campbell, G., & Hulme, L. (2016). Kumamoto earthquake: Cascading geological hazards and compounding risks. *Frontiers in Built Environment, 2*, 19.

Gong, Z.H. (2007). *Study on the seismic safety evaluation of Huangcaoping Tunnel of National Road No. 318* [Dissertation]. Chengdu University of Technology.

Goto, Y., Matsuda, Y., Ejiri, J., & Ito, K. (1988). Influence of distance between juxtaposed shield tunnels on their seismic responses. In *Proceedings of the 9th World Conference on Earthquake Engineering* (pp. 569–574). Japan Association for Earthquake Engineering.

Graff, K.F. (1973). *Wave motion in elastic solids.* University of Ohio Press.

Grasselli, G. (2001). *Shear strength of rock joints based on quantified surface description* [PhD Thesis]. Swiss Fed Inst Tech.

Grasselli, G. (2006). Manuel rocha medal recipient shear strength of rock joints based on quantified surface description. *Rock Mechanics and Rock Engineering, 39*, 295–314.

Grasselli, G., Wirth, J., & Egger, P. (2002). Quantitative three-dimensional description of a rough surface and parameter evolution with shearing. *International Journal of Rock Mechanics and Mining Sciences, 39*, 789–800.

Griffith, A.A. (1921). The phenomena of rupture and flows in solids. *Philosophical Transactions of the Royal Society of London, Series A, 221*, 163–167.

Griffith, A.A. (1924). Theory of rupture. In: *Proceedings of the 1st International Congress for Applied Mechanics* (pp. 53–63). Gauthier-Villars et Cie.

Hallbauer, D.K., Wagner, H., & Cook, N. G. W. (1973). Some observations concerning the microscopic and mechanical behaviour of quartzite specimens in stiff, triaxial compression tests. *International Journal of Rock Mechanics, 10*, 713–726.

Han, F., & Liu, D.K. (1997). Scattering of plane SH-wave on semi-canyon topography of arbitrary shape with lining in anisotropic media. *Applied Mathematics and Mechanics, 18*(8), 753–761.

Hashash, Y.M., Hook, J.J., Schmidt, B., John, I., & Yao, C. (2001). Seismic design and analysis of underground structures. *Tunnelling and Underground Space Technology, 16*(4), 247–293.

Hazzard, J.F., Young, R.P., & Maxwell, S.C. (2000). Micro-mechanical modeling of cracking and failure in brittle rocks. *Journal of Geophysical Research (Solid Earth), 105*(7), 1978–2012.

He, C., Li, L., & Zhang, J. (2014). Seismic damage mechanism of tunnels through fault zones. *Chinese Journal of Geotechnical Engineering, 36*(3), 427–434. (in Chinese)

Higdon, R.L. (1986). Absorbing boundary conditions for difference approximations to the multidimensional wave equation. *Mathematics of Computation, 47*(176), 437–459.

Higdon, R.L. (1987). Numerical absorbing boundary conditions for the wave equation. *Mathematics of Computation, 49*(179), 65–90.

Hoek, E. (1965). The design of a centrifuge for the simulation of gravitational fields in mine models. *Journal of the South African Institute of Mining and Metallurgy, 65*, 455–487.

Hoek, E. (1990). Estimating Mohr-Coulomb friction and cohesion values from the Hoek-Brown failure criterion. *International Journal of Rock Mechanics and Mining Sciences & Geomechanics Abstracts, 27*, 227–229.

Hoek, E., & Brown, E.T. (1980a). *Underground excavations in rock*. Institution of Mining and Metallurgy.

Hoek, E., & Brown, E.T. (1980b). Empirical strength criterion for rock masses. *ASCE—Journal of the Geotechnical Engineering Division, 106*(9), 1013–1035.

Hoek, E., Wood, D., & Shah, S. (1992). A modified Hoek-Brown criterion for jointed rock masses. In *Proceedings of the International ISRM Symposium on Rock Characterization—EUROCK' 92* (pp. 209–214). British Geotechnical Society.

Honarvar, F., & Sinclair, A.N. (1998). Nondestructive evaluation of cylindrical components by resonance acoustic spectroscopy. *Ultrasonics, 36*(8), 845–854.

Hou, Y.J. (2006). Centrifuge shakers and testing technique. *Journal of China Institute of Water Resources and Hydropower Research, 4*(1), 15–22 (in Chinese)

Hudson, J.A. (1980). Overall properties of a cracked solid. *Mathematical Proceedings of the Cambridge Philosophical Society, 88*, 371–384.

Hudson, J.A. (1981). Wave speeds and attenuation of elastic waves in material containing cracks. *Geophysical Journal of the Royal Astronomical Society*, 64(1), 133–150.

Hudson, J.A. (1986). A higher order approximation to the wave propagation constants for a cracked solid. *Geophysical Journal of the Royal Astronomical Society*, 87(1): 265–274.

Hudson, J.A. (1990a). Overall elastic properties of isotropic materials with arbitrary distribution of circular cracks. *Geophysical Journal International*, 102(2), 465–469.

Hudson, J.A. (1990b). Attenuation due to second-order scattering in material containing cracks. *Geophysical Journal International*, 102(2), 485–490.

Hudson, J.A. (1994). Overall properties of anisotropic materials containing cracks. *Geophysical Journal International*, 116(2), 279–282.

Hudson, J.A., & Harrison, J.P. (1997). *Engineering rock mechanics: An introduction to the principles* (1st ed.). Elsevier Science Ltd.

Hung, C., Monsees, J., Munfah, N., & Wisniewski, J. (2009). *Technical manual for design and construction of road tunnels–civil elements* (FHWA-NHI-10–034). US Department of Transport Federal Highway Administration, 702pp.

Hushmand, B., Scott, R. F., & Crouse, C. B. (1988). Centrifuge liquefaction tests in a laminar box. *Geotechnique*, 38(2), 253–262.

Hwang, J.H., & Lu, C.C. (2007). Seismic capacity assessment of old Sanyi railway tunnels. *Tunnelling and Underground Space Technology*, 22(4), 433–449.

Iai, S. (2005). International Standard (ISO) on seismic actions for designing geotechnical works-an overview. *Soil Dynamics and Earthquake Engineering*, 25(7–10), 605–615.

Isago, N., & Kusaka, A., 2018. Road tunnel damage caused by the 2016 Kumamoto earthquake and future aseismic measures. *Tunnel*, 49(2), 131–141. (in Japanese)

ISO 23469, 2005. *Bases for design of structures–seismic actions for designing geotechnical works*. ISO International Standard. ISO TC 98/SC3/WG10.

ISRM. (1978). Suggested methods for the quantitative description of discontinuities in rock masses. *International Journal of Rock Mechanics and Mining Sciences & Geomechanics Abstracts*, 15, 319–368.

Itasca Consulting Group, Inc. (ICG). (2004). *PFC2D version 3.0: Theory and background*. ICG.

Iwatate, T., Kobayashi, Y., Kusu, H., & Rin, K. (2000). Investigation and shaking table tests of subway structures of the Hyogoken-Nanbu earthquake. In *Proceedings of the 12th World Conference on Earthquake Engineering* (Paper No. 1043). International Association for Earthquake Engineering.

Jaeger, J.C., Cook, N. G. W., & Zimmerman, R.W. (2007). *Fundamentals of rock mechanics*, 4th ed. Blackwell Publishing Ltd.

Japan Road Association. (2003). *Technical standard for structure design of road tunnel*. Japan Road Association. (in Japanese)

Japan Road Association. (2015). *Road tunnel maintenance handbook*. Japan Road Association. (in Japanese)

Japan Society of Civil Engineers. (1989). *Dynamic analysis and seismic design—part 1: Seismic motion and its properties*. Japan Gihodo Press. (in Japanese)

Jeffreys, H. (1926). The reflection and refraction of elastic waves. Monthly Notices of the Royal Astronomical Society. *Geophysical Supplement, 1,* 321–334.

Jia, C., Wang, Z., & Zhu, W. (2013). Study on influencing factors of dynamic response of underground cavern under earthquake. *Chinese Journal of Underground Space and Engineering, 4,* 017. (in Chinese)

Jiang, L., Chen, J., & Li, J. (2010a). Seismic response of underground utility tunnels: Shaking table testing and FEM analysis. *Earthquake Engineering and Engineering Vibration, 9*(4), 555–567.

Jiang, S.P., Wen, D., & Zheng, S. (2011). Large-scale shaking table test for seismic response in portal section of Galongla tunnel. *Chinese Journal of Rock Mechanics and Engineering, 30*(4), 649–656. (in Chinese)

Jiang, Y.J., Wang, C.X., & Zhao, X.D. (2010b). Damage assessment of tunnels caused by the 2004 mid Niigata prefecture earthquake using Hayashi's quantification theory type II. *Nature Hazards, 53*(3), 425–441.

Jiang, Y.J., Xiao, J., Tanabashi, Y., & Mizokami, T. (2004). Development of an automated servo-controlled direct shear apparatus applying a constant normal stiffness condition. *International Journal of Rock Mechanics and Mining Sciences, 41,* 275–286.

Jiang, Y.J., & Zhang, X.P. (2019). Structure of longitudinal shock absorber and tunnel lining with the longitudinal shock absorber. Chinese Patent. ZL201611050285.5. (in Chinese)

Jin, F., Wang, G.L., & Jia, W.W. (2001). Application of distinct element-boundary element coupling model in underground structure dynamic analysis. *Journal of Hydraulic Engineering, 1*(2), 24–28.

Jing, L. (2003). A review of techniques, advances and outstanding issues in numerical modelling for rock mechanics and rock engineering. *International Journal of Rock Mechanics and Mining Sciences, 40,* 283–353.

Jing, L., Stephansson, O., & Nordlund, E. (1993). Study of rock joints under cyclic loading conditions. *Rock Mechanics and Rock Engineering, 26*(3), 215–232.

Jones, J.P., & Whittier, J.S. (1967). Waves at a flexibly bonded interface. *Journal of Applied Mechanics, 40,* 905–909.

Joseph, P.J., Einstein, H.H., &Whitman, R.V. (1988). *A literature review of geotechnical centrifuge modeling with particular emphasis on rock mechanics* (Report No. AD-A213–793). Massachusetts Institute of Technology.

Kawamoto, S., Hiyama, Y., & Kai, R. (2016a). Crustal deformation caused by the 2016 Kumamoto earthquake revealed by GEONET. *Bulletin of the Geospatial Information Authority of Japan, 64,* 27–33.

Kawamoto, S., Hiyama, Y., & Ohta, Y. (2016b). First result from the GEONET real-time analysis system (REGARD): The case of 2016 Kumamoto earthquake. *Earth, Planets and Space, 68*(1), 190.

Kawashima, K. (2000). Seismic design of underground structures in soft ground: A review. In F.M. Kusakabe (Ed.), *Geotechnical aspects of underground construction in soft ground.* A.A. Balkema.

Kendall, K., & Tabor, D. (1971). An ultrasonic study of the area of contact between stationary and sliding surfaces. *Proceedings of the Royal Society of London, Series A, 323,* 321–340.

Kimura, T., Takemura, J., & Saitoh, K. (1988a). Development of a simple mechanical shaker using a cam shaft. *Proceedings of Centrifuge, 88*, 107–110.

Kimura, T., Takemura, J., & Saitoh, K. (1988b). Development of an electrohydraulic centrifuge earthquake simulator. *Proceedings of Centrifuge, 88*, 103–106.

Kiyomiya, O. (1995). Earthquake-resistant design features of immersed tunnels in Japan. *Tunnelling and Underground Space Technology, 10*(4), 463–475.

Ko, H. (1994). Modelling seismic problems in centrifuge. *Proceedings of the International Conference, Centrifuge, 94*, 3–12.

Konagai, K., Takatsu, S., Kanai, T., Fujita, T., Ikeda, T., & Johansson, J. (2009). Kizawa tunnel cracked on 23 October 2004 Mid-Niigata earthquake: An example of earthquake-induced damage to tunnels in active-folding zones. *Soil Dynamics and Earthquake Engineering, 29*(2), 394–403.

Krauthammer, T., & Chen, Y. (1989). Soil-structure interface effects on dynamic interaction analysis of reinforced concrete lifelines. *Soil Dynamics and Earthquake Engineering, 8*(1), 32–42.

Krinitzsky, E.L. (1995). Deterministic versus probabilistic seismic hazard analysis for critical structures. *Engineering Geology, 40*(1–2), 1–7.

Krinitzsky, E.L. (1998). The hazard in using probabilistic seismic hazard analysis for engineering. *Environmental and Engineering Geoscience, IV*(4), 425–443.

Kubenko, V.D., & Cherevko, M.A. (1978). Diffraction of elastic waves. *Soviet Applied Mechanics, 14*(8), 789–798.

Kuesel, T.R. (1969). Earthquake design criteria for subways. *Journal of the Structural Division, 95*(ST6), 1213–1231.

Kumamoto River and National Highway Office, Kyushu Regional Development Bureau, Ministry of Land, Infrastructure, Transport and Tourism. (2017). *Recovery study of road subjected to Kumamoto earthquake (the Takamori Line of Kumamoto Prefectural Route): Tunnel damage condition and recovery plan.* Kumamoto River and National Highway Office, Kyushu Regional Development Bureau, Ministry of Land, Infrastructure, Transport and Tourism.

Kunita, M., Takemata, R., & Iai, Y. (1994). Restoration of a tunnel damaged by earthquake. *Tunnelling and Underground Space Technology, 9*(4), 439–448.

Kutter, B. L. (1983). Deformation of centrifuge models of clay embankments due to 'bumpy road' earthquakes. *International Journal of Soil Dynamics and Earthquake Engineering, 2*(4) 199–205.

Lambert, C., & Coll, C. (2014). Discrete modeling of rock joints with a smooth-joint contact model. *Journal of Rock Mechanics and Geotechnical Engineering, 6*, 1–12.

Lanzano, G., Bilotta, E., Russo, G., Silvestri, F., & Madabhushi, S.G. (2012). Centrifuge modeling of seismic loading on tunnels in sand. *Geotechnical Testing Journal, 35*(6), 854–869.

Lee, C.J., Wei, Y.C., & Kuo, Y.C. (2012). Boundary effects of a laminar container in centrifuge shaking table tests. *Soil Dynamics and Earthquake Engineering, 34*, 37–51.

Lee, H.S., Park, Y.J., & Cho, T.F. (2001). Influence of asperity degradation on the mechanical behavior of rough rock joints under cyclic shear loading. *International Journal of Rock Mechanics and Mining Sciences, 38*(7), 967–980.

Lee, V.W., & Trifunac, M.D. (1979). Response of tunnels to incident SH-waves. *Journal of Engineering Mechanics-ASCE*, *105*(4), 643–659.

Li, C.C. (2010). Field observations of rock bolts in high stress rock masses. *Rock Mechanics and Rock Engineering*, *43*(4), 491–496.

Li, H.B., Feng, H.P., & Liu, B. (2006). Study on strength behaviors of rock joints under different shearing deformation velocities. *Chinese Journal of Mechanical Engineering*, *25*(12), 2436–2440.

Li, L., Hagan, P.C., Saydam, S., & Hebblewhite, B. (2016a). Shear resistance contribution of support systems in double shear test. *Tunnelling and Underground Space Technology*, *56*, 168–175.

Li, L., Hagan, P.C., Saydam, S., Hebblewhite, B., & Li, Y. (2016b). Parametric study of rockbolt shear behaviour by double shear test. *Rock Mechanics and Rock Engineering*, *49*(12), 4787–4797.

Li, T. (2012). Damage to mountain tunnels related to the Wenchuan earthquake and some suggestions for aseismic tunnel construction. *Bulletin of Engineering Geology and the Environment*, *71*(2), 297–308.

Li, Z.L., & Liu, D.Z. (1987). Ray method of SH wave propagation in an anisotropic medium with a circular cavity. *Earthquake Engineering and Engineering Vibration*, *7*(1), 1–8.

Lillesand, T.M., Kiefer, R.W., & Chipman, J.W. (2007). *Remote sensing and image interpretation* (5th ed.). John Wiley & Sons.

Lin, A., Satsukawa, T., Wang, M., Mohammadi, Asl, Z., Fueta, R., & Nakajima, F. (2016). Coseismic rupturing stopped by Aso volcano during the 2016 Mw 7.1 Kumamoto earthquake, Japan. *Science*, *354*(6314), 869–874.

Lin, C.H., Lee, V.W., Todorovska, M.I., & Trifunac, M.D. (2010). Zero-stress, cylindrical wave functions around a circular underground tunnel in a flat, elastic half-space: Incident P-waves. *Soil Dynamics and Earthquake Engineering*, *30*(10), 879–894.

Lin, M.L., & Wang, K.L. (2006). Seismic slope behavior in a large-scale shaking table model test. *Engineering Geology*, *86*, 118–133.

Liu, D.K. (1988). Dynamic stress concentration around a circular hole due to SH-wave in anisotropic media. *Acta Mechanica Sinica*, *4*(2), 146–155.

Liu, D., Gai, B., Tao, G. (1982). Applications of the method of complex functions to dynamic stress concentrations. *Wave Motion*, *4*(3), 293–304.

Liu, G.L., Song, E.X., Liu, H.B., & Gong, C.L. (2008). Dynamic centrifuge tests on seismic response of tunnel in saturated sandy foundation. *Rock and Soil Mechanics*, *29*(8), 2070–2076. (in Chinese)

Liu, J., & Li, B. (2005). A unified viscous-spring artificial boundary for 3-D static and dynamic applications. *Science in China Series E Engineering & Materials Science*, *48*(5), 570–584.

Liu, Q., & Wang, R. (2012). Dynamic response of twin closely-spaced circular tunnels to harmonic plane waves in a full space. *Tunnelling and Underground Space Technology*, *32*, 212–220.

Liu, Q., Zhang, C., & Todorovska, M.I. (2016). Scattering of SH waves by a shallow rectangular cavity in an elastic half space. *Soil Dynamics and Earthquake Engineering*, *90*, 147–157.

Liu, Q., Zhao, M., Wang, L. (2013). Scattering of plane P, SV or Rayleigh waves by a shallow lined tunnel in an elastic half space. *Soil Dynamics and Earthquake Engineering*, 49, 52–63.

Liu, S.W., Datta, S.K., Khair, K.R., & Shah, A.H. (1991). Three-dimensional dynamics of pipelines buried in backfilled trenches due to oblique incidence of body waves. *Soil Dynamics and Earthquake Engineering*, 10(4), 182–191.

Liu, X.S., Tan, Y.L., & Ning, J.G. (2018). Mechanical properties and damage constitutive model of coal in coal-rock combined body. *International Journal of Rock Mechanics and Mining Sciences*, 110, 140–150.

Lombard, B., & Piraux, J. (2006). Numerical modeling of elastic waves across imperfect contacts. *SIAM Journal on Scientific Computing*, 28(1), 172–205.

Low, K. L. (2004). *Linear least-squares optimization for point-to-plane ICP surface registration*. Technical Report, Department of Computer Science, University of North Carolina at Chapel Hill, TR04-004.

Luco, J.E., & De Barros, F. C. P. (1994). Seismic response of a cylindrical shell embedded in a layered viscoelastic half-space. I: Formulation. *Earthquake Engineering & Structural Dynamics*, 23(5), 553–567.

Matula, M., & Holzer, R. (1978). Engineering topology of rock masses. In *Proceedings of Felsmekanik Kolloquium, Grundlagen ung Andwendung der Felsmekanik* (pp. 107–121). Wilhelm Ernst & Sohn Verlag für Architektur und technische Wissenschaften.

Maugeri, M., & Soccodato, C. (2014). *Earthquake geotechnical engineering design*. Springer.

McGarr, A. (1981). Analysis of peak ground motion in terms of a model of inhomogeneous faulting. *Journal of Geophysical Research: Solid Earth*, 86(B5), 3901–3912.

McGarr, A. (1982). Upper bounds on near-source peak ground motion based on a model of inhomogeneous faulting. *Bulletin of the Seismological Society of America*, 72(6A), 1825–1841.

McGarr, A. (1983). Estimating ground motions for small nearby earthquakes. In *Seismic design of embankments and caverns* (pp. 113–127). ASCE.

McGarr, A., Green, R. W. E., & Spottiswoode, S.M. (1981). Strong ground motion of mine tremors: Some implications for near-source ground motion parameters. *Bulletin of the Seismological Society of America*, 71(1), 295–319.

McHugh, E., & Signer, S.D. (1999). Roof bolt response to shear stress: Laboratory analysis. In *Proceedings of 18th International Conference on Ground Control in Mining* (pp. 232–238). WV University.

Milne, D. (1990). Standardized joint descriptions for improved rock classification. In *Proceedings of the 31st US Symposium on Rock Mechanics* (pp. 35–41). American Rock Mechanics Association (ARMA).

Mirzaghorbanali, A, Nemcik, J., & Aziz, N. (2014). Effects of cyclic loading on the shear behaviour of infilled rock joints under constant normal stiffness conditions. *Rock Mechanics and Rock Engineering*, 47, 1373–1391.

Moghadam, M.R., & Baziar, M.H. (2016). Seismic ground motion amplification pattern induced by a subway tunnel: Shaking table testing and numerical simulation. *Soil Dynamics and Earthquake Engineering*, 83, 81–97.

Mononobe, N., & Matsuo, H. (1929). On the determination of earth pressure during earthquakes. In *Proceedings of the World Engineering Conference* (pp. 177–185). University of Michigan Press.

Monsees, J.E., & Merritt, J.L. (1988). Seismic modelling and design of underground structures. *Numerical Methods in Geomechanics, 1988*, 1833–1842.

Moradian, Z. A., Ballivy, G., Rivard, P., Gravel, C., & Rousseau, B. (2010). Evaluating damage during shear tests of rock joints using acoustic emissions. *International Journal of Rock Mechanics and Mining Sciences, 47*, 590–598.

Moreland, L.W. (1974). Elastic response of regularly jointed media. *Geophysical Journal of the Royal Astronomical Society, 37*(3), 435–446.

Moya, L., Yamazaki, F., & Liu, W. (2017). Calculation of coseismic displacement from lidar data in the 2016 Kumamoto, Japan, earthquake. *Natural Hazards and Earth System Sciences, 17*(1), 143.

Mukoyama, S., Sato, T., Takami, T., & Nishimura, T. (2017). *Estimation of coseismic surface displacement in the Aso Caldera area before and after the 2016 Kumamoto earthquake by topographical data analysis from differential LiDAR DEM analysis* (Report of the 2016 Kumamoto). Oita Earthquake Disaster Research Mission, Japan Society of Engineering Geology, pp. 55–63.

Muller, J.R., & Harding, D.J. (2007). Using LIDAR surface deformation mapping to constrain earthquake magnitudes on the Seattle fault in Washington state, USA. *Urban Remote Sens Joint Event, IEEE*, 1–7.

Muskat, M., & Meres, M.W. (1940). Reflection and transmission coefficients for plane waves in elastic media. *Geophysics, 5*, 115–148.

Myer, L.R., Pyrak-Nolte, L.J., & Cook, N. G. W. (1990). Effects of single fractures on seismic wave propagation. In *Proceedings of ISRM Symposium on Rock Joints* (pp. 467–474). A.A. Balkema.

Nakano, T., Kobayashi, T., Yoshida, K., & Fujiwara, S. (2016). Field survey of non-tectonic surface displacements caused by the 2016 Kumamoto earthquake around Aso valley. *Bulletin of the Geospatial Information Authority of Japan, 64*, 47–54.

Newmark, N.M. (1968). Problems in wave propagation in soil and rock. In *Proceedings of the International Symposium on Wave Propagation and Dynamic Properties of Earth Materials* (pp. 7–26). National Academy of Sciences, Washington, DC, USA.

Nissen, E., Maruyama, T., & Arrowsmith, J.R., Elliott, J. R., Krishnan, A. K., Oskin, M. E., & Saripalli, S. (2014). Coseismic fault zone deformation revealed with differential lidar: Examples from Japanese M_w~7 intraplate earthquakes. *Earth and Planetary Science Letters, 405*, 244–256.

Nuttli, O.W. (1979). *The relation of sustained maximum ground acceleration and velocity to earthquake intensity and magnitude* (Miscellaneous Paper S-73-1, Report 16). US Army Corps of Engineers Waterways Experiment Station, p. 74.

O'Connell, R.J., & Budiansky, B. (1974). Seismic velocities in dry and saturated cracked solids. *Journal of Geophysical Research, 79*(35), 5412–5426.

Oda, K., Hattori, S., & Takayama, T. (2016). Detection of slope movement by comparing point clouds created by SfM software. *International Archives of the Photogrammetry, Remote Sensing and Spatial Information Sciences, XLI-B5*, 553–556.

Oda, M., Yamabe, T., Ishizuka, Y., Kumasaka, H., & Tada, H. (1992). Elastic analyses on excavation of an intersecting tunnel with emphasis on the presence of geological discontinuities. In *Rock Characterisation, ISRM Symposium, EUROCK'92* (pp. 324–329). A.A. Balkema.

Ogawa, Y., & Koike, T. (2001). Structural design of buried pipelines for severe earthquakes. *Soil Dynamics and Earthquake Engineering, 21*(3), 199–209.

Ohtomo, K., Suehiro, T., Kawai, T., & Kanaya, K. (2001). Research on streamlining seismic safety evaluation of underground reinforced concrete duct-type structures in nuclear power stations-Part-2. Experimental aspects of laminar shear sand box excitation tests with embedded RC models. *Transactions, SMiRT, 16,* Paper no. 1298.

Ohtomo, K., Suehiro, T., Kawai, T., & Kanaya, K. (2003). Substantial cross section plastic deformation of underground reinforced concrete structures during strong earthquakes. *Doboku Gakkai Ronbunshu, 724,* 157–175.

Okamoto, S. (1984). *Introduction to earthquake engineering* (2nd ed.). University of Tokyo Press.

Okano, N., Kawagoe, K., Kojima, Y., & Tsuru, T. (2018). Tunnel damage on the Minami-Aso railway caused by the 2016 Kumamoto earthquake. *Tunnel, 49*(2), 123–130. (in Japanese)

Oreste, P. (2007). A numerical approach to the hyperstatic reaction method for the dimensioning of tunnel supports. *Tunnelling and Underground Space Technology, 22*(2), 185–205.

O'Rourke, M.J., Hmadi, K.E. (1988). Analysis of continuous buried pipelines for seismic wave effects. *Earthquake Engineering & Structural Dynamics, 16*(6), 917–929.

Ortiz, L.A., Scott, R.F., & Lee, J. (1983). Dynamic centrifuge testing of a cantilever retaining wall. *Earthquake Engineering and Structural Dynamics, 11,* 251–268.

Osaki, Y. (1984). *Introduction to spectral analysis of ground motion.* Kajima Institute Publishing. (In Japanese)

Owen, G.N., & Scholl, R.E. (1981). *Earthquake engineering of large underground structures* (Report FHWA/RD-80/195). Federal Highway Administration and National Science Foundation.

Palmström, A. (1995). *RMi – A rock mass characterization system for rock engineering purposes* [PhD thesis]. Oslo University.

Palmström, A. (1996). Characterizing rock masses by the RMi for use in practical rock engineering–Part 1: The development of the Rock Mass index (RMi). *Journal of Tunnelling and Underground Space Technology, 11*(2), 175–188.

Pao, Y.H., Ku, G. C. C., & Ziegler, F. (1983). Application of the theory of generalized rays to diffractions of transient waves by a cylinder. *Wave Motion, 5*(4), 385–398.

Pao, Y.H., & Mow, C.C. (1973). *Diffraction of elastic waves and dynamic stress concentrations.* Crane Russak.

Park, J., & Song, J. (2009). Numerical simulation of a direct shear test on a rock joint using a bonded-particle model. *International Journal of Rock Mechanics and Mining Sciences, 46,* 1315–1328.

Park, S.C., Yang, H., Lee, D.K., Park, E. H., & Lee, W. J. (2018). Did the 12 September 2016 Gyeongju, South Korea earthquake cause surface deformation?. *Geosciences Journal, 22*(2), 337–346.

Paterson, M.S. (1978). *Experimental rock deformation: The brittle field.* Springer.

Penzien, J. (2000). Seismically induced racking of tunnel linings. *Earthquake Engineering & Structural Dynamics, 29*(5), 683–691.

Phillips, F.C. (1971). *The use of the stereographic projection in structural geology* (3rd ed.). Edward Arnold.

Phillips, J.S., & Luke, B.A. (1991). Tunnel damage resulting from seismic loading. In *International Conferences on Recent Advances in Geotechnical Earthquake Engineering and Soil Dynamics* (pp. 207–217). Missouri University of Science and Technology.

Pitilakis, K., & Tsinidis, G. (2014). Performance and seismic design of underground structures. In M. Maugen & C. Soccodato (Eds.), *Earthquake geotechnical engineering design* (pp. 279–340). Springer.

Potyondy, D., & Cundall, P. (2004). A bonded-particle model for rock. *International Journal of Rock Mechanics and Mining Sciences, 41*, 1329–1364.

Power, M.S., Rosidi, D., & Kaneshiro, J.Y. (1998). Seismic vulnerability of tunnels-revisited. In L. Ozedimir (Ed.), *Proceedings of the North American Tunneling Conference.* Elsevier.

Priest, S.D. (1985). *Hemispherical projection methods in rock mechanics.* Allen & Unwin.

Priest, S.D. (1993). *Discontinuity analysis for rock engineering.* Chapman & Hall.

Pyrak-Nolte, L.J. (1988). *Seismic visibility of fractures* [PhD Thesis]. Department of Materials Science and Mineral Engineering, University of California.

Pyrak-Nolte, L.J. (1996). The seismic response of fractures and the interrelations among fracture properties. *International Journal of Rock Mechanics and Mining Sciences & Geomechanics Abstracts, 33*(8), 787–802.

Pyrak-Nolte, L.J., Myer, L.R., & Cook, N.G.W. (1990). Transmission of seismic waves across single natural fractures. *Journal of Geophysical Research, 95*(B6), 8617–8638.

Reiter, L. (1991). *Earthquake hazard analysis issues and insights.* Columbia University Press.

Ren, G., Smith, J.V., Tang, J.W., & Xie, Y.M. (2005). Underground excavation shape optimization using an evolutionary procedure. *Computers and Geotechnics, 32*(2), 122–132.

Riquelme, A.J., Abellán, A., & Tomás, R. (2015). Discontinuity spacing analysis in rock masses using 3D point clouds. *Engineering Geology, 195*, 185–195.

Riquelme, A.J., Abellán, A., Tomás, R., & Jaboyedoff, M. (2014). A new approach for semi-automatic rock mass joints recognition from 3D point clouds. *Computers and Geosciences, 68*, 38–52.

Roscoe, K.H. (1968). Soils and models tests. *Journal of Strain Analysis for Engineering Design, 3*(5), 57–64.

Roy, N., & Sarkar, R. (2017). A review of seismic damage of mountain tunnels and probable failure mechanisms. *Geotechnical and Geological Engineering, 35*(1), 1–28.

Saito, T., Mukouyama, M., & Taguchi, Y. (2007). Damages to railroad tunnels in the Niigataken Chuetsu-oki Earthquake Shinetsu line Yoneyama to Kashiwazaki. *Tunnelling and Underground Space Technology, 38*(12), 891–900.

Sang-Eun, L., Man-II, K., Jae-Hyeon, P., Chang-Kun, P., Meea, K., & Gyo-Cheol, J. (2006). Damage process of intact granite under uniaxial compression: Microscopic observations and contact stress analysis of grains. *Geosciences Journal*, *10*(4), 457–463.

Schnabel, P.B., Lysmer, J., & Seed, H. B. (1972). *Shake: A computer program for earthquake response analysis of horizontally layered sites* (Report no. UCB/EERC-72/12). University of California.

Schoenberg, M. (1980). Elastic wave behavior across linear slip interfaces. *Journal of Acoustics Society of America*, *68*(5), 1516–1521.

Schoenberg, M. (1983). Reflection of elastic waves from periodically stratified media with interfacial slip. *Geophysical Prospecting*, *31*(2), 265–292.

Schoenberg, M., & Douma, J. (1988). Elastic wave propagation in media with parallel fractures and aligned cracks. *Geophysical Prospecting*, *36*(6), 571–590.

Schoenberg, M., & Muir, F. (1989). A calculus for finely layered anisotropic media. *Geophysics*, *54*(5), 581–589.

Schoenberg, M., & Sayers, C. M. (1995). Seismic anisotropy of fractured rock. *Geopyisics*, *60*(1), 204–211.

Severn, R.T. (2011). The development of shaking tables-A historical note. *Earthquake Engineering and Structural Dynamics*, *40*, 195–213.

Sharma, S., & Judd, W.R. (1991). Underground opening damage from earthquakes. *Engineering Geology*, *30*(3–4), 263–276.

Shen, M. (1998). *Mechanics of rock mass*. Tongji University Press. (in Chinese)

Shen, Y.S., Gao, B., Wang, Z.Z., & Wang, X.Y. (2008). Model test for a road tunnel in the region of high seismic intensity. *Modern Tunnelling Technology*, *45*(5), 38–42.

Shen, Y.S., Gao, B., Yang, X., & Tao, S. (2014). Seismic damage mechanism and dynamic deformation characteristic analysis of mountain tunnel after Wenchuan earthquake. *Engineering Geology*, *180*, 85–98.

Shi, S.X., Han, F., Wang, Z.Q., & Liu, D.K. (1996). The interaction of Plane SH-wave and non-circular cavity surfaced with lining in anisotropic media. *Applied Mathematics and Mechanics*, *17*(9), 855–867.

Shih, K.T., Balachandran, A., Nagarajan, K., Slatton, C., & George, A. (2008, 14–17 July). Fast real-time LIDAR processing on FPGAs. DBLP, 2008. In *Proceedings of the 2008 International Conference on Engineering of Reconfigurable Systems & Algorithms*. ERSA.

Shimizu, M., Kurisu, M., & Katou, S. (2005). Damages of railway tunnels by Niigata Chuetsu earthquake. *Tunnelling and Underground Space Technology*, *36*(5), 421–428.

Shimizu, M., Saito, T., Suzuki, S., & Asakura, T. (2007). Results of survey regarding damages of railroad tunnels caused by the Mid Niigata Prefecture Earthquake in 2004. *Tunnelling and Underground Space Technology*, *38*(4), 265–273.

Slob, S. (2010). *Automated rock mass characterization using 3-D terrestrial laser scanning* [Dissertation]. Delft University of Technology.

Smerzini, C., Aviles, J., Paolucci, R., & Sánchez-Sesma, F.J. (2009). Effect of underground cavities on surface earthquake ground motion under SH wave propagation. *Earthquake Engineering & Structural Dynamics*, *38*(12), 1441–1460.

Son, M., & Cording, E. J. (2007). Ground-lining interaction in rock tunneling. *Tunnelling and Underground Space Technology*, 22(1), 1–9.

Spottiswoode, S. M., & McGarr, A. (1975). Source parameters of tremors in a deep-level gold mine. *Bulletin of the Seismological Society of America*, 65(1), 93–112.

Srivastava, L. P., & Singh, M. (2015). Effect of fully grouted passive bolts on joint shear strength parameters in a blocky mass. *Rock Mechanics and Rock Engineering*, 48, 1197–1206.

Sternberg, E. (1960). On the integration of the equations of motion in the classical theory of elasticity. *Archive for Rational Mechanics and Analysis*, 6, 34–50.

St. John, C. M., & Zahrah, T. F. (1987). Aseismic design of underground structures. *Tunnelling and Underground Space Technology*, 2(2), 165–197.

Structure and Planning Institute of Japan. (2005a). *User manual of the seismic response analysis program for stratified soil layers: k-SHAKE*. Structure and Planning Institute of Japan. (in Japanese)

Sturzenegger, M., & Stead, D. (2009). Close-range terrestrial digital photogrammetry and terrestrial laser scanning for discontinuity characterization on rock cuts. *Engineering Geology*, 106(3–4), 163–182.

Sun, G. Z. (1983). *The foundation of rock mechanics*. Science Press.

Suzuki, M., Hayashi, Y., Imai, J., Tada, M., & Shiga, M. (2001). Strange state survey and disaster recovery of the Touya tunnel associated with the mount USU eruption. *Japan Society of Civil Engineers*, 11, 233–238. (in Japanese)

Synge, J. L. (1956). Elastic waves in anisotropic media. *Journal of Mathematics and Physics*, 35(1–4), 323–334.

Takai, S., Date, H., Kanai, S., Niina, Y., Oda, K., & Ikeda, T. (2013). Accurate registration of MMS point clouds of urban areas using trajectory. *ISPRS Annals of the Photogrammetry, Remote Sensing and Spatial Information Sciences*, II-5/W2, 277–282.

Tao, L., Hou, S., Zhao, X., Qiu, W., Li, T., Liu, C., & Wang, K. (2015). 3-D shell analysis of structure in portal section of mountain tunnel under seismic SH wave action. *Tunnelling and Underground Space Technology*, 46, 116–124.

Taylor, R. N. (1995). *Geotechnical centrifuge technology*. Blackie Academic & Professional.

Teymur, B., & Madabhushi, S. P. G. (2003). Experimental study of boundary effects in dynamic centrifuge modeling. *Geotechnique*, 53(7), 655–663.

Ting, T. C. T. (2010). Existence of anti-plane shear surface waves in anisotropic elastic half-space with depth-dependent material properties. *Wave Motion*, 47(6), 350–357.

Tistel, J., Grimstad, G., & Eiksund, G. (2018). Testing and modeling of cyclically loaded rock anchors. *Journal of Rock Mechanics and Geotechnical Engineering*, 9(6), 1010–1030.

Towhata, I. (2008). *Geotechnical earthquake engineering*. Springer Science & Business Media.

Tse, R., & Cruden, D. M. (1979). Estimating joint roughness coefficients. *International Journal of Rock Mechanics and Mining Sciences & Geomechanics Abstracts*, 16(5), 303–307.

Turan, A., Hinchberger, S. D., & Naggar, H. M. (2009). Design and commissioning of a laminar soil container for use on small shaking tables. *Soil Dynamics and Earthquake Engineering*, 29, 404–414.

United States Geological Survey. (2010). *Earthquake facts.* https://prod-earthquake.cr.usgs.gov/learn/facts.php.

Vrettos. (2005). Design issues for immersed tunnel foundations in seismic areas. In *Proceedings of the 1st Greece-Japan Workshop: Seismic Design, Observation, and Retrofit of Foundations* (pp. 257–266). Vrettos.

Wang, F., Jiang, X., & Niu, J. (2017). The large-scale shaking table model test of the shallow-bias tunnel with a small clear distance. *Geotechnical and Geological Engineering, 35*(3), 1093–1110.

Wang, G., Jiang, Y. J., Wang, W.M., & Li, T. C. (2009a). Development and application of an improved numeric control shear-fluid coupled apparatus for rock joint. *Rock and Soil Mechanics, 30*(10), 3200–3208.

Wang, G., Wu, X., Jiang, Y., Huang, N., & Wang, S. G. (2013). Quasi-static laboratory testing of a new rock bolt for energy-absorbing applications. *Tunnelling and Underground Space Technology, 38*, 122–128.

Wang, J.H., Zhou, X.L., & Lu, J.F. (2005). Dynamic stress concentration around elliptic cavities in saturated poroelastic soil under harmonic plane waves. *International Journal of Solids and Structures, 42*(14), 4295–4310.

Wang, J.N. (1993). *Seismic design of tunnels: A simple state of the art design approach.* Parsons Brinckerhoff Inc.

Wang, M.N., Lin, G.J., Yu, L., & Cui, G.Y. (2012). *Tunnel seismic and shock absorption.* Science Press.

Wang, T.T., Hsu, J.T., Chen, C.H., & Huang, T.H. (2014). Response of a tunnel in double-layer rocks subjected to harmonic P-and S-waves. *International Journal of Rock Mechanics and Mining Sciences, 70*, 435–443.

Wang, Z., Gao, B., Jiang, Y., & Yuan, S. (2009b). Investigation and assessment on mountain tunnels and geotechnical damage after the Wenchuan earthquake. *Science in China Series E: Technological Sciences, 52*(2), 546–558.

Wang, Z., Li, W., Bi, L., Qiao, L., Liu, R., & Liu, J. (2018). Estimation of the REV size and equivalent permeability coefficient of fractured rock masses with an emphasis on comparing the radial and unidirectional flow configurations. *Rock Mechanics and Rock Engineering, 51*(5), 1457–1471.

Wang, Z.Z., & Zhang, Z. (2013). Seismic damage classification and risk assessment of mountain tunnels with a validation for the 2008 Wenchuan earthquake. *Soil Dynamics and Earthquake Engineering, 45*, 45–55.

Wasantha, P., Ranjith, P., & Shao, S. (2014). Energy monitoring and analysis during deformation of bedded-sandstone: Use of acoustic emission. *Ultrasonics, 54*, 217–226.

Weissman, K., & Prevost, J.H. (1989). Centrifugal modeling of dynamic soil-structure interaction. *Earthquake Engineering and Structural Dynamics, 18*, 1145–1161.

White, J.A. (2014). Anisotropic damage of rock joints during cyclic loading: Constitutive framework and numerical integration. *International Journal for Numerical and Analytical Methods in Geomechanics, 38*, 1036–1057.

White, W., Lee, I.K., & Valliappan, S. (1977). Unified boundary for finite dynamic models. *Journal of the Engineering Mechanics Division, 103*(5), 949–964.

Whitman, R. V., & Lambe, P. C. (1986). Effect of boundary conditions upon centrifuge experiments using ground motion simulation. *Geotechnical Testing Journal, 9*, 61–71.

Wikipedia. (2018). *Earthquakes (M6.0+) since 1900 through 2017*. https://commons.wikimedia.org/wiki/File:Map_of_earthquakes_1900-.svg.

Wong, K.C., Datta, S.K., & Shah, A.H. (1986). Three-dimensional motion of buried pipeline I: analysis, II: numerical results. *Journal of Engineering Mechanics*, *112*(12), 1319–1337.

Wu, D., Gao, B., Shen, Y., Zhou, J., & Chen, G. (2015). Damage evolution of tunnel portal during the longitudinal propagation of Rayleigh waves. *Natural Hazards*, *75*(3), 2519–2543.

Wu, X., Jiang, Y., & Li, B. (2018a). Influence of joint roughness on the shear behaviour of fully encapsulated rock bolt. *Rock Mechanics and Rock Engineering*, *51*(3), 953–959.

Xia, C.C., Song, Y., & Tang, Z. (2012). Particle flow numerical simulation for shear behavior of rough joints. *Chinese Journal of Rock Mechanics and Engineering*, *31*, 1545–1552. (in Chinese)

Xia, C.C., Tang, Z.C., & Xiao, W.M. (2014). New peak shear strength criterion of rock joints based on quantified surface description. *Rock Mechanics and Rock Engineering*, *47*(2), 387–400.

Xie, H., Peng, R., & Ju, Y. (2004). Energy dissipation of rock deformation and fracture. *Chinese Journal of Rock Mechanics and Engineering*, *23*, 3565–3570. (in Chinese)

Xin, C.L., Gao, B. Wang, Y, Zhou, J., & Shen, Y.S. (2015). Shaking table tests on deformable aseismic and damping measures for fault-crossing tunnel structures. *Rock and Soil Mechanics*, *36*(4), 1041–1049. (in Chinese)

Xin, C.L., Wang, Z.Z., & Gao, B. (2018). Shaking table tests on seismic response and damage mode of tunnel linings in diverse tunnel-void interaction states. *Tunnelling and Underground Space Technology*, *77*, 295–304.

Xu, H., Li, T., & Wang, D. (2013). Study of seismic responses of mountain tunnels with 3D shaking table model test. *Chinese Journal of Rock Mechanics and Engineering*, *32*(9), 1762–1771. (in Chinese)

Xu, X., Ling, D., & Chen, Y. (2010). Correlation of microscopic and macroscopic elastic constants of granular materials based on linear contact model. *Chinese Journal of Geotechnical Engineering*, *32*(7), 991–998. (in Chinese)

Xue, S., Liu, Y., & Li, X. (2013). Review of some problems about research on soil-structure dynamic interaction. *World Earthquake Engineering*, *29*(2), 1–9.

Yakovlevich, D.I., & Borisovna, M.J. (1978). Behavior of tunnel liner model on seismic platform. In *VI Sump on Earthquake Engineering* (pp. 379–382). University of Roorkee.

Yamagiwa, A., Hatanaka, Y., Yutsudo, T., & Miyahara, B. (2006). Real-time capability of GEONET system and its application to crust monitoring. *Bulletin of the Geographical Survey Institute*, *53*, 27–33.

Yamazaki, F., & Liu, W. (2016). Remote sensing technologies for post-earthquake damage assessment: A case study on the 2016 Kumamoto earthquake. In *6th Asia Conference on Earthquake Engineering*. Asian-Pacific Network of Centers for Earthquake Engineering Research.

Yashiro, K., Kojima, Y., & Shimizu, M. (2007). Historical earthquake damage to tunnels in Japan and case studies of railway tunnels in the 2004 Niigataken-Chuetsu earthquake. *Quarterly Report of RTRI*, *48*(3), 136–141.

Yasuda, N., Tsukada, K., & Asakura, T. (2014). A consideration of seismic damage mechanism of mountain tunnel based on three-dimensional elastodynamic. *Journal of Japan Society of Civil Engineers (Tunnel Engineering)*, 70(1), 1–14.

Yeh, G. C. K. (1974). Seismic analysis of slender buried beams. *Bulletin of the Seismological Society of America*, 64(5), 1551–1562.

Yi, C., Zhang, P., Johansson, D., & Nyberg, U. (2014). Dynamic response of a circular lined tunnel with an imperfect interface subjected to cylindrical P-waves. *Computers and Geotechnics*, 55, 165–171.

Yoshikawa, K. (1981). Investigation about past earthquake disasters of railway tunnels. *Quarterly Reports of RTRI*, 22(3).

Yu, B., Zhang, Z., Kuang, T., & Liu, J. R. (2016a). Stress changes and deformation monitoring of longwall coal pillars located in weak ground. *Rock Mechanics and Rock Engineering*, 49, 3293–3305.

Yu, C.W., & Dravinski, M. (2010). Scattering of plane harmonic P, SV and Rayleigh waves by a completely embedded corrugated elastic inclusion. *Wave Motion*, 47, 156–67.

Yu, H.T., Chen, J.T., Bobet, A., & Yuan, Y. (2016b). Damage observation and assessment of the Longxi tunnel during the Wenchuan earthquake. *Tunnelling and Underground Space Technology*, 54, 102–116.

Yu, H.T., Chen, J.T., Yuan, Y, & Zhao, X. (2016c). Seismic damage of mountain tunnels during the 5.12 Wenchuan earthquake. *Journal of Mountain Science*, 13(11), 1958–1972.

Yu, H.T., Yuan, Y., Liu, X., Li, Y. W., & Ji, S. W. (2013). Damages of the Shaohuoping road tunnel near the epicenter. *Structure and Infrastructure Engineering*, 9(9), 935–951.

Yu, Y.Z., & Chen, Z.F. (2005). A review on development of shaking table system for geotechnical centrifuge. *Water Resources and Hydropower Engineering*, 36(5), 19–21.

Zelikson, A., Leguay, P., & Pascal, C. (1983). Centrifugal model comparison of pile and raft foundations subjected to earthquakes. *International Journal of Soil Dynamics and Earthquake Engineering*, 2(4), 222–227.

Zeng, X. (1998). Seismic response of gravity quay walls, I: centrifuge modeling. *ASCE-Journal of Geotechnical and Geoenvironmental Engineering*, 124(5): 406–417.

Zeng, X., & Schofield, A.N. (1996). Design and performance of an equivalent shear-beam container for earthquake centrifuge modeling. *Geotechnique*, 46(1), 83–102.

Zhang, J.M., Yu, Y.J., Pu, J.L., Yin, K.T., Huang, S.Y., & Zhang, X.Y. (2004). Development of a shaking table in electro-hydraulic servo-control centrifuge. *Chinese Journal of Geotechnical Engineering*, 26(6), 843–845. (in Chinese)

Zhang, W.M., Lai, Z.Z., & Xu, G.M. (2002). Development of an electrohydraulic shake table for the centrifuge. *Hydro-science and Engineering*, 1, 63–66.

Zhang, X., Jiang, Y., & Maegawa, K. (2020). Mountain tunnel under earthquake force: A review study on possible causes of damage and restoration methods. *Journal of Rock Mechanics and Geotechnical Engineering*, 12(2), 414–426.

Zhang, X., Jiang, Y., & Sugimoto, S. (2018). Seismic damage assessment of mountain tunnel: A case study on Tawarayama tunnel due to the 2016 Kumamoto Earthquake. *Tunnelling and Underground Space Technology*, 71, 138–148.

Zhang, X.M., Yang, J.S., & Liu, B.C. (2006). Study on blasting seismic dynamic effect for tunnels with close interval in bedded rock mass. *Journal of Hunan University of Science & Technology (Natural Science Edition)*, 21(4), 70–74.

Zhang, X.P., Wang, G., Jiang, Y.J., Wu, X.Z., Wang, Z., & Huang, N. (2014). Simulation research on granite compression test based on particle discrete element model. *Rock and Soil Mechanics*, 35(Supp.1), 99–105. (in Chinese)

Zhao, J. (1998). A new JRC-JMC shear strength criterion for rock joint. *Chinese Journal of Rock Mechanics and Engineering*, 17, 349–357. (in Chinese)

Zhao, J., & Cai, J.G. (2001). Transmission of elastic P-waves across single fractures with a nonlinear normal deformational behavior. *Rock Mechanics and Rock Engineering*, 34(1), 3–22.

Zhao, J., Cai, J.G., Zhao, X.B., & Li, H.B. (2008a). Dynamic model of fracture normal behaviour and application to prediction of stress wave attenuation across fractures. *Rock Mechanics and Rock Engineering*, 41(5), 671–693.

Zhao, J., Cai, J.G., Zhao, X.B., & Song, H.W. (2003). Transmission of elastic P-wave across single fracture with nonlinear normal deformation behavior. *Chinese Journal of Rock Mechanics and Engineering*, 22(1), 9–17. (in Chinese)

Zhao, J., Zhao, X.B., & Cai, J.G. (2006a). A further study of P-wave attenuation across parallel fractures with linear deformational behaviour. *International Journal of Rock Mechanics and Mining Sciences*, 43, 776–788.

Zhao, X.B., Zhao, J., & Cai, J.G. (2006b). P-wave transmission across fractures with nonlinear deformational behaviour. *International Journal for Numerical and Analytical Methods in Geomechanics*, 30, 1097–1112.

Zhao, X.B., Zhao, J., Cai, J.G., & Hefny, A.M. (2008b). UDEC modelling on wave propagation across fractured rock masses. *Computers and Geotechnics*, 35, 97–104.

Zhao, X.B., Zhao, J., Hefny, A.M., & Cai, J.G. (2006c). Normal transmission of S-wave across parallel fractures with Coulomb slip behavior. *Journal of Engineering Mechanics*, 132(6), 641–650.

Zhou, D.P. (1998). Dynamic behavior of portal of tunnel subjected to strong ground motion. *Earthquake Engineering and Engineering Vibration*, 18(1), 124–130.

Zhou, Y., Misra, A., Wu, S.C., & Zhang, X. (2012). Macro-and meso-analyses of rock joint direct shear test using particle flow theory. *Chinese Journal of Rock Mechanics and Engineering*, 31, 1245–1256. (in Chinese)

Zhu, J.B., Zhao, X.B., Li, J.C., Zhao, G.F., & Zhao, J. (2011). Normally incident wave propagation across a joint set with the virtual wave source method. *Journal of Applied Geophysics*, 73, 283–288.

Zimmerman, C., & Stern, M. (1993). Boundary element solution of 3-D wave scatter problems in a poroelastic medium. *Engineering Analysis with Boundary Elements*, 12, 223–240.

Index

For Product Safety Concerns and Information please contact our EU
representative GPSR@taylorandfrancis.com
Taylor & Francis Verlag GmbH, Kaufingerstraße 24, 80331 München, Germany

www.ingramcontent.com/pod-product-compliance
Lightning Source LLC
Chambersburg PA
CBHW060329220326
41598CB00023B/2654